Management of Global Construction Projects

Edward Ochieng

Senior Lecturer, Liverpool John Moores University

Andrew Price

Professor of Project Management, Loughborough University

David Moore

Reader, Robert Gordon University, Aberdeen

First published 2013 by
PALGRAVE MACMILLAN

Palgrave Macmillan in the UK is an imprint of Macmillan Publishers Limited, registered in England, company number 785998, of Houndmills, Basingstoke, Hampshire RG21 6XS.

Palgrave Macmillan in the US is a division of St Martin's Press LLC, 175 Fifth Avenue, New York, NY 10010.

Palgrave Macmillan is the global academic imprint of the above companies and has companies and representatives throughout the world.

Palgrave® and Macmillan® are registered trademarks in the United States, the United Kingdom, Europe and other countries.

ISBN 978–0–230–30321–8

This book is printed on paper suitable for recycling and made from fully managed and sustained forest sources. Logging, pulping and manufacturing processes are expected to conform to the environmental regulations of the country of origin.

A catalogue record for this book is available from the British Library.

A catalog record for this book is available from the Library of Congress.

Short contents

Contents

Figures and Tables

Figures

Tables

About the Authors

Edward Ochieng PhD, PGCertHELT, MSc, BSc (Hons), FHEA is a Senior Lecturer in project management. Edward has a PhD from Loughborough University. Edward's research is focussed on construction project management. He has presented at national and international conferences, such as Association of Researchers in Construction Management (ARCOM), Australian Universities Building Educators Association (AUBEA), CIB World Congress, American Society for Engineering Education (ASEE) and International World of Construction Project Management where he has shared his knowledge on 'Project Complexity, Performance, Capital Effectiveness and Team Integration'. Edward has published one book and over 40 refereed papers. He has extensive research experience relating to people and organisational challenges and solution development for managing large capital and heavy engineering projects. He has 2 PhD completions, and is currently supervising 4 PhD students on project related topics (Risk Management, Agile, Knowledge Management and Sustainability).

Andrew Price DSc, PhD, BSc (Civil Eng), FCIOB, FICE, CEng is a Professor of Project Management with over 30 years design, construction and industry-focused research experience including several large collaborative research projects and centres. He has graduated 50+ PhD/EngDoc students from 25 countries and acted as External Examiner at 12 Universities as well as a Visiting Professor at four overseas Universities. Andrew has published 6 Books and over 300 papers in refereed journals and conferences. He has been a principal or co-investigator on 24 completed research projects. He was a Co-investigator on SUEMoT Consortia (£1.6M) and Loughborough's IMCRC. Andrew is currently Co-Director of the EPSRC funded (£13M over 7 years) HaCIRIC. He has been a member of 11 DETR/DoE/DTI/CIRIA Steering Groups and 9 European Construction Institute Steering Groups and Task Forces dealing with major project performance in the heavy engineering sector.

 **David Moore PhD, BSc, MCIOB** is a reader in project management at Robert Gordon University, Aberdeen. David has been involved in funded research examining the behaviours of superior-performing project managers, and which was rated as tending to international significance by EPSRC. His research activity covers areas ranging from buildability, through sustainable design and the use of solar technologies, to perception and cognition in a construction industry context. David has authored, co-authored and contributed to five books and over 80 papers. His published work reflects his research interests, including sustainable development (in terms of encouraging design acceptance of indigenous materials and methods), use of solar technologies for material production and wealth generation, low carbon construction skills generally, factors in the visual perception of information embedded in complex charts, culture and conflict in design and build teams, and project management competences and performance level.

Publisher's acknowledgements

The publisher and authors would like to thank the organizations listed below for permission to reproduce material from their publications:

Shutterstock for permission to reproduce figures 2.2, 2.3, 6.9 and 12.7.

Arvasya and the Marmaray Project for permission to reproduce figure 2.4

FIATECH for permission to reproduce the Capital Projects Technology Roadmap in figures 12.4, 12.5 and 12.6. FIATECH is an operational unit of CII (The Construction Industry Institute) based at the Cockrell School of Engineering, The University of Texas at Austin, TX.

Acknowledgements

The authors wish to acknowledge many individuals whose help and cooperation aided in the completion of this book.

Edward Ochieng: A huge thank you is owed to my wonderful parents Nelson Wilfred Ochanji Ochieng and Pheobe Omollo Amade Ochieng (Rest in Peace Mum). You are both my heroes. Thank you for your undying support and encouragement you have given me throughout my studies. I am thankful to my sister Dr Bertha Ochieng for her superb job of providing me with constructive criticisms, useful suggestions, encouragement and support. A special thank you is owed to my brothers, sisters, nephews, and nieces and cousin (Anthony Awimbo) for all their help and encouragement. Thanks are also extended to my many friends (Dr Ximing Ruan, Chris Bird, Graham Robinson) and colleagues at Liverpool John Moores University, especially to my late friend Yassine Melaine (Rest in Peace Yassine), Dr Raymond Abdulai and Dr Wilfred Matipa. They all contributed so much to the book through informal discussions, advice and criticisms. Many thanks go to Helen Bugler, for giving us the opportunity to publish this book. I would also like to thank Jenny Hindley-appreciate your on-time guidance and coordination at each stage of the publication process. My final thanks are to the Lord, for guiding me through the trials and tribulations that I never thought that I would see my way through.

Andrew Price: I would like to take this opportunity to thank those who made this book possible. First, I would like to thank my wife, Sandra, for the love, help and encouragement she has given me since we first met, and my parents, Frank and Jean, for their advice and guidance over the years. I have been in the privileged position of being able to supervise many high-quality postgraduate research students and research staff from many different countries; they have been a pleasure to work with and I have enjoyed watching their progress and significant achievements. Loughborough University has been an exciting place to work and I am grateful to my colleagues who have provided a tremendous amount of support. Most of my research projects have been highly collaborative and depended upon contributions made by industrial partners and academic colleagues from other universities: the time and support that they have provided is highly valued and much appreciated.

David Moore: I would like to acknowledge the support of my wife, Christine, through her ability to stand back and cut through the complexity (mainly added by her husband!) and offer something positive; a priceless skill in short supply and just one of the many things that make her special. I would also like to acknowledge the life and work of Dr Douglas Hague, who recently passed away – a man of learning and dry humour who was a good friend, research collaborator, and who will be missed on those frustrating days when there is a need to simply talk rubbish!

Preface

Brief description of the book

Construction is incontestably one of the most important contributors to the global economy in terms of employment and GDP. However, there are a number of well-recognised issues facing the global construction industry, including new approaches to project management, procurement, advances in technology, market uncertainty and changes in the structure of the economy; and the adoption of new environmental policies. At the same time, the challenges of project management have proliferated. Clients, designers, contractors, subcontractors, policymakers and researchers must respond positively. New challenges require new thinking and approaches. This book examines some of the key issues relevant for students and professionals interested in and working with organisations in delivering global construction projects. It takes into account the many important developments that have taken place and offers recommendations for successfully completing projects that present challenges such as managing culturally diverse and globally dispersed construction teams. A key focus in the text is how cultural complexity and uncertainty can be managed on global construction projects. *Management of Global Construction Projects* takes a decision-making, business-oriented approach to the management of projects, an approach reinforced throughout the twelve chapters with current case studies and vignettes. The traditional project management process and ways of managing global construction projects are increasingly becoming obsolete. Rigid construction project management processes can no longer cope with the demands of the new global economy. Managing sustainability is also crucial to project success in today's globally connected business world. This book covers this important issue amongst many contemporary global project management themes.

The need for this book

The theory and practice of global construction project management has changed fundamentally since 2007, in particular because of the increasingly demanding performance expectations of clients, new contractual agreements and greater understanding of national and regional issues in the internationalised global construction environment. One of the effects of globalisation is that international global construction projects are serviced by diverse project teams of designers, contractors, subcontractors and suppliers. In addition, the workforce is recruited both locally (regional/national) and internationally. This means that organisations with an international potential now have to deliver their projects using multicultural project teams. The challenge is how to equip a global project manager with the skills to effectively manage soft and hard project issues in order to deliver these demanding projects. In Chapter 7, the authors identify a number of multi-dimensional factors that either facilitate or limit the effectiveness of multicultural team working. These have been synthesised into a framework capturing eight key dimensions that must be taken into account in multicultural team working. Such factors include leadership style; team selection and composition process; team development process; cultural communication; cultural collectivism; cultural trust; cultural management and cultural uncertainty. These factors are particularly significant because the existing body of research into poor performance and people management within global construction project management does not take into account the relevant challenges faced by clients, suppliers, contractors, subcontractors and designers.

Key features of the book

This book offers a coherent approach to the subject matter including up-to-date data in its support. We have tried to keep its organisation clear and rational, moving from the general to the specific and from the historical, through the contemporary and into the future. It covers topics of national and international relevance in global project structure and organisation, global project management process, cultural complexity, global stakeholders in projects, global partnering and alliancing, global project finance, the future of project management and strategic issues in global construction projects. We believe it to be the first inclusive reference text in this field, highlighting the reality of current global construction project management and looking at future trends which are often disregarded within the 'traditional' project management perspective. We have aimed to provide an accessible entry point to learning the fundamentals of global construction project management, with clear pointers to further detailed information. We present contemporary appraisals of the literature for each chapter and examine the multiple aspects of global construction project management.

Intended audience

This book will appeal to students working on project management programmes, construction management, civil engineering, business management, MBA and international development studies. In addition, the book will appeal to practitioners involved in managing global projects, all those who work with diverse project teams and, in particular, anyone who studies or works with global construction organisations. Given that globalisation has led to the cultural issues in projects becoming more complex than ever, academic researchers and PhD students may also use this book to pursue new directions in cross-cultural project management. Naturally, the book will be helpful for a wide audience in project management.

The structure of the book

The book is divided into twelve chapters:

- **Chapter 1:** The introductory chapter briefly presents a broad introductory account of global project construction management.
- **Chapter 2:** This chapter considers some of the recent global developments in construction project management.

- **Chapter 3:** This chapter examines the role of stakeholders in global construction projects.
- **Chapter 4:** To facilitate an understanding of the nature of global construction challenges, this chapter focuses on successes and failures in construction projects.
- **Chapter 5:** This chapter examines three main traditional organisational structures for global project management and organisational issues in global construction projects.
- **Chapter 6:** The objective of this chapter is to provide advice and methodologies appropriate for managing the project management process in a global environment.
- **Chapter 7:** This chapter examines the nature of global construction project teams and their place in the global construction environment. The main thrust of this chapter is to explore the cultural complexity that exists in global construction project teams.
- **Chapter 8:** This chapter takes the reader through all aspects of global partnering and alliancing in construction projects. It examines drivers for, and pre-requisites of, partnering and alliancing in global construction projects.
- **Chapter 9:** This chapter examines the nature of global project finance and how construction organisations work with global financial institutions.
- **Chapter 10:** The aim of this chapter is to examine what gives rise to uncertainty in global construction projects.
- **Chapter 11:** This chapter lays out specific actions for industry which will contribute to the achievement of overarching targets in the next 20 years.
- **Chapter 12:** This chapter considers some of the more important differences in project delivery and the implications for practical global construction organisations.

Abbreviations

AfDB	African Development Bank
AKDN	Aga Khan Development Network
APM	Association of Project Management
AsDB	Asian Development Bank
ATP	Accredited Training Personnel
BIM	Building Information Modelling
BIS	Department for Business Innovation and Skills
BOT	Build–Operate–Transfer
BRICS	Brazil, Russia, India, China, South Africa
CE	Conformité Européenne
CIB	International Council for Research and Innovation in Building and Construction
CIOB	Chartered Institute of Building
CMFN	Canadian Model Forest Network
CoST	Construction Transparency Initiative
CR	Corporate Responsibility
CRM	Customer Relationship Management
CSF	Critical Success Factor
CSR	Corporate Social Responsibility
CVM	Contingent Valuation Method
DBFO	Design, Build, Finance and Operate
DEED	Detailed Engineering Design
DFID	Department for International Development
DSCR	Debt Service Cover Ratio
EA	Environmental Assessments
EBITDA	Earnings before Interest, Tax, Depreciation and Amortisation
EIB	European Investment Bank
EPSRC	Engineering and Physical Sciences Research Council
ERG	Efficiency and Reform Group
ERP	Enterprise Resource Planning
ESG	Environmental, Social and Governance
EU	European Union
FEED	Front End Engineering Design
GCO	Global Construction Organisation
GDP	Gross Domestic Product
GRI	Global Reporting Initiative
HSE	Health and Safety Executive
IA	Impact Assessment
IBRD	International Bank for Reconstruction and Development
ICCPM	International Centre for Complex Project Management

ICE	Import, Conversion, Export
ICSID	International Centre for Settlement of Investment Disputes
ICT	Information Communications Technology
IDA	International Development Association
IFC	International Finance Corporation
ILO	International Labour Organization
IPO	Initial Public Offering
IRR	Internal Rate of Return
IS	Information Systems
IT	Information Technology
JCT	Joint Contracts Tribunal
LIBOR	London Interbank Offered Rate
KM	Knowledge Management
KMS	Knowledge Management System
KPI	Knowledge Performance Indicator
LEED	Leadership in Energy Design
LLI	Local Level Indicators
LRV	Law of Requisite Variety
MeTA	Medicines Transparency Alliance
MIGA	Multilateral Investment Guarantee Agency
NGO	Non-Government Organisation
NIHL	Noise-Induced Hearing Loss
OECD	Organisation for Economic Co-Operation and Development
OGC	Office of Government Commerce (UK)
PDP	Prospect Development Plan
PDRI	Project Definition Rating Index
PESTLE	Political, Economic, Social, Technical, Legal, Environmental
PFI	Private Finance Initiative
PMI	Project Management Institute
PPP	Public Private Partnerships
PRINCE2	Projects in Controlled Environments
PRM	Partner Relationship Management
ROCE	Return on Capital Employed
ROI	Return on Investment
SARCI	Saudi Responsible Competitiveness Index
SBCI	Sustainable Buildings and Climate Initiative
SCM	Supply Chain Management
SDS	Sustainable Development Strategy
SIBOR	Singapore Interbank Offered Rate
SMART	Specific, Measurable, Achievable, Realistic and Time-Scaled
SME	Small to Medium Enterprise
SOA	Self-Organising Application
SPV	Special-Purpose Vehicle
SR	Social Responsibility
SROI	Social Return on Investment
SRSG	Special Representative on Business and Human Rights (United Nations)
SSCM	Sustainable Supply Chain Management
TQM	Total Quality Management

TTM	Transtheoretical Model of Change
VAPH	Value Added per Human (employee)
VISTA	Vietnam, Indonesia, South Africa, Turkey and Argentina
WHO	World Health Organization
WTO	World Trade Organization
WTP	Willingness to Pay
ZAMSIF	Zambia Social Investment Fund

1

Introduction

1.1 Chapter objectives

In this chapter you will learn about:

>> the aims of this book;
>> global construction project management;
>> what we mean by 'globalisation';
>> recent global project developments;
>> regional global differences;
>> global construction policies;
>> the structure of the book.

1.2 Primary aims of this book

The aim of the book is threefold:

- to examine how contractors, subcontractors, suppliers and governments can improve project performance; and
- to examine how these actors can increase awareness of how international differences can be managed in global construction projects; and
- to address managerial intercultural and organisational challenges in global construction projects.

1.3 Global construction project management

'Global projects' is a term that is often used imprecisely. Very few empirical studies have been made on the role of such projects in construction project management. It is apparent that a project is called global when it involves key stakeholders that represent national, regional and international differences, separated by great geographical distance and potentially significant cultural and organisational distances.[1] For a global construction project to merit that name there must be both substantial funds available and a real vision for a future built environment. Global projects, as we describe them in this book, are complex: they place considerable demands upon the supply chain, the design team and the project management team. Global projects do not come about simply by scaling-up a typical construction project – building 3000 houses instead of 300 houses does not turn the latter into a global construction project. The global nature of construction projects can be due to the presence of a contractor in various locations or due to the fact that a project manager has to manage a multicultural construction team. In an era of globalisation, project managers in the construction industry face unique challenges in coordinating clients, financiers, developers, designers and contractors from different countries.

It is worth noting that the global construction industry has been under pressure to evolve into a sector that is constantly changing to fit the needs of the broader context in which operations are executed. The global construction industry generates employment and income for a significant percentage of the population, and covers an extensive variety of technologies and practices of scale.[2] About one-tenth of the global economy is dedicated to constructing and operating homes and offices. The industry consists of construction and repair and therefore serves to maintain the infrastructure. The industry builds public and private housing, non-residential public buildings (e.g. hospitals and schools), industrial facilities (e.g. factories and processing plants) and commercial properties. Factories and plants, which contractors in this industry design and construct, contribute to the manufacturing of many different consumer products. The global construction industry also consumes a significant amount of natural resources, using up at least one-sixth of the world's wood, minerals, water and energy supplies.[2]

Construction activity takes place wherever there is human settlement. In most countries, the government is a major client for the construction industry. A comprehensive study by the International Labour Organization (ILO) verified that there is a correlation between disposable income and any government investment in construction work.[3] The 1998 statistics indicated that government expenditure in construction varied from US$5 per head in Ethiopia to almost US$5000 in Japan, suggesting a concentration of construction output in the rich, developed nations. Europe accounted for as much 30 per cent of global output while the USA accounted for 21 and Japan 20 per cent.[3] Despite its huge growth in recent years, China absorbed only 6 per cent and India 1.7 per cent. Turning to the distribution of construction employment, 77 per cent of the construction workforce comes from high-income nations whilst 23 per cent is from low-income nations. Global construction projects vary in size, complexity, form and repeatability but are generally classified as large, medium or small. Classification is based, primarily, on the estimated total cost and the thresholds of the delegated international guidelines financial authority. Where the project cost is close to a classification boundary, risk and sustainability factors are considered in determining the classification. Generally, global construction projects with high risk will move into the next higher classification, but for this book, the projects we have considered are large and complex, and in addition they:

- have a finite life;
- require inputs from many disciplines;
- progress through phases from definition to completion; and
- are complicated, changing dynamic situations.

The costs involved in global construction projects are huge – in some cases the budget for a single project can be more that a small country's annual gross domestic product (GDP). Not surprisingly, there are insufficient numbers of wealthy individuals to form a viable global construction project market. There have to be a variety of clients willing to spend billions on projects across a range of product types in different regions of the globe. It is not uncommon for consortia to be formed specifically for the purpose of bringing together the funding required for a large project. These consortia may comprise individuals, individuals and companies, government and companies; a considerable diversity of clients is possible when individuals, companies or governments see either an opportunity or a need for some form of construction product at the scale of a global project.

While the diverse nature of such consortia can bring uncertainty (frequently, consortia will have some players that have never previously worked together and there may be uncertainty regarding their abilities or financing), it is also an inevitable characteristic for the majority of global projects. Without the diversity of players there will simply be insufficient funds for the project to proceed. Where enough clients exist there is a real incentive for the development of global construction contractors. Examples of such companies are Skanska, headquartered in Sweden, AMEC, headquartered in London, and Bouygues, in Paris. As part of entering the market, the contractors need to develop expertise in a number of areas, including the development of project management expertise in the global context.

Global construction contractors also face uncertainty. The concept of uncertainty is addressed in detail in Chapter 10, but here it is relevant to consider the impact of uncertainty in terms of those key issues of the size and nature of the global construction project market. Uncertainty in a global market environment can be both a threat and an opportunity. Global construction projects can be considered as a response to a perceived need that may be largely or indeed entirely humanitarian in nature. Such projects may also be a response to a purely commercial opportunity. In the case of the former kind of project the 'threats' will typically be social in nature, such as loss of life, loss of livelihood or environmental degradation. There is little opportunity in this sector of the construction market for the making of a commercial profit. Indeed, such projects are typically funded by a combination of government aid money and aid money flowing from either not-for-profit organisations (charities) or philanthropic individuals. Nonetheless, global construction contractors may seek an involvement in such projects simply to evidence their social, ethical and moral credentials. However, at some point they have to be involved in projects that will allow them the opportunity to make a commercial profit. For such projects, there has to be a level of certainty in the wider economy before clients will decide to proceed. This is evidenced by the 2012 report which outlined that the London office development sector could have projects up to the value of £12 billion at risk over a five-year period.[4] The threat to the sector is suggested as being uncertainty on the part of potential tenants – who are unsure whether or not they should move to new premises or remain in their current premises – but largely because of uncertainty about the wider economy and debt concerns within the Eurozone. While the report identifies 150 potential projects delivering a total of 53 million square feet of new office space, there is also around

70 million square feet of office space covered by current leases due to expire by 2017. Without sufficient certainty that enough tenants currently holding leases within that 70 million square footage will decide to move (and the leaseholders may decide not to renew their leases even if they decide to stay put), the projects to build the new office space will not be given the go-ahead by their investors, and the larger construction contractors will find themselves operating within a reduced marketplace. In this case contractors will need to decide if they are going to compete within the reduced marketplace or move to a different one – to relocate.

Global construction contractors are, by their very nature, capable of relocating; construction companies have always had to be mobile in that their product has to be constructed in a specific but new location every time they find a client. However, the true extent of that mobility is, for the majority of construction contractors, considerable. Chapter 3 considers the nature of stakeholders in global projects and discusses the problems of identifying stakeholders and the needs to which a contractor may have to react. The greater the mobility of a contractor, the more diverse is the stakeholder community. This factor has tended to act against contractors moving outside a relatively well-defined geographical location; many contractors have found that such a move can be problematic and performance is not always at the level required or expected. It is worth noting that global construction contractors have developed the management expertise required to move across regional, national and even continental boundaries in order to exit falling markets and enter rising markets. This kind of movement simply reflects the manner in which the global construction clients operate their businesses. For example, a large UK supermarket chain, Tesco, announced in February 2012 that it was launching its initial public offering (IPO) of a Thai property fund with the intention of raising around US$585 million to fund further expansion.[4] While Tesco may not be at the top of everyone's list when identifying global construction clients, it is the world's third largest retailer with over 5300 stores in 14 countries and has sufficient retail property to form the foundation of funds such as its initial public offering, the Tesco Lotus Retail Growth Freehold and Leasehold Property Fund.

1.4 What is globalisation?

One of the most significant discussions in construction management research is globalisation. The past decade has seen the rapid growth of construction companies operating in more than one country. As revealed in this book, the process of globalisation has significant business, social, cultural and economic implications. A number of factors have influenced the process of globalisation in the construction sector:

- **Enhancement of communications**: mobile technology, adoption of building information modelling (BIM), and the use of the internet to improve team productivity has led to greater communication between contractors in different countries.
- **Work availability and skills:** nations such as Poland and India have cheap labour costs and also high skill levels.

- **Freedom of trade:** organisations like the ILO and World Trade Organization (WTO) have helped to eliminate barriers between nations.
- **Transnational corporations:** globalisation has resulted in many construction organisations merging or buying operations in other nations.[5]

According to DeWit and Meyer,[6] globalisation can be defined as the process of becoming more global; but in fact no single definition exists. A number of organisations and authors have used the term globalisation to refer to diverse multicultural project teams, universal similarity and international scope.[7,8] For example, if a contractor decides to sell the same low carbon material in all its international markets, the product will be referred to as a universal product, as opposed to a locally tailored low carbon product. In this book, the term 'universal similarity' has been used to depict the variance dimension. Construction organisations with project operations around the world can be branded international. From the reviewed literature, authors have used the term international to describe firms that are local or regional in scope.[7,8] It worth noting that the construction industry is local by its very nature in terms of many variables such as social, political, economic, regulatory and procurement conditions.[9] The globalisation of the construction sector has become a reality under today's globalisation trend. According to Chinowsky and Songer,[10] from the supply side, a number of construction organisations have joined the global market because the construction sector shares similar attributes with other sectors that facilitate globalisation, such as transportation, funds and reduced costs of communication. From the demand side, the collapse of trade barriers has led to more opportunities for organisations with competitive advantages. Discussions on globalisation tend to focus on three levels:

- globalisation of organisations;
- globalisation of businesses; and
- globalisation of economies.[6]

Current project management research has extended theories and concepts that decontextualise global projects from their cultural, organisational and international settings. Global projects can be large scale, and complex construction projects involve project members from more than one country, and require project teams to negotiate great geographical distances, and cultural, regional, organisational and international differences.[1]

1.5 Recent global project developments

Recent global project trends in construction indicate that in the year 2020, seven nations – Indonesia, Canada, Australia, Russia, the US, China and India – will account for 65 per cent of the projected $4.5 trillion increase in global construction productivity.[11] In 2010, the infrastructure provision accounted for 32 per cent of global construction output. Both residential and non-residential sectors accounted for 68 per cent of the global output. From 2015 to 2020 comparable rates of economic growth will take place; this will be an average of 5.6–5.7 per cent per annum. Globally, it has been suggested that the infrastructure sector will have an annual growth rate of 5.4 per cent per annum. In order to achieve future growth in Africa, Asia, Europe and the US, governments are going to have to invest in infrastructure projects. Most governments in the four continents will do so, but a number of countries are facing huge challenges in funding infrastructure projects. On a national scale, construction activity will be greatly determined by wider economic factors, including wider GDP growth potential, population growth, specific developments in employment patterns and urbanisation. These four variables will, for example, influence growth in world cities over the next ten years.[11]

1.6 Regional global differences

By 2020, the global construction market will be worth US$12 trillion, compared with US$7.5 trillion at present.[11] The value of global construction in new emerging economies will more than double. Global construction productivity is expected to increase by 70 per cent, with developed countries' output rising 35 per cent, compared with the 110 per cent rise forecast in developing countries. The key driver to this growth is expected to come from infrastructure projects in developing countries and in non-residential construction in emerging economies. According to the *Global Construction Report*,[11] China will overtake the US as the world's largest construction sector in 2018. As a result of this forecast, the Chinese construction industry is expected to be worth US$2.5 trillion – a staggering 19.1 per cent of the global production. In Western Europe a number of countries will experience slow growth, with the UK and Greece showing the least growth, and countries such as France, Spain and Italy not far behind. On the other hand, Eastern Europe will experience higher growth, with a 100 per cent rise in construction productivity expected in the next eight years. Economists see Poland as one of the ten fastest growing construction countries in the world, and, further east, Russia also shows great potential. Together with China and Russia, India, Nigeria and Vietnam will experience fast growth in the new global construction economy. In Pacific Asia, Japan will show the weakest growth up to 2022. In 2003, China surpassed Japan as the world's largest construction market. Economists predict that the situation for Japan will worsen, and that it will lose its position as the world's third largest construction market as India rises in significance.[11,12] As Figure 1.1 shows, amongst the other major nations, Indonesia is expected to be one of the fastest growing up to 2020. Canada is expected to move to rank fifth largest. In each region the three major long-term drivers will include the rate of change in population growth, the level of economic activity and the degree of urbanisation.

1.7 Global construction policy issues

Some of the long-term policy issues driving the global construction industry are particularly significant to construction clients and governments. Current issues and implications for the global construction industry are noted in what follows.[11]

1.7.1 Population growth

A number of developed economies are likely to experience little or no population growth over the next decade. Some major developed economies such as

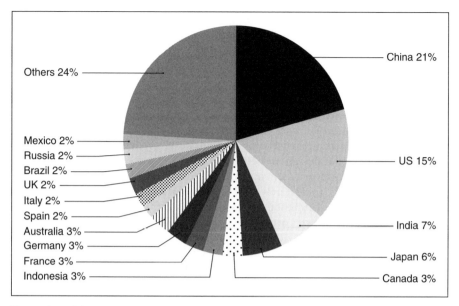

Figure 1.1   Fifteen largest construction markets in 2010–2020 $Billion at 2010 average exchange rates
(*Source*: Adapted from Global Construction Report 2010[11])

Japan, Germany and Italy are likely to experience a decline in population. This trend will have major implications for economic growth and demand for housing, and will counterbalance other factors such as rising income levels. On the whole, the demographic trends are likely to have a positive impact on the construction industry. In China, due to the hold-up effects of the one-child policy, population growth is expected to slow. All the same, the impact of demographic changes on emerging economies is likely to be strongly positive. This will be strengthened by other variables that will encourage housing demand such as increasing workforce mobility. In emerging economies, the continuous population growth has led to demand for better quality housing, higher quality utilities and higher living standards. These trends have further maximised demand for the construction 'product' in emerging economies.

1.7.2 Budgetary policy

In the Eurozone, governmental budgetary policy has had major implications for global construction activities. The recent economic recession has left long-term scars on the global economy. For instance, in the developed world, a number of countries have accumulated large financial shortfalls. In the medium term, this is likely to limit government spending in major infrastructure construction projects being delivered in developing countries. This makes the picture look dreary for construction organisations operating in Europe. In the UK, major heavy engineering construction projects have helped the government to counter the predicted depression. These include the London 2012 Olympics, the upgrade and maintenance of the major orbital motorway around Greater London (the M25), and the Thameslink and Cross London Rail

Links. Interestingly, emerging economies will not experience strict budgetary limitation, although there will be some major exemptions to this, as in Russia.

1.7.3 Environmental policy

Recent changes in environmental policy will have an increasing influence on how global construction projects are to be delivered. The global construction industry acknowledges that environmental and life cycle models are needed to help decision makers in the design and construction of heavy engineering construction projects. The notion of environmental sustainability is explored in more detail in Chapter 11. In the new global economy, stakeholders have become more sensitive to environmental issues, but it is difficult to assess the impact because a number of policies are still unclear. In addition, it is difficult to determine how committed both developed countries and emerging economies are to having fully integrated environmental policies in their construction activities. In order to address the above issues, clients will have to take a more proactive and educated leadership role. Direct engagement at both national and international level between institutions and governments will help stimulate research in global environmental sustainability. The primary role of secondary stakeholders in the delivery of large heavy engineering construction projects is becoming increasingly recognised. For instance, the UK government recently announced a new £32.7 billion HS2 high-speed rail project.[12] The project has been met with massive resistance on environmental grounds – and not only just from those people who will find themselves neighbouring the proposed link, but from politicians as well. All the same, a number of business leaders, unions and economists have come out in favour of the project. As shown in Figure 1.2, the emerging significance of the environmental policy to the construction industry will be based on three main drivers – increased awareness, cost pressures, and laws and regulation.

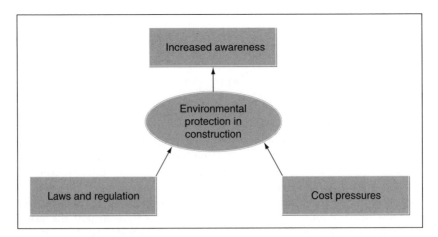

Figure 1.2 Environmental policy drivers for the construction industry
(*Source:* Adapted from European Powers Construction 2008 Report[13])

1.8 The structure of the book

The book is divided into 12 chapters:

- **Chapter 2** considers some of the recent global developments in construction project management.
- **Chapter 3** examines the role of stakeholders in global construction projects.
- **Chapter 4** focuses on project success and failures in construction projects in order to facilitate an understanding of the nature of global construction challenges.
- **Chapter 5** examines three main traditional organisational structures for global project management and organisational issues in global construction projects.
- **Chapter 6** provides advice and methodologies appropriate for managing the project management process in a global environment.
- **Chapter 7** examines the nature of global construction project teams and their place in the global construction environment, exploring the cultural complexity present in global construction project teams.
- **Chapter 8** takes the reader through all aspects of global partnering and alliancing in construction projects. It examines drivers for, and prerequisites of, partnering and alliancing in global construction projects.
- **Chapter 9** explores the nature of global project finance and how construction organisations work with global financial institutions.
- **Chapter 10** examines the issues that give rise to uncertainty in global construction projects.
- **Chapter 11** lays out specific actions undertaken by industry which will contribute to the achievement of overarching targets in the years to 2032.
- **Chapter 12** considers some of the more important differences in project delivery and the implications for practical global construction organisations.

1.9 References

1. Editorial (2010). Global projects: Strategic perspectives, *Scandinavian Journal of Management*, **26**, pp. 343–351.
2. United Nations Environmental Programme (1996). The construction industry and environment. *Industry and Environment*, **19** (2), pp. 1–14.
3. WIEGO (2007). Construction workers. http://wiego.org/informal-economy/occupational-groups/construction-workers [Accessed October 2007].
4. Longbottom, W. (2011). World's most expensive house lies abandoned... because billionaire owners believe it would be bad luck to move in. http://www.daily mail.co.uk/news/article-2053231/Worlds-expensive-house-Antilia-Mumbai-lies-abandoned.html [Accessed June 2012].
5. BBC (2012). Transnational corporations. http://www.bbc.co.uk/schools/gcsebite size/geography/globalisation/globalisation_rev3.shtml [Accessed July 2012].
6. De Wit, B. and Meyer, R. (2010). *Strategy Process, Content, Context: An International Perspective*. 4th edn. Hampshire, UK: Cengage Learning.
7. Ochieng, E. G. and Price, A. D. F. (2010). Managing cross-cultural communication in multicultural construction project teams: The case of Kenya and UK, *International Journal of Project Management*, **28** (5), pp. 449–460.
8. Young, B. R. and Javalgi, R. G. (2007). International marketing research: A global project management perspective, *Business Horizons*, **50**, pp. 113–122.

9. Ofori, G. (2003). Frameworks for analysing international construction, *Construction Management and Economics*, **21** (4), pp. 379–391.

10. Chinowsky, P. S. and Songer, A. D. (2011). *Organisational Management Construction*. London: Spon Press.

11. *Global construction 2020*. Oxford Economics, 3 March 2011, London.

12. Gwyn, T. and Stratton, A. (2012). HS2 high-speed rail project gets green light. http://www.guardian.co.uk/uk/2012/jan/10/hs2-rail-project-green-light [Accessed January 2012].

13. Kelly, J. (2008). *European powers of construction 2008: Analysis of key players and markets in construction*. Deloitte, UK.

2

Strategic Issues in Global Project Management

2.1 Introduction

As more construction organisations have accepted project management as a way of life, change in construction project management practices have taken place at an astounding rate. Critically, construction organisations that compete with one another on a day-to-day basis are sharing their achievements with other construction organisations through benchmarking activities. This chapter considers some recent global strategic issues, contemporary developments in project management, project governance and global developments in construction project management. It covers:

- global strategic issues;
- corporate strategic issues;
- tools and techniques for the management of strategic issues;
- project governance;
- current approaches to project management.

2.2 Learning outcomes

> The chapter's specific learning outcomes are to enable the reader to gain an understanding of:
>
> >> strategic issues in global construction projects;
> >> application of project governance in global construction projects;
> >> contemporary developments in project management.

2.3 Global strategic issues

As observed from the reviewed literature, multinational construction organisations have already found opportunities for growth through the establishment
of operations in the BRICS countries of Brazil, Russia, India, China and
South Africa.[1,2] The BRICS countries have witnessed robust growth in trade
over the past decade. From 2001 to 2010, trade between the five countries
grew at an annual rate of 28 per cent to reach nearly 230 billion US dollars.
According to Wilson, if this trend continues, in less than 40 years, the BRICs
countries could surpass the G6 in US dollar terms.[2] Interestingly, as multinational construction organisations seek to relocate their business strategies to
meet new demands, clients are now searching for opportunities in a second
wave of emerging countries such as the VISTA nations of Vietnam, Indonesia,
South Africa, Turkey and Argentina. Evidence shows that construction is boom-
ing in Indonesia, where the economy is rapidly expanding, but improvements
to infrastructure are required in order to sustain growth.[2] In South Africa,
there has been an augment in both public and private sector infrastructure
investment, partly as a result of the 2010 World Cup tournament. Multinational construction organisations have momentous opportunities to win work
as governments progressively open doors to inbound investment, through
joint ventures, foreign direct investment, private–public partnerships, partnering and private finance initiatives.[1] It should be noted that the challenges
are substantial, as multinational construction organisations need to administer business culture, regulatory issues, procurement strategies and the local
market, which vary widely globally. To be successful, today's construction organisations must appreciate the emerging global engineering construction sector
and be in a position to take advantage of emerging markets. Additionally,
as the BRIC nations continue to be drenched by multinational corporations,
the VISTA nations are an alternate option for consideration. The weak but
unwavering currencies in a number of VISTA and African nations also affords
them competitive merit and opportunities to supply developed nations with
materials and products.[3] Vietnam, Turkey and Indonesia have introduced tax
breaks for capital investments, subsidies for new businesses, and low-cost
financing to attract new foreign investments. The three countries proffer free
trade zones, where normal trade barriers such as tariffs and quotas have been
abolished and officious requirements have also been lowered in the hope of
attracting foreign businesses globally. A report by PricewaterhouseCoopers
(PwC) suggests that VISTA country governments view foreign businesses as
an important source of capital for their economies.[3] By comparison, China
and Russia prefer to give the advantage to local construction contractors over
outsiders. Both sanctioned policies that confine foreign contractors from investing in 'strategic' sectors and provide preferential grants to local contractors.
As today's developed economies become a shrinking part of the global economy, investing in the newly developing nations could prove very worthwhile.
According to Wilson, advance identification of, and investment in, the 'right'
emerging economies, particularly the up-and-coming emerging economies,
may be an increasingly vital strategic choice for multinational construction
organisations.[2]

2.4 Concept of strategic issues: global construction industry

Since the 1960s the process of developing strategy has undergone a number of name changes from 'long-range planning to corporate planning to strategic planning'.[4] As a result, there are strongly differing viewpoints on most of the key themes within the field; the deviations run so deep that a universal definition of the term 'strategy' is not possible.[5] As summarised below, there are a number of definitions of strategy, and the more there are, the more they tend to mystify rather than elucidate. De Wit and Meyer state that: *'strategy can be broadly conceived as a course of action for achieving an organisation's purpose'*;[5] Quinn: *'a strategy is the pattern or plan that integrates an organisation's major goals, policies and action sequences into a cohesive whole'*;[6] Johnson and Scholes: *'strategy is the direction and scope of an organisation over the long term: which achieves competitive advantage for the organisation through its configuration of resources within a changing environment to meet the needs of markets and to fulfil stakeholder expectations'*.[7] Interestingly, Mintzberg examined the process of strategy configuration as seen by different schools of thought, thus signifying the many fascinating and varied standpoints.[8] According to Price, a key feature of strategic management is strategic planning, that is, the process of setting missions, goals and objectives; clarifying policies and principles;[4] searching for opportunities and threats while preparing to take advantage of the first and avoiding the second. Depending on the project environment and the organisation strategic mission, this procedure can be formal or informal.

A number of multinational construction organisations utilise a diverse business model, with multiple divisions operating in a number of different countries. In recent years, it has been suggested that multinational construction organisations have a wide-ranging, complicated organisational structure of legal entities supporting them. Construction strategic issues lie at the heart of construction organisations. The most dramatic change to affect them over the last decade has been the globalisation of economic activity. The scale of the change over the period 2002 to 2012 has been notable as construction organisations have grown considerably in scale and scope. The largest multinational construction organisations are still mainly based in the developed world but there has been a strong shift in the development of international investment by organisations from Russia, China and India.[1] Globalisation has taken place in almost every construction segment. For instance, architecture, heavy engineering construction, building services, civil and structural engineering all now operate at least in part at an international level. In recent years, there has been a strong activity surge in the commercial sector and new public housing sectors, with larger multinational construction organisations the main beneficiaries of the increased workload. Globalisation is a lively process and in many construction divisions it has only just begun to have an impact. Mobility, for instance, is something taken for granted in the context of developed nations but has increasingly become an economic possibility for contractors in Asia and Africa.

In this context, what then are the strategic issues for tomorrow's global construction organisations? In order for construction organisations to be successful, it is of paramount importance that some key strategic laws are utilised – otherwise doing business in newly emerging economies will become very expensive because of the higher business risks. Political stability and the level of socioeconomic development will determine the structure of investment that clients can make in newly emerging economies.

2.5 Corporate strategic issues

Technology, communication, economic challenges and market advances have changed the global outlook of time, distance and spatial boundaries.[9] Rapid technological, social and economical improvements in the last decade have provided many threats, as well as opportunities for multinational construction organisations. During the period 1992 to 2012 construction organisations could classify themselves as being local, regional, national or international in scope, and expect that the above definitions were clearly distinct.[10] Nonetheless, with the introduction of technological modernisations, these boundaries have been distorted to the point where any construction organisation (irrespective of their supposed scope) can hypothetically play a part in a design or construction project of any scale in any location. These changes experienced by construction organisations remain unprecedented. Hence, the need to plan more tactically and better anticipate future possibilities, opportunities and threats is more essential than ever before. Improving their capacity to help forecast possible alternative futures, and plan for them, is important if construction organisations are to prepare for, and become accustomed to, the emerging developments and prospects that lie ahead.[10] The concepts of conventional competitors, organisation loyalty and employee progression have changed at a pace that has not formerly been encountered since the Industrial Revolution. Of particular interest is the introduction of several issues that form the need for a strategic management viewpoint by global construction organisations. These issues relate to business model implementation, appreciating cultural diversity, legal and tax settings, national technical features and material rights, corporate responsibility, and new emerging markets.

2.5.1 Implementing an appropriate business model

Construction organisations are accustomed to legal and regulatory lucidity. Currently, however, there is no global framework to match the global nature of construction business operations. As has been witnessed with the response to climate change, international agreements can be very hard to negotiate and implement. Some construction contractors are operating at a level of integration unmatched by the legal systems of the various countries that they operate in. This means that a contractor based in one country must comply with the rules of its 'home' country but it must also and concurrently conform to the law in any other country in which it operates. One way in which home construction organisations can address this issue is to set up a joint venture with one or more local contractors from the target country or zone. The appropriate local contractor(s) will be networked within their sphere of activity and have the necessary legal awareness to operate effectively within the local or regional market. Joint venturing can be a creative process, and one in which both contractors harmonise their individual strengths so as to create a more effective organisation. One contractor, for example, can focus on the stipulation of personnel, while the other concentrates on implementation of construction technology and know-how. One of the preconditions for a medium-term affiliation in a global joint venturing is a basis of trust that is able to endure tension. Both contractors must value cultural differences and should certify that their partners are capable and have sufficient authority to make assessments and resolutions. Irrespective of whether organisations access new markets via a joint venture or alone, two basic prerequisites must be met: sufficient network in the target nation, and complete loyalty to the home construction organisation.[11] This can be achieved by appointing two executive senior managers with equal authority – a representative from the home construction organisation and a local manager from the target country.

2.5.2 Appreciation of cultural diversity

The biggest challenge facing multinational construction organisations is managing cultural diversity stemming from religion, time, language, geography, climate and politics. It is worth mentioning that a number of multinational construction organisations preparing for entry into new markets do not pay sufficient attention to cultural diversity. Meticulous preparation for any cross-border move is essential.[11] This is particularly important with regard to assumptions concerning communication, especially the relative sophistication of information communications technologies (ICTs) allowing people to be geographically disparate but working together, and the mix of people who in working together are being exposed to a wide range of cultural norms.

There has been a change, in response to ICT developments, in the way that many construction engineering projects are delivered. This is especially noticeable in Western Europe where local levels of investment have dropped and many project management contractors are now working on projects in other parts of the world.[11] The increased application and development of rapid worldwide electronic communications has led to a number of construction engineering projects being designed and developed in dispersed locations many thousands of miles away from the actual construction sites. In addition, there has been a tendency by clients to develop and undertake such projects in partnership with other companies as joint ventures, often collaborating with local companies based in the territory where the assets will be built. This has resulted in projects with team members from different cultures and backgrounds working together. A number of authors, including Weatherly, agree that project success is difficult enough to accomplish where the project team is located close to the construction project environment,[12] and the situation is made considerably complex for multicultural project teams that are widely separated geographically and have dissimilar organisational and regional cultures. The geographical division of multicultural project teams poses its own communication challenges.

The management and development of construction project teams within a global context unavoidably leads to a consideration of diversity and related challenges.

Within overseas construction projects, organisations will have to help their project managers to appreciate the international context and develop the ability to understand everyday issues from different cultural perspectives. Bartlett and Goshal identified the main challenge facing organisations intending to work overseas as the introduction of practices, which balance global competitiveness,[13] multinational flexibility and the building of global learning capability. Multinational construction organisations must develop the cultural sensitivity and ability to manage and build future capabilities if they are to achieve this balance. Current thinking on diversity requires organisations to value explicitly multicultural teamwork, to adapt to it and use it to generate improvements in project work performance and team effectiveness. At this juncture, it is worth highlighting that the colonial approach to construction project delivery is now an artefact of history. Multinational construction organisations need to accommodate individuals who can bridge cultural diversity and learn how to work in a global environment. As the global environment is becoming more complex and change occurs at an increasingly rapid rate, contractors must improve their ability to address cultural strategic diversity issues.

2.5.3 Integration of legal and tax settings

In order for construction organisations to survive global financial uncertainty, they will have to look into the whole range of legal, tax and economic settings. Global construction project delivery is becoming flawless, evolving towards a single global marketplace that surpasses national borders. In spite of this, tax establishments across the global market still administer income on nationwide, jurisdictional region bases. With continuous regulatory, legislative and judicial adjustments, construction organisations operating across borders will be challenged to follow and appreciate ever-changing global developments in the market. A notable feature to this is that strategic adjustments, globalisation, economic realisms and operational change will require construction organisations to follow and identify with strategic internal proposals. The integration of the above streams will require construction organisations to be flexible and well-versed in global economic developments. They will need to understand the tax impact on global business operations. Simplifying organisational structures will bring significant benefits from a cost viewpoint, considerably reducing statutory audits, tax audits and administration costs. As well as the fact that direct cost savings may be significant, they will also save substantial indirect costs by becoming more lithe firms that benefit from a universal corporate culture and approach across assorted regions.

2.5.4 Familiarity of a country's technical features and supplying material rights

According to Kelly, the national and regional specialities in technical construction work should not be undervalued.[14] From a global perspective, there are major differences in national and regional standards – from formats to materials, and planning to construction project delivery. Global construction organisations (GCOs) will have to be aware of regional and national practices when supplying materials. In Germany, for example, planning work has to be supplied in packages so as to be compatible to the fee structure for architects and engineers. The growth of global economic volatility, trade and the configuration of supranational economic and political trading blocks have all obviously broadened the presence on the market of construction materials of different regional origins. Consequently, GCOs will have to pay attention to factors such as tax policies that negatively affect their global project delivery in international and domestic markets. Contractors with good reputations will have a competitive advantage in positioning their construction materials in global markets, due to their financial strength and specialist expertise.

2.5.5 Corporate responsibility

GCOs will need to take a view as to the legal limits within their working environment and to be aware when they are approaching the margin line. They will have to understand their role and impact on society and, as they enter new markets, ensure that they do not disrupt local economies. Recent global economic shakiness and international corporate activities have affected local labour markets and currency values. GCOs that want to be sustainable will have to recognise the reality of the local impact of their activity and work hard to apply skills and technology so as to make a positive impact on the society. As well-established global construction organisations move into developing countries, there is a warranted role in helping local construction organisations through training, education and encouragement of establishments to form a civil society. GCOs have a direct long-term role in contributing inventiveness to the development of strong societies and supporting national authorities, and doing so within clearly defined legal boundaries. They must take great care not to transgress the boundary of legality, but also because the activity goes beyond the traditional strategic planning. The new global scene means that construction organisations will have a direct long-term role in the advancement of the environments in which they work. For instance, investment in education in developing countries will help to protect and augment their long-term position whilst also having a permanent positive impact on societies.

2.5.6 New emerging markets

Traditionally, the construction industry has been divided into industrial, commercial, heavy and residential sectors. This classification has allowed construction organisations to establish themselves into narrowly defined competitive markets. In so doing, each sector had its leaders, challengers and followers, thus establishing a field of intense but ordered competition to bid for projects.[9] Competition from outside organisations has been a secondary concern, but it is worth highlighting that this stability is slowly changing in Western markets. With pressure increasing on turnovers and market boundaries, construction organisations are being forced to seek alternatives to traditional markets. Boundaries accepted as the limits of market focus can no longer hold back construction organisations from delving into substitute income opportunities. The entire construction project represents opportunities for qualified professionals. The knowledge of how to classify, find and pursue these opportunities must be integrated as part of an expanded construction organisation strategy.[9]

The six strategic issues outlined above are issues that tomorrow's GCOs will have to deal with. They will have to make use of advisers who can impart the essential know-how in both the home country and investment country. In addition, they will have to learn how to work in alliances with state entities whose goals do not start with turnover and the interests of investors. Construction organisations will have to understand how to maximise the benefits of such alliances. The unique set of challenges facing the industry today will require most organisations to rethink their business operations and refine their business models.

2.6 Tools and techniques for the management of strategic issues

According to Price, there are a number of activities that can lead to the process of developing a strategy.[4] In adopting the most suitable activities companies should consider what the outputs of the strategy process are intended to be and the conditions in which projects will have to be executed. In the new global economy, the strategic management process has become more multifaceted and costly. A thorough understanding of strategic tools and techniques within the global construction sector is a prerequisite to improving the techniques. Several of the most extensively used tools are: SWOT, Porter's five forces analysis, PESTLE and value chain analysis.

2.6.1 SWOT analysis

A SWOT analysis is a project tool that can provide prompts to clients, project managers, stakeholders and teams involved in the analysis of what is effective or less effective in the project delivery process. An acronym of strengths, weaknesses, opportunities and threats, SWOT analysis is one of the most widely used tools in appraising a project based on each of these factors. In order for this technique to be utilised effectively, construction organisations will have to ensure project objectives are well defined, and that both internal and external factors have been identified. The S and W can relate to the internal factors of the project and organisation, whilst the O and T can relate to the external factors of the project and organisation. There are three ways in which SWOT can be utilised in global construction projects:

- **SWOT identification:** this involves the classification of strengths, weaknesses, opportunities and threats.
- **SWOT comparison:** a logical decision support tool that uses the variables identified during the SWOT process. Based on the outcome, project managers can then make a simple evaluation of the positive features against the negatives ones.

- **SWOT analysis:** once the above two processes have been carried out, SWOT analysis can then be performed to ensure that the client, project manager and stakeholders are in agreement about what is to be done.

Before commencing any analysis process, project managers need to have clear and specific, measurable, achievable, realistic and time-scaled (SMART) objectives. Project managers need to ensure that the analysis is not carried out in isolation. When the project objective is shared then a brainstorming session can be run by the project manager, who needs to encourage group participation and ensure that all team members are well aware of or informed about the project issue. There are two basic methods that the project manager could use: pull and push. The push method entails forming an opinion and then arguing for it; the pull technique depends on seeking the opinions of others. Project managers applying the push technique need to form an opinion first and then utilise skilful questioning to encourage project teams to form the same view. Thus it can take longer to achieve a decision using the pull technique. A skilful project manager needs to utilise both methods, depending on the team or project situation. SWOT analysis can be used to audit the overall strategic position of a global construction project and its environment – and it is worth highlighting that SWOT analysis can be used in conjunction with other tools and analysis (e.g. PESTEL analysis, Porter's five forces and value analysis).

2.6.2 Porter's five forces

In this technique, there are five forces that ascertain the competitive intensity and attractiveness of a market: supplier power, buyer power, competitive rivalry, threat of substitution and threat of new entry. Porter's five forces can help GCOs to identify where power lies in a business situation. This will be useful in evaluating the strength of an organisation's current competitive position, and the strength of a position into which it may look to move. Understanding the dynamics of competitors within a global environment is critical for several reasons. First, it can help clients to evaluate the potential opportunities for their projects. Second, clients can also use Porter's five forces to differentiate themselves from their competitors. Clients can also use Porter's five to identify the root causes of the level of competitiveness of a given venture, and thus the drivers of its return on investment. For instance, any UK construction client intending to invest in new emerging markets would need to have a tremendous awareness of market growth, product characteristics, competitors, exit barriers, strategic alliances, diversification and legal requirements. The changing global environment provides construction clients with both opportunities and threats. There is intense competition on all fronts in the industry.

2.6.3 PESTLE analysis

PESTLE analysis is frequently used with SWOT. This tool is mainly used to identify external political, economic, social, technological, environmental and legislative factors that might impact on the project. The best way to execute a PESTLE analysis in projects is through the use of thought association in relation to macro-environment factors. Listing one variable under a particular factor, such as economic, may generate others under the linked factors of political, legislative and environmental. A key feature of this tool is that stakeholders can have a brainstorming session when formulating project strategies. GCOs can use this tool to understand the political, economic, socio-cultural and technological environment. In addition, this tool can be used to assess market growth or decline, and as such the position, potential and direction for a new business venture.

2.6.4 Value analysis

Before making a global strategic decision, construction organisations will have to understand how project activities within a project life cycle can create value for clients. One way to affect this will be to carry out a value chain analysis. In the context of the construction sector, this is based on the principle that contractors exist to create value for their clients. The three main steps for conducting a value chain analysis are: (1) splitting the project management office operations into primary and support activities; (2) allocating cost to each project task; (3) identifying project tasks that are significant to a client's approval and project success. A notable feature to value analysis is that it will assist contractors in effectively utilising an operational project model that delivers organisational and project value. A logical strategy will help GCOs respond to both internal and external pressures. The creation of an effective strategy usually requires three distinct phases: strategic analysis, strategic choice and strategy implementation. GCOs can utilise the above tools to assess their competitive environment and policy direction.

2.7 Project governance

This can be defined as an internal model introduced to help safeguard the interests of clients, and mitigate risks throughout the project life cycle.[14] As Renz noted, governance is about 'checks and balances', about 'direction and control'.[15] From an organisational standpoint, it is usually correlated with top leadership of an organisation. For whichever governance structure that is put in practice, careful consideration should be given as to how each of the following issues will be addressed:

- active client involvement;
- technology-enabled project management and controls;
- organisational structure;
- SMART goals;
- accountability and transparency.

Tomorrow's GCOs will have to utilise project governance models so as to manage projects that come with institutional lending requirements; to administer risks linked with taxpayers; to take account of stakeholder interests; to cost and schedule the project and anticipate the expected return on investment. A well-governed project will recognise and follow a few guiding principles that exemplify leading practices. Appreciation of risks at an early phase is significant.[14] Having a framework with robust controls around costs, scheduling and key performance variables will help to organise and manage the project more successfully and economically. In order for the project to be managed well, GCOs will need to have competent project staff in place. The challenging global economic climate (possibly continuing into 2018), acting in combination with continuous tax adjustments, has made the hiring and retaining of capable project staff increasingly difficult. It is in fact a major concern to multinational construction organisations, a number of which are selectively hiring today and, in good shape financially, are poised to regain significant market share when the global economy rebounds. However, their approach towards hiring construction project staff has changed dramatically and the sense of urgency to filling project positions has decreased greatly.

One of the most significant discussions in project governance literature is that well-governed projects will feature a formal monitoring procedure that starts at the management project level and goes all the way up to operational and strategic level. Effective operation of a project governance framework relies on stakeholders, construction project team members and representatives of various contractors involved in project delivery all having a clear understanding of what is required of them.[16] The following points highlight key areas that GCOs will need to consider before introducing a project governance framework:[14,15,16]

- identification of stakeholders and their roles;
- a statement of requirements;
- responsibility and authority;
- a formal reporting structure and feedback system;
- a project management structure and procedures that are fit for purpose;
- the support given to project managers;
- how independent project reviews will be carried out;
- ways in which post-project evaluations will be instigated.

Getting the right level of global project governance, and understanding how it will be implemented, is inexorably time-consuming. Difficulties arise in the project life cycle when priorities are not defined at an early phase of the project. Clients, programme managers and senior project managers have to ensure that the required project governance principles highlighted above are carried out at the right time. In the wake of any global turbulence in the markets, clients will seek greater assurance that arrangements are in place to keep projects on track with the risks being managed. Multinational construction organisations will need to develop models for project governance that are different to those involving traditional organisation structures, in that a model defines accountabilities and responsibilities for strategic decision making across the project. This will be particularly useful to project management processes such as change control and strategic (project) decision making. When employed well, it can have a considerably positive effect on the operation, quality and speed of decision making on important issues affecting projects.

To ensure project governance success, GCOs will have to focus on the entire life cycle of a project, start the process early and hold high standards surrounding cost, schedule and execution. What is emerging from the industry is that corporate boards, global joint venture partners, shareholders and stakeholders are expecting their projects to be administered with a governance framework

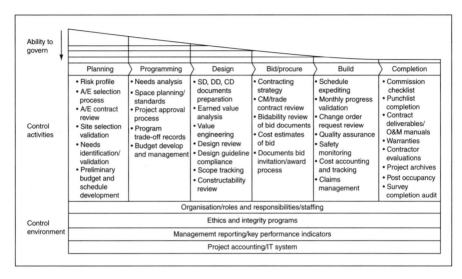

Figure 2.1 Capital program governance model
(*Source:* Adapted from Kelly 2007[14])

that facilitates transparency and dependable reporting. Clients must insist on the same. Processes, controls and procedures must be practically introduced by identifying, monitoring, averting and reporting any activities that risk the discretion, veracity and availability of project data. Essentially, systems must be able to classify, manage and mitigate risks and appraise project performance, including the appropriate utilisation of resources.[14,15,16]

One possible model (see Figure 2.1) addressing the previous issues also exemplifies how important it is that the control environment and the governance framework be instituted in the early phases of the project. The top of the model displays a curve that decreases to the right, signifying that the ability to monitor costs, or to introduce a control environment, decreases swiftly through the project life cycle. Several processes cut across the life cycle of any construction project (as observed at the bottom of the model). As noted in a previous section, clients need to introduce processes and controls at the early phase of the project. All project team members need to identify with the processes, control, roles, responsibilities, ethics, integrity requirements, management reporting and key performance indicators. The aim is for everyone, from the client to the construction project manager, to have an overall health audit of the project at several key project phases. This also permits instigation of an early warning scheme, so that when a crisis arises, senior management's attention is timely allowing quick decisions to be made. Finally, the project accounting and information technology coordination allows project staff to monitor project costs precisely back to an accounting system and ensure that financial statements are accurate. The model also makes explicit six key segments of a project that can be utilised to classify governance issues and address the risks.

1. **Planning:** at this phase it is essential to have clear accountability, which should include the thoughts of senior managers involved in project planning during the original scoping of the project.

2. **Programming and design:** one of the main challenges in this phase is managing the outsourcing of design and engineering services. The client and project manager have to ensure that there is tight control and management of outside engineers and architects so that their views, deliverables and schedules are well managed. At this phase the project manager has to work closely with the client so as to steer clear of any dramatic modifications to the plan's progress during the construction phase. One strategy is to freeze the design once key stakeholders have signed off the scope of work. A critical success factor here is to ensure that all key stakeholders review the design of the project.
3. **Procurement:** during the procurement phase, organisations need to certify that there is an open, competitive and fair process, and that there is a sensible amount of bidding around the project.
4. **Construction:** at the construction phase, there need to be timely project updates on the planned schedule and progress, and key performance indicators so that key stakeholders involved with the project can monitor these.
5. **Completion:** a project is considered complete when the final product meets all of the initial stated requirements and is accepted by the client. It is therefore important for the client to carry out a project review so as to ensure costs are documented and in compliance with the contract.

From this it can be seen that the application of project governance is crucial to the success of any construction project. In particular, the right organisational culture is essential, because it can help construction organisations to cultivate a sense of accountability and an ownership approach.[9]

2.8 Current approaches to project management

Project management and related research continues to grow and develop. In response to projects being developed in new sectors, countries and application areas, the demands on project management continue to change. These changes have altered the way project management is viewed and practiced and this is reflected by the literature. In recent years, the discipline of project management has changed dramatically in its application, to accommodate emerging management processes and philosophies related to organisational development. Numerous studies have already examined changes to the field (including those by Lewis *et al.*, Soderlund and Winter *et al.*) using many different approaches, resulting in diverse and at times contradictory findings.[17,18,19] Turner articulated the fundamentals of project management theory and addressed a hugely important challenge for the research community,[20] which is that much of project management research lacks explicit statements of its theoretical underpinning.[20] Since theory helps to direct researchers into productive lines of enquiry, the lack of explicit theory has hampered the pursuit of research and the development of a cumulative foundation in the advancement of knowledge.

One of the main weaknesses of normative theory in project management is that it treats failures according to the theory aberrations. It offers no reason as to why deviation has occurred, nor how to correct it other than to say, '*Do it right next time*'. One could suggest, therefore, that knowing what normative theory prescribes is not enough to secure the right behaviour. In construction/project management research, a theory is needed that will help to understand the conditions and drivers that lead functional behaviour so that it can be influential in addressing the root causes of these failures. As Turner stated, theories that are descriptive in this way are often referred to as positive – as opposed to normative.[20] One of the key questions that have occupied the research community is whether or not project management is a profession. The issue of identity has been a controversial and much disputed subject within the field of project management. The answer seems to be at best '*not yet*'. Before it can become a profession it needs to be recognised as an academic discipline, and that has not yet been achieved.[20] There are a number of reasons for this. One reason is that there is no second theory of project management. It is empirical knowledge rather than theoretical knowledge. Another reason is that it is not clear from the literature where it sits in the academic community. One could suggest that it could sit in either the management faculty, the engineering faculty, the faculty of the built environment or the computer sciences faculty. From publications in the *International Journal of Project Management* (*IJPM*), *Journal of Construction Engineering and Management* (*CEM*), *Journal of Construction Engineering and Project Management* (*JCEPM*), *Construction Management Economics* (*CME*), *Engineering Construction and Architectural and Management* (*ECAM*), and research carried out to date, it could be suggested that it sits most appropriately in the faculty of the built environment.

It is arguable whether project management is used consistently and generically. Results of work by Crawford found variations in project management knowledge and practices between sectors, nations and application areas.[21] Organisational developments in recent years have seen the emergence of project management across a number of industries and sectors. According to Whittington *et al.*, there has been an increase in new developments and new initiatives being pursued through projects and programmes.[22] KPMG confirmed the growing adoption of project management standards and practices across large numbers of organisations.[23] Despite the number of developments in practice, several authors highlighted that the current conceptual base of project management continues to attract disparagement for its lack of relevance to practice, and as a result, to improved performances of projects across a number of industrial sectors (Hobday;[24] Kloppenborg and Opfer;[25] Koskela and Howell;[26] Meredith;[27] Morris[28]).

There is no single theoretical base from which to explain and guide the management of projects. There are instead a number of theoretical approaches, many of which overlap. According to Winter *et al.*, these operate both for individual aspects of project management (e.g. control, risk, leadership) and for the discipline as a whole.[19] From the strands discussed, the most dominant strand of project management thinking is the rational, universal, deterministic model, which emphasises the planning and control dimensions of project management.[27,28] It is the first strand arrived at by critical path analysis and scheduling. It has, however, been criticised for failing to deal adequately with the emergent nature of front-end work.[29,30] The second strand of thinking was more theoretically based and emerged in the late 1960s and 1970s from the literature on organisational design.[31] The third strand appeared in the 1980s and was concerned, for example, with major projects.[32] These studies emphasised a wider view of projects, recognising the importance of the front-end issues and managing exogenous factors, as well as the more traditional execution-focused endogenous ones. As claimed by Williams, issues facing both practitioners and researchers now seem to be well beyond the hard systems perspective so often related with project management.[33]

In 2003, the UK's Engineering and Physical Sciences Research Council (EPSRC) funded a research network (Rethinking Project Management) to define a research agenda aimed at inspiring and broadening the field of project management beyond its current conceptual foundations. The principal finding of the network was the need for new thinking in the areas of project complexity, social process, value creation, project conceptualisation and practitioner development. The above-mentioned five research areas are not new to academics and experienced practitioners but this is not what the network aimed to achieve.[19] As illustrated in Table 2.1, the five directions shown are the key principal areas in which new ideas and approaches are needed.

Table 2.1 Directions for future research in project management

Theory ABOUT Practice   **Direction 1**

The Lifecycle Model of Projects and PM ⟶	Theories of the Complexity of Projects
From: the simple lifecycle-based models of projects, as the dominant model of projects and project management.	**Towards:** the development of new models and theories, which recognise and illuminate the *complexity* of projects and project management, at all levels.
And **from:** the (often examined) assumption that the lifecycle model is (assumed to be) the actual 'terrain' (i.e. the actual reality 'out there' in the world).	And **towards:** new models and theories which are explicitly presented as only *partial* theories of the complex 'terrain'.

Implication

The need for *multiple images* to inform and guide action at all levels in the management of projects, rather than just the classical lifecycle model of project management, as the main guide to action (with all its codified knowledge and techniques). Note: theories ABOUT practice can also be used as theories FOR practice.

Theory FOR Practice   **Direction 2**

Projects as Instrumental Processes ⟶	Projects as Social Processes
From: the instrumental lifecycle image of projects as a linear sequence of tasks to be performed on an objective 'out there', using codified knowledge, procedures and techniques, and based on an image of projects as temporary apolitical production processes.	**Towards:** concepts and images which focus on social interaction among people, illuminating: the flux of events and human action, and the framing of projects (and the profession) within an array of social agenda, practices, stakeholder relations, politics and power.

Table 2.1   (Continued)

Direction 3

Product Creation as the Prime Focus ⟶	Value Creation as the Prime Focus
From: concepts and methodologies, which focus on: *product creation*- the temporary production, development, or improvement of a physical product, system, or facility etc – and monitored and controlled against specification (quality) cost and time.	**Towards:** concepts and frameworks, which focus on: *value creation* as the prime focus of projects, programmes, and portfolios. Note, however, 'value' and 'benefit' as having multiple meanings linked to different purposes: organisational and individual.

Direction 4

Narrow Conceptualisation of Projects ⟶	Broader Conceptualisation of Projects
From: concepts and methodologies which are based on the narrow conceptualisation that projects start from a well-defined objective 'given' at the start, and are named and framed around single disciplines e.g. IT projects, construction projects, HR projects etc.	**Towards:** concepts and approaches which facilitate: broader and ongoing conceptualisation of projects as being multidisciplinary, having multiple purposes, not always pre-defined, but permeable, contestable and open to renegotiation throughout.

Theory IN Practice   ### Direction 5

Practitioners as Trained Technicians ⟶	Practitioners as Reflective Practitioners
From: training and development which produces: practitioners who can follow detailed procedures and techniques, prescribed by project management methods and tools, which embody some or all of the ideas and assumptions of the 'from' parts of I to 4.	**Towards:** learning and development which facilitates: the development of reflective practitioners who can learn, operate and adapt effectively in complex project environments, through experience, intuition and the pragmatic application of theory in practice.

(*Source*: Adapted from Winter et al. 2006[19])

To illustrate fully the different types of concepts and approaches that are required, the five directions are presented under three particular headings: theory about practice, theory for practice, and theory in practice. From the three categories, the network found that there were significant differences in theory and knowledge constructed on the basis of studying projects and project management processes. In essence, the network suggested that there needs to be much more emphasis on research focusing on concepts and theories closely reverberating project complexity realities so as to provide practitioners with practical concepts and approaches more in alignment with contemporary thinking. The final heading in Table 2.1, 'theory in practice', covers how practitioners learn their craft and practise their craft using relevant theory from the already published literature. As shown in Table 2.1, a 'from' position is highlighted for each direction, the dominant position (as we perceive it) and a 'towards' position representing the new direction of thought, vis-à-vis, a new direction for future research.[19]

The clearest pattern observed from all the practitioner inputs to the network is the total complexity of projects and programmes across all industries at all levels. It includes all manner of aspects, such as: including an array of stakeholders; the different agendas, theories, practices and discourses operating at different levels within different groups in the ever-changing fluctuation of events. Therefore, there is a need to introduce new theories about actual project management practice, which would recognise and clarify the complexity of projects and project management. A second pattern that emerges from the network is the need to introduce new theories for practice, which would include new images, concepts, frameworks and approaches to help practitioners deal with complexity in the midst of practice. Future research in the field needs also to focus on the area of theory 'in' practice, that is, the actual application of theory in the midst of action. Two important implications for research arise: Directions 1–5 (see Table 2.1) mirror the concerns of practitioners in the areas of project complexity, social process, value creation, project conceptualisation and practitioner development. In addition, Directions 1–5 highlight the need for an interdisciplinary application to conceptualise and theorise project management practice, and a cautious consideration of the methodological issues by academics in order to facilitate the creation of knowledge professed as useful by practising senior managers in organisations and projects.

The advancement of project management can be exemplified through a series of mega projects, which are large-scale ventures that require enormous financial and physical resources. They encompass power plants, highways, dams, bridges, pharmaceutical plants, airports and tunnels. Cicmil *et al.* noted that mega construction projects such as the Hoover Dam in the 1930s near Las Vegas in the US, and the Manhattan Project were among the earliest to be managed through contemporary principles of project management.[34] Since then, mankind has continued to deliver mega projects such as Hong Kong International Airport, Abraj Al Bait Towers in Saudi Arabia, Aswan Dam in Egypt and Beijing National Stadium in China. In recent decades, the increase in interest in projects and project management has its roots in notions of contemporary civilisation typified by government systems, standardisation and large-scale operations.[34] According to Soderlund, the word 'project' has been used to describe contemporary project organisations.[18] Evidence shows that the term 'project' is used normatively: to support what project-based organisations must develop into if they are to be competitive in today's market.[18] It has been suggested that three key features of contemporary firms and civilisation are characteristically cited in the rise of projects:

1. increase of complexity in products;
2. rapidly changing markets;
3. the equivalent of knowledge in production processes.[18]

Ekstedt *et al.*, showed that projects have been regarded as suitable ways to control ventures in turbulent environments.[35] Projects have been used to stimulate a learning environment and augment creativity so as to deliver complex products.[24] Organisations are ever more realising that the ability to manage, and to respond quickly, economically and strategically to global market dynamics is heavily dependent on project management models. It has been highlighted that a number of models and techniques used in the project management domain do not elucidate or amplify our knowledge about the behaviour of project-based organisations.[36] In spite of this, the growing demand to invest in emerging markets is also increasing demands to manage and prioritise a growing list of projects. The way in which projects are delivered has been forced to change because of the economic, social, political, technological and legislative environments. There is a suggestion that a primary reason for these changes has been the never-ending drive for more efficiency and greater competition, nationally and internationally. In this respect, there is a need to streamline the global project delivery process. Successful global project delivery needs the ability of construction contractors to respond agilely to sudden changes in client requirements. The rapid change and greater demands for accountability, and high-quality projects supported by rigorously applied professional behaviours, will be the standard that operates for the future. The inference is that global-based construction organisations will have to alter their culture from a reactive one to a proactive one.

Greater demands from clients for a better-quality construction project delivered on lower unit cost and reduced standard deviation across the portfolio on out-turn costs are now part of the sector ethos. Current thinking on global project delivery requires contractors to integrate cost, time, quality, risk, safety, economic sustainability, social sustainability, environmental sustainability and value at the front end of projects. The inclination towards '*projectification*' has significantly changed the application of, and also, most likely, the value added by, project management.[37] The focal point on the value added by the project management domain is at the centre of comprehending the increased application of project management frameworks on a universal level. Results of work by Whittington *et al.* showed that project-based organising is at the top of the strategic programme of a number of organisations in Europe.[22] As the global market is becoming more complex and changes are taking place faster, contractors will have to improve their ability to manage project management models. Success in managing global projects will depend on: managing the project in a turbulent environment, from a diverse multidisciplinary team, managing functional plans consecutively and interdependently, maintaining a high level of project communication, and monitoring concurrently the utilisation of resources. For construction contractors this means managing a portfolio of complex projects. The issue of dealing with complexity is the prime focus in today's global-based construction organisations.

Project complexity can be found in two dimensions: outside the project and inside the project. Li and Guo, present quite a different perspective on project complexity by suggesting that complexity in mega construction projects can be generated from three different levels: technical, social and managerial.[38] Social complexities of mega construction projects are determined by unintended impact of mega projects on the environment and social systems within their location of implementation; managerial complexities originate from business and governance facets of projects including financial arrangement, scheduling, resource classification and decision management; technical complexities transpire from the design and technologies utilised in the construction process.[38] More recently, Morris highlighted that most project managers grapple with technical substantive issues, within the context of commercial and schedule, budget, requirement, specification, and health and safety constraints.[36]

One of the key reasons why project complexity varies is that clients have different goals, interests and expectations of the construction project, which can originate from different levels of the project. A second reason is that the project process is often focused on the key aspects of project phases. Moving from one phase to another probably means that the focus is on the project result rather than its process. A third reason is that in different project phases you find different driving factors and deliverables. Successful project management delivery will require analysis of how project complexity affects the social, economical, technical and managerial project constraints. Clients and senior construction managers will need this knowledge in order to manage the complexity of global construction projects. It is crucial that throughout the project life cycle clients and senior construction managers develop plans and standardise the project with the purpose of managing project complexity. It is not surprising that complex construction projects require an outstanding level of management and that the use of conventional systems developed for ordinary projects have been found to be unsuitable for complex ones.[38]

The continuous need in global construction engineering projects for speed, cost and quality control, safety in the working environment and avoidance of disputes, together with technological advances, environmental issues and fragmentation of the global construction industry, have resulted in a spiralling and hasty increase in the complexity of projects. It has today reached a level where senior construction managers must very seriously consider its influence on the success of global construction engineering projects. There is a need to introduce new theories about *actual* project management practice, which would recognise and clarify the complexity of global projects and project delivery in sectors. To date there has been little agreement on what constitutes project management theory, nonetheless there is *A Guide to the Project Management Body of Knowledge*, which is an epistemology of the field, (the PMBOK Guide). It is worth noting that there are a number of theoretical project management models that might be suitable for the delivery of projects. In addition, the management of projects is now a vibrant area of academic interest, attracting a number of new researchers in management education. While there is some increase in research activity, the research contribution remains detached from the problems that practitioners face in industry. It has been suggested that much of project management research is being done by social scientists driven by an interest in projects that are focused on organisational phenomena rather than examining the needs of individuals who are trying to manage these organisations.[36] The research community will need to focus on the role of senior managers in their value creation activities and delivery of projects. Recent research showed that a number of studies have acknowledged the increased role of project managers and project directors in modern organisations.[18] In addition, the research community ought to examine the following question: what are the consequences to global project models of complexity? Morris suggested that project management should be devised so that its significance is clear.[36] Recent evidence suggests that the prospect of project management rests with the research community being able to comprehend and interact successfully with eight allied disciplines:

- Operations research, decision sciences, operation management and supply chain.
- Quality management and process improvement.
- Performance management, earned value management, project finance and accounting.
- Strategy, integration, portfolio management and value of project management.
- Engineering construction, contracts and legal aspects.
- Technology, innovation, research and development.
- Information technology and information systems.
- Organisational behaviour and human resources management.[39]

One of the most significant current discussions in project management is that it has the potential of integrating different fields to focus on a marvel of study, i.e. global projects. It is becoming increasingly difficult to ignore the development of project management (both the practical and theoretical parts of it). The main assertion addressed in this section is that there are openings for more research in the global construction industry. Still, the industry lacks in-depth analysis of front-end uncertainty, value creation, complexity and capital effectiveness. While there is some increase in project management research, the field lacks comprehensive case studies, studies of processes, and concurrent research that would be valuable for understanding elementary issues of projects and project organisations.[36,37,38] It would be beneficial to carry out research studies that would build theories for understanding fundamental issues of global construction projects.

2.9 Infrastructures for global project management

Over the past decade, construction clients have witnessed a momentous alteration of their organisations. Globalisation, economic interdependence between countries and financial adversity has forced governments and organisations to act.[40] Considering the speed of change since a decade ago, what will the typical construction organisations of the near future look like? And what can the sector clients do to prepare the labour force for change? Over the period to 2022, changes in the way that construction organisations function will not be revolutionary or disruptive; they will be an annexe of the evolution already discernible in many organisations today. The emergence of new markets, the global financial debacle commencing around 2010, and demographic pressures are among the forces making organisations invest in new markets.[40,41,11] The inclination is especially strong among large construction firms that have the resources to grow. Changes to organisational models and labour force profile will spell new infrastructures for global project management. Among them are:[40,11]

2.9.1 More diversity

Global construction teams will come from a range of backgrounds. Those with national and regional knowledge of an emerging market, a global viewpoint and an instinctive sense of the corporate culture will particularly be appreciated. Global construction organisations will be sending their workforce abroad more frequently, often for short project-based assignments. The theory of diversity and soft project skills is examined in more detail in Chapter 7.

2.9.2 Appreciation of soft project skills

Construction organisations will have to focus on enhancing communication skills, cultural awareness and corporate values. The labour force will be larger and spread over more continents, making cross-cultural communication more important and challenging. Senior managers will be required to have the ability to work across cultures and utilise best business practices that can be transmitted to other nations. Cross-border integration is expected to become more concentrated, and a global viewpoint will just be as important as national knowledge.

2.9.3 Automated project roles

More project roles will be automated and outsourced. This may allow construction organisations to influence global resources more economically, but it will also increase complexity at the strategic, operations and project levels of organisations. Today's work environment has evolved from being skill-based to one that is knowledge-based. According to Chinowsky and Meredith, the construction industry has witnessed the introduction of knowledge-based tasks as a focal point of organisation operations.[9] The focus is now on technology, automation, economics and market advancement. The developments in human resources and emerging markets demand that construction firms respond to changing conditions in the employee and client marketplace. Information communications technologies (ICT) have had an impact on all features of the construction sector. Current information technologies have provided construction professionals with access to information repositories and evolving communication pathways.[9] This has had a profound impact on the construction sector in several segments including data management, site management and intra-contractors' communications. Recent developments in information technology have heightened the need for building information modelling (BIM), and many construction professionals have pigeonholed BIM as the answer to poor interoperability and data management. For instance, BIM is now widespread in the UK's engineering, architecture and construction segments. Over the past century there has been an increase in hyperarchical information access and transfer. Hyperarchical information allows any construction professionals to access data repositories anywhere in the world. These rapid advancements are having a positive effect in the construction sector as organisations increasingly recognise the need to integrate technologies as a component to their long-term strategies. This issue is further explored in Chapter 12. The notion of automated project roles is also explored in detail in Chapter 12.

2.9.4 Interpretation of environmental sustainability

In a global environment where energy costs, energy security and climate change are in the headlines every second day, it is easy to comprehend why governments are putting together policies to persuade organisations to commission ever more energy-efficient buildings. There is some increase in environmental sustainability initiatives. For instance, the European Union's newly restructured Energy Building Performance Directive, China's Building Energy Efficiency programme and the newly instigated Better Building programme launched by the Obama administration.[1] This topic will be examined further in Chapter 11.

2.9.5 Impact of austerity measures

Risks and uncertainties increased to high levels as a result of the debt crisis facing Eurozone countries during 2011–12.[1] A key issue is that financial markets have lost confidence in the ability of EU political leaders to bestow long-term structural solutions. Our view is that due to this lack of confidence, construction organisations in Eurozone countries will face much greater financial challenges as, for example, the share of government spending on buildings and structures has decreased. As banks have lost trust in the ability of governments and construction organisations to pay, global construction organisations will have to utilise stringent austerity measures so as to provide long-term stimulus measures. Robust governance has always been critical to project success, but in the wake of recent turbulence in the markets it seems likely that lending criteria will tighten, and that funders will seek greater assurance that arrangements are in place to keep projects on track with the risks being managed. On a brighter note, however, a reasonable growth in real terms is expected in 2015–20.[1] Global construction organisations that make the changeover successfully will have to move towards a vibrant balance of power between local operations and headquarters, utilise technology that facilitates communication across continents and diversify talent on all levels of the organisation (i.e. strategic, operations and project).

2.10 Chapter summary

This chapter has shown that there has been a strong surge in activity in the commercial and new public housing sectors, with larger multinational construction organisations the main beneficiaries of the increased workload. The way in which projects are delivered has been forced to change because of the economic, social, political, technological and legislative environments. Rapid technological, social and economical improvements in the last decade have provided many threats, as well as opportunities for multinational construction organisations. The global construction environment is becoming more complex and changes are occurring at a rapid rate, so that multinational contractors must improve their ability to address strategic issues. Flourishing project management will require analysis of how project complexity affects the project constraints of social, economical, technical and managerial.

2.11 Case study: the Marmaray project in Turkey

The Marmaray project comprises an upgrading of the local commuter rail system and the construction of the Istanbul Strait Crossing in Turkey, connecting Halkalı on the European side with Gebze on the Asian side, through an uninterrupted, modern, high-capacity rail system. It consists of a tunnel under the Istanbul Strait with connections either side, and is one of the world's largest current infrastructure projects.

The total length of the project is approximately 76 kilometres, of which the immersed tunnel part under the Istanbul Strait is 1.4 kilometres; the approaching tunnels are 12.2 kilometres, and the suburban rail sections nearly 63 kilometres. Three new underground stations and one surface station will be constructed and 38 existing surface stations will be upgraded and renovated. The project will have three tracks. Two of the tracks will serve as a high-capacity commuter rail system, and the third track will be used by intercity passenger and freight trains to allow an uninterrupted railway connection between the continents of Asia and Europe. The project is also significant for Turkey's connection to the Trans-European Network.

The key features of the project consist of: an immersed tube tunnel, bored tunnels, cut-and-cover tunnels, grade structures, the stations and tracks and the procurement of modern rolling stock.

2.11.1 Background to the case study

The project was first put forward in 1860. However, contemporary techniques could not allow for the tunnel under the Strait to be on or under the seabed, and therefore the design that evolved was a 'floating' type of tunnel placed on pillars constructed on the seabed. In the years since that first initiative, several proposals have been tested and different technologies have been considered so as to allow flexibility in the design.

Figure 2.2 The Istanbul Strait – the new rail system will connect the land either side of the Strait through an underground tunnel

Figure 2.3 Istanbul, the largest city in Turkey – a historic centre at the heart of the Marmaray project

In recent times a scientific study for crossing the Istanbul Strait with a railway tunnel was made by Istanbul Rail Tunnel Consultants (IRTC) via the Turkish Ministry of Transportation in 1985–1987. In 1996 a new transport and feasibility study was made by Yüksel Project International-Louis Berger Int. Inc-De Consult via the Ministry of Transportation into upgrading the existing commuter railway and integrating the railway tunnel. In the Istanbul Master Plan prepared by Istanbul Technical University, the upgrading of the commuter railway and the crossing were shown to be one of the most important railway projects of modern times.

The technique chosen to cross the Istanbul Strait is the immersed tube tunnel. This technique was developed in the nineteenth century, with the first immersed tube tunnel being built in North America for sewerage purposes. In the current Eurozone, Holland was the first nation to utilise an immersed tube tunnel technique, and in 1942 the Maas Tunnel in Rotterdam was opened. In Asia, Japan was the sole nation to embrace this technique, and in 1944, the two-tube tunnel Aji River Tunnel was opened. It seems that such a type of project remained uncommon until a vigorous and well-proven industrial technique for project delivery was introduced in the 1950s.

2.11.2 Objectives of the project

The Marmaray project will influence not only the daily traffic pattern of Istanbul, the largest city in Turkey, but will also influence the development of the city and the region. The objectives of the project are:

- To provide a long-term solution to Istanbul's current urban transport problems
- To relieve existing operating problems on the mainline railway services

- To provide direct railway connections between the continents of Asia and Europe
- To increase capacity, reliability, accessibility, punctuality and safety on the commuter rail services
- To reduce travel time and increase comfort for a large number of commuter train passengers
- To reduce air pollution resulting from the exhaust gases and thereby improve the air quality of Istanbul
- To reduce adverse effects on historical buildings and heritage sites by potentially reducing the number of cars in the old centre of Istanbul.

2.11.3 Contracts

There are three major contracts for the project: the Engineering and Consultancy Services Contract including the BC1 Contract for the tunnelling works, the deep stations and related Electro-Mechanical works; the Commuter Rail Upgrading Contract (CR3) which covers upgrading the existing commuter rail system on both ends of the BC1 Contract including a new third track and the installation of a completely new electrical and mechanical system for whole of the project length including the BC1 part; and the Procurement of Rolling Stock Contract CR2. The Consultancy Services were started in 2001 and the Avrasya Joint Venture (Oriental Consultants of Japan, Yüksel Project of Turkey and JARTS of Japan) was subsequently awarded the tender and started work in March 2002. Contract CR3 was signed in late 2011 and this contract is currently in progress with phased completion from end 2013 until mid-2015. The Contract CR2 was awarded to the Hyundai-Rotem Company (HRC) and was signed in November 2008 with the Contractor starting work on 25 December 2008.

The Istanbul Strait crossing section of the Marmaray project should start operating on 29 October 2013. When the Marmaray project is finished, the most important part of the railway network from the East (Beijing) to the West (London) will have been completed and an uninterrupted connection will be in place.

2.11.4 Financial

A loan agreement between the Japan International Cooperation Agency (JICA) and the Republic of Turkey was signed under Official Development Assistance (ODA). The loan is to cover the costs for the engineering and consulting services, including supervision of and the construction costs for the Istanbul Strait Crossing phase of the project.

The total cost of the project is expected to be roughly 4.6 billion USD including the engineering and consulting services and other costs such as land acquisitions and archeological excavations.

2.11.5 Organisation

Figure 2.4 represents the organisation of the Marmaray project.

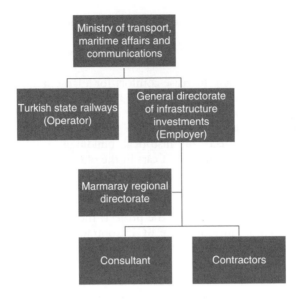

Figure 2.4 Marmaray project organisation

2.11.6 Disapproval of the Marmaray project

Criticism of the project has come from politicians, railway workers and railway aficionados. This criticism focuses on:

- The decommissioning of Haydarpasa port.
- The closing of two main railway terminals, Haydarpasa Terminal and Sirkeci Terminal. The two terminals have been historically important as a railway stations.
- Other disapproval includes replacement of major railway stations and railway bridges, such as Goztepe railway station.
- All intercity trains will be terminating at Gebze and Hakali stations. This means that passengers will have to transfer to local commuter trains which are overcrowded.

2.12 Discussion questions

1. What are some of the key strategic issues that affected this project?
2. Discuss how you would implement a project governance model in this project.
3. Which strategic tool was utilised for this project, and why?
4. Classify some of the strategic issues which are likely to affect global construction operations and propose measures that could be utilised to manage the strategic issues.

5. Out of the six strategic issues discussed in this chapter, which ones apply to this project? What are the potential advantages and disadvantages of the strategic issues you have identified?
6. How does familiarity with a country's technical features help contractors to deliver global construction projects?

2.13 References

1. Global construction 2020. *Oxford Economics*, 3 March 2011, London.
2. Wilson, D. (2003). Dreaming with BRICs: *The Path to 2050, GS Global Economics*, paper no. 99. Goldman Sachs.
3. PwC Report (2010). Engineering growth: fourth-quarter 2010 global engineering and construction mergers and acquisitions analysis.
4. Price, A.D.F. (2003). The strategy process within large construction organisations, *Engineering Construction and Architectural Management*, **10** (4), pp. 283–296.
5. DeWit, B., and R. Meyer. (1998). Strategy, process, content and context, 2nd ed. London: Thomson Learning.
6. Quinn, J.B. (1980). *Strategies for Change: Logical Incrementalism*, Homewood, IL: Irwin.
7. Johnson, G. and Scholes, K. (1997). *Exploring Corporate Strategy*. 4th ed. London, Prentice Hall.
8. Mintzberg, H. (1994). The fall and rise of strategic planning, *Harvard Business Review*, 72, 107–114.
9. Chinowsky, P.S. and Meredith, J.E. (2000). Strategic management in construction, *Journal of Construction Engineering and Management*. January/February 2000.
10. Soetanto, R., Goodier, C.I., Austin, S.A. Dainty, A.R.J. and Andrew, A.D.F. (N.D.). Enhancing strategic planning in the UK construction industry. (2007). Enhancing strategic planning in the UK construction industry. In G. Burt (ed.). Proceedings of 3rd International Conference on Organizational Foresight: Learning the Future Faster: University of Strathclyde, Glasgow, 16 August.
11. Ochieng, E.G. and Price, A.D.F. (2009). Framework for managing multicultural project teams, *Engineering, Construction and Architectural Management*, **16**(6), pp. 527–543.
12. Weatherley, S. (2006). ECI in partnership with engineering construction industry training board (ECITB), *ECI UK 2006 Master class Multi-cultural Project Team Working*: <http://www.gdsinternational.com/infocentre/artsum.asp?lang=en&mag=182&iss=149&art=25863> [Accessed December 2006].
13. Bartlett, C.A. and Goshal, S. (1989). *Managing across Borders*. Boston, MA: Harvard Business School Press.
14. Kelly, J. (2007). European powers of construction 2007: Surveying the landscape, analysis of key players and markets of construction. Deloitte, London.
15. Renz, P.S. (2007). *Project Governance. Implementing Corporate Governance and Business Ethics in Nonprofit Organizations*. Berlin: Physica.
16. HM Treasury (2007). Project governance: a guidance note for public sector projects: http://www.hm-treasury.gov.uk/d/ppp_projectgovernanceguidance231107.pdf [Accessed June 2011].
17. Lewis, M. L., Welsh, M. A., Dehler, G. E., and Green, S. G. (2002). Product development tensions: exploring contrasting styles of project management, *Academy of Management Journal*, **45**(3), pp. 546–564.

18. Söderlund, J., 2003. Building theories of project management: past research. *International Journal of Project Management*, **22**, pp. 183–191.
19. Winter, M., Smith, C., Morris, P., and Cicmil, S. (2006). The main findings of UK government-funded research network, *International Journal of Project Management*, **24**, pp. 638–649.
20. Turner, J. R. (2006). Towards a theory of project management: the nature of the project, *International Journal of Project Management*, **24**(1), pp. 1–3.
21. Crawford, L. (2002). Profiling the competent project manager. In Slevin, D. *et al.*, eds. *The Frontiers of Project Management Research*. Newton Square (PA): Project Management Institute, pp. 151–176.
22. Whittington, R. *et al* (1999). Change and complementarities in the new competitive landscape: an European panel study. 1992–1996. *Organisation Science*, **5**, pp. 583–600.
23. KPMG (2000). *Programme Management Survey*, UK.
24. Hobday, M. (2000). The project based organisation: an ideal form for managing complex products and systems, *Res Policy*, **29**(7–8), pp. 871–893.
25. Kloppenborg, T. J. and Opfer, W. A. (2000). Forty years of project management research: trends, interpretations and predictions: In *Project Management Institute Research Conference*. Paris: PMI PG, pp. 41–59.
26. Koskela, L. and Howell, G. (2002). The underlying theory of project management is obsolete. In *Proceedings of PMI Research Conference*, Seattle, Project Management Institute.
27. Meredith, J. (2002). Developing project management theory for managerial applications: the view of a research Journal's editor. *In Proceedings of PMI Research Conference*, Seattle, Project Management Institute.
28. Morris, P. W. G. (2000). Researching the unanswered questions of project management. In: *Conference Proceedings of Project Management Institute Research Conference*, Paris.
29. Checkland, P. (1989). Soft systems methodology. Chapter 4 in *Rational Analysis for a Problematic World*. New York: Wiley, ch. 4.
30. Morris, P. W. G. (2002). Science, objective knowledge and the theory of Project management, *Civil Engineering Proceedings ICE*, **150**, pp. 82–90.
31. Lawrence, P. R. and Lorsch, J. W. (1967). Organisation and Environment, managing differentiation and integration. Harvard University, Division of Research, Graduate School of Business Administration: Boston, MA.
32. Morris, P. W .G. and Hough, G. H. (1987). *The Anatomy of Major Projects: A Study of the Reality of Project Management*. Chichester: Wiley.
33. Williams, T. M. (1999). The need for new paradigms for complex projects, *International Journal of Project Management*, **17**(5), pp. 269–273.
34. Cicmil, S., Hodgson, D.E., Lindgren, M. and Packendorff, J. (2009). Project management behind the facade ephemera: theory and politics in organization, *Journal of Ephemera: Theory and Politics in Organization*, **9**(2), pp. 78–92.
35. Ekstedt, E., Lundin, R.A., Söderholm, A. and Wirdenius, H. (1999). *Neoindustrial Organizing*, London: Routledge.
36. Morris, P.W.G., (2010). Research and the future of project management, *International Journal of Managing Projects in Business*, **3**(1), pp. 139–146.
37. Midler, C., 1995. Projectification of the firm: the Renault Case, *Scandinavian Management Journal*, **11**(4), 363–375.
38. Li, H. and Guo. H.L. (2011). Complexities in managing mega construction projects, *Editorial for International Journal of Project Management*, **29**(7), pp. 795–796.

39. Kwak, Y.H. and Anbari, F.T. (2008). Analysing project management research: Perspectives from top management journals, *International Journal of Project Management*, **27**, pp. 435–446.

40. The Economist (2010). Closing the gap: the link between project management excellence and long-term success. A report from the *Economist Intelligence Unit*.

41. Eurofer the European Steel Association (2011). Economic and steel market outlook 2011–2012, *Q4 Report from Eurofer's Economic Committee*.

3

Stakeholder Management

3.1 Introduction

For global construction projects to be successful a number of key factors need to be 'managed'. In the best projects this management is near invisible and it may seem that the project was actually quite a simple one rather than the hugely complex project that was perhaps anticipated. As with any activity, a high level of expertise can make implementation seem deceptively easy; the trick is to identify where expertise is needed and then apply what is referred to as *encapsulation*: the ability to increasingly do less and less (focus), but do it increasingly better and better (in other words, to become expert). A useful illustration of this ability to identify where expertise may raise the standard can be found in a quotation by George Trevelyan: 'Education ... has produced a vast population able to read but unable to distinguish what is worth reading.'[1] A good project manager is able to 'read' project stakeholders and then decide which ones are worth (or require) giving most attention to.

Any area of human activity, such as stakeholder management, inevitably evolves its own language and terminology, and one of the first steps in developing expertise in any new area of activity is to define and understand the key terms and concepts within the relevant language. Encapsulation can be regarded as a form of structure that helps the development of expertise to progress. This is particularly effective if it is kept in mind that the essence of encapsulation is the placing of elaborate knowledge, gained in the context of a specific scenario, into a 'stripped-down' or simplified model of the key relationships within that knowledge. For example, there is quite a high probability that you are able to ride a bicycle or drive a car and, in the case of the latter, may even have a licence to prove your level of expertise. If so, how did you develop your current level of expertise regarding either of these two skills? Essentially you will have started by constructing elaborate knowledge; all the information concerning the nature of a bicycle or car (what a steering wheel looks like, how it moves, etc.) which is then applied to give an *understanding* of how the bicycle or car functions in use.

The range of information included in this activity will initially be wide but will then be narrowed (as in doing increasingly less and less) as the activity is repeated. The narrowing is achieved through establishing causal relationships between pieces of information (press this – the accelerator – and you go faster) that allow some of the detail to be discarded; eventually you cease consciously thinking about the relationship between the pressure from your foot, the movement of the accelerator and the increasing speed of the car. This in turn allows you to become better at applying those causal relationships that allow you to actually drive the car or ride the bicycle (do it increasingly better and better) until such point as you become 'expert'. Reading a single chapter of information concerning stakeholder management will not turn anyone into an expert; that is simply not a realistic outcome. A more realistic outcome would be the development of an understanding of what comprises stakeholder management and how key causal relationships can be identified and used to develop expertise. In other words, this chapter will allow you to start doing less and less but increasingly better and better. However, before you can start doing less and less, you have to have an overview of what you will then begin to do less and less of; you have to start by building up the aforementioned elaborate knowledge.

3.2 Learning outcomes

This chapter's specific learning outcomes are to enable the reader to gain an understanding of:

>> global stakeholders and sustainability;
>> stakeholders in projects;
>> stakeholder perspectives.

3.3 Global stakeholders and sustainability

Within the context of your present and future projects there will be any number of experts, and also a few novices, who can make your life as a project manager either easier or more difficult. When placed in the context of achieving a sustainable project (not to be confused with a *sustainability* project; the word sustainable is in danger of becoming overused in its environmental context), the need to engage with (manage) stakeholders in an effective manner becomes particularly important. For those projects that do incorporate specific sustainability targets, these represent an additional emphasis on using resources efficiently. While sustainability targets are typically presented in terms related to conserving relatively scarce resources such as water, and also of saving energy, reducing CO_2 emissions, etc., the project manager should not forget that the project also comprises human resources. This particular resource brings with it knowledge and expertise, both of which are valuable resources for construction projects given their relatively high level of 'manual' input, but also brings a degree of autonomy (unlike mechanical resources such as plant and equipment) that can result in it deciding to 'protect' that which makes it potentially valuable; its knowledge and expertise.

The protecting of expertise and knowledge can be achieved in a variety of different ways and this is a potential problem that global construction projects can suffer significantly from, largely due to differing culture norms with regard to how expertise (or knowledge, wisdom, sagacity, insight, or any one of the possible labels that can be applied) is valued. The cultures of many developed countries have changed greatly in this regard since around the time of the Industrial Revolution. As such societies moved away from being largely agrarian, the ability to buy an increasing range of goods produced by others (on a scale much larger than that of traditional craft 'industries') has led to an emphasis on consumerism. The connection to the natural environment typical of agrarian societies has gradually diminished and knowledge has become something that, if it is the right kind of knowledge, brings financial rewards to an individual. Knowledge is no longer focused on that which allows survival within a physical environment; it has become a financial commodity where value is determined on the basis of what others are willing to pay for. When dealing with a global project, particularly one based in a developing country, a project manager may find it difficult to determine valuable knowledge in the same manner as much of the project's workforce (given that the workforce could vary between being almost entirely locally recruited, and therefore relatively homogenous, and almost entirely 'externally' recruited, and therefore possibly heterogeneous). It is not uncommon, for example, for cultures to believe in a direct link between age and wisdom; the older an individual, the wiser he or she is believed to be. In the event that the project manager is considerably younger than the majority of the local 'leaders', he or she is likely to encounter difficulty in establishing any authority based on 'wisdom' (expertise). Another area where cultural values may diverge is the value and importance placed on society, community and family. If a particular group of stakeholders see a construction project as being detrimental to their community's values and culture, they will simply not support it and the project may lose a valuable resource as a result.

Overall, construction projects tend to rely heavily on people, even though mechanisation has increased significantly in the majority of construction projects over the last fifty years or so. While there are examples of highly automated construction projects, they tend to be in countries such as Japan, where the construction industry has its own specific problems in recruiting people, thereby driving a need to develop more automated processes. If the very small number of highly automated projects are discounted, the 'normal' situation is one where projects rely heavily on human input in terms of knowledge and expertise. Moore[2] suggested that projects also rely on the manner in which the human resource makes available its knowledge and expertise; the human resource is unique amongst project resources in that it can make decisions (has autonomy). The project manager needs stakeholders to make available or release knowledge and expertise in support of the project. However, stakeholders may decide to become critics, rather than supporters (or allies), of the project. Those critics who are particularly vocal may well be the ones that a project manager regards as being most damaging to the project and therefore requires to be marginalised in an attempt to achieve damage limitation by 'sidelining' them. This traditional approach to dealing with project critics, based as it is very much on the positional authority (rather than sapiential or expertise-based authority) of the project manager, is not now regarded as an example of good stakeholder management. A 2007 conference focused on sustainability stakeholder engagement identified a number of 'should do's', and one worth noting at this point is:

> Engage even your most vocal critics to try to find common ground. They impact consumer and media perceptions and ultimately sales – positively or negatively. [3]

Two key aspects of this suggested engagement with critics are that the project manager should not listen *only* to supporters or allies of the project – the apparently fiercest of critics can actually prove to be the most useful stakeholders to engage with, particularly if their 'passion' can be refocused – and the need to base engagement on any common ground that can be identified; focusing only on the areas where you disagree with your critic will not help to change their behaviour so that they become less critical of the project. Do not forget that this critic may be in the project's external environment (nowhere within the official project supply chain) and thus you will have little or no control (no punitive or rewarding actions can be implemented) over their actions. In this situation, the more traditional project managers could do well to consider adopting a contrary approach; either do what everyone else is doing but do it better, or do it differently.[4] When dealing with vocal critics, doing confrontation (usually what everyone else, possibly including your critic, is doing) better will not help. Where confrontation becomes expressed as brinkmanship (do this or else!) success becomes judged on not being the first one to back down, as this person will lose everything; brinkmanship leads to an 'all-or-nothing' attitude that is not ultimately beneficial to the project.

Taking the contrary approach of doing something different (such as engaging rather than confronting) will change the environment and thus also give an opportunity for behaviour change. The Pareto Criterion can be used to give a clearer picture of what the project manager is aiming for when adopting a 'doing something different' approach. Various aspects of Pareto (analysis, distribution, efficiency, etc.) are referred to in a variety of subject areas (economics, social welfare, etc.) but the essence of the Pareto Criterion is that if one individual can be made better off (not always in a financial sense) without any other individual being made worse off, then there is no justifiable reason for not making that individual better off.[5] If the project manager can give the critic(s) both inside and outside their project something that makes them feel more positive about the project (and thus move to support it) without making anyone else within the project worse off, then why not do it? It would, after all, be a win-win situation; no one loses anything and the project gains more support. Unfortunately, not everyone is capable of adopting this approach, due to their having become locked into behaviours that are more focused on 'winning' a disagreement rather than on achieving a sustainable outcome.

One different approach would be to envision dealing with a Norwegian! Norwegians have a reputation for being very principled, direct and honest. They do not generally respond positively to a confrontational approach based on alpha male behaviours:

> Hard-selling techniques will get you nowhere in Norway. Avoid bragging and exaggerations and make a well-documented presentation that gets

your counterpart involved and lets him/her buy from you rather than you selling through one-way communication. Norwegians are usually not tactical negotiators. If they say your product is too expensive they probably mean it.[6]

Also, in terms of stakeholder management, the following Norwegian characteristic is worth considering:

> The Norwegian management style is characterised as 'participative' and a manager is seen as a coach and very much as part of the team. Communication is open and consultative and goals are aimed for by the team as a whole.[6]

Taking the perspective that each human resource is a stakeholder in the project can help to reinforce the positive behaviours of those experts who have chosen to be helpful. It can also help to change or modify for the better the negative behaviours of those experts who have chosen to be difficult. Both of these management actions contribute towards achieving a project that is sustainable, in that it is capable of being successfully completed and in terms of its impact on the wider (external) environment. On the latter point, it is worth noting that the size of the project is becoming less of an issue – as large, medium and small companies all recognise the significance of sustainability even if they are unsure how exactly they should respond to it:

> Sustainability has become important to companies of all sizes . . . they understand its importance, but aren't quite sure how to engage in it.[7]

Engagement is a key consideration in the context of stakeholder management, particularly if it is positive engagement in support of the project. For many cultures in developing countries, the act of engagement is a critical aspect of deciding if an individual or community is going to 'do business' with a project or not. However, the nature of that engagement can vary considerably. Indians, for example, value long-term relationships and typically prefer to get to know possible business partners in detail before reaching any agreement. Thus, the act of engagement may involve long discussions about family; many businesses are family-run and there is a degree of suspicion of those outside of the family. In addition, the respect given to an individual is based on a combination of age (old age equals wisdom), status and rank. It also helps if you have a university degree. Throughout the engagement period, the project manager should try to consistently evidence personal traits that are seen as being positive within Indian business culture: friendliness (but not loud and boisterous as this is taken as indicating someone of a dishonest nature), compassion, humility and a willingness to find common ground (zones of accordance). In comparison to India, Thailand may initially seem to be very similar in terms of the value placed on respect, but there are some key differences and these mean that building a relationship based on trust may take even longer to achieve than in India. In Thailand respect flows from an individual's ability to maintain emotional restraint and harmony at all times; never openly exhibit annoyance or be overly assertive, as a loss of face will result and the engagement will come to a halt. While there is nothing wrong with championing your particular cause (sincerity is an admired personal trait), do so with politeness, modesty and honesty.

Research has indicated that if a company is to be successful in pushing the sustainability message there needs to be an internal 'champion' that is capable of not only being a supporter of the cause but of also working in a particular manner:

> Companies need an internal champion who aligns interests of high priority stakeholders with internal decision makers.[8]

There is no reason to believe that projects cannot benefit from a champion who is capable of aligning stakeholders and decision makers, even when a particular project may not have an explicit sustainability aspect to it. The approach used by sustainability champions does not comprise any actions that could not be applied in any kind of project. Given that large global construction projects are increasingly being judged against sustainability-related success criteria (along with the more traditional time, cost and quality criteria), there is no real reason why project managers cannot learn from the championing approach when engaging with stakeholders. This should not be taken as automatically meaning the project manager is the person to act as the champion – there are good arguments to indicate that the champion will be more successful if they are seen as being credible in the context of the actions being proposed; the project manager may simply lack the expertise to be seen by others as credible within a particular set of circumstances.

In order to be successful, a champion first needs to be able to identify stakeholders and then evaluate which ones are the high-priority ones. When working within your own culture this activity may be a relatively simple one, particularly if the project involves few resources and is not focused on either a complex product or complex processes. If, for example, you are an SME (small to medium enterprise) with some expertise in retrofitting traditional houses so as to reduce their level of heat loss, and have built up relationships with suppliers and subcontractors that you can trust in a defined locality (e.g. Aberdeenshire, Scotland), you will pretty much know the majority of the stakeholders (but you will almost certainly never have referred to them as such). But what if you decide to step outside of that defined local market? Perhaps you happen to be browsing the Internet one day and come across an article on FISH (FISH Norway, to be precise). On further reading, it transpires that FISH Norway actually has nothing to do with fishing: 'The overall vision of the FISH program is to provide a new and innovative business model that allows Danish companies and organizations to enter the Norwegian market for sustainable buildings and urban development.'[9] As you read more of the article you become aware that the driver for this programme is the requirement that by 2015 all new Norwegian houses will meet Passive standards; just the kind of work you have been involved in for some time. Could FISH Norway provide an opportunity for your company to move into a new market? Obviously, you do not meet the requirement to be a Danish company (as you are based in Aberdeenshire) but the programme does indicate that there is a market in Norway, and the Norwegians are generally regarded as being a wealthy nation (in 2009 Norway had the world's largest sovereign wealth fund, outside of the Middle East, of £259 billion;[10] by 2011 the fund had increased to US$570 billion[11] or £355 billion) so there may well be opportunities there for your company. However, a useful action before packing your suitcase and booking a flight to Oslo would be to identify possible Norwegian and/or Danish stakeholders in the kind of sustainable construction that you have been championing in Aberdeenshire.

3.4 Identifying stakeholders in global projects

Anyone acting as a champion, or indeed any other key role within the project, needs to have an understanding of what is meant by the term 'stakeholder'. While there are quite a few definitions available, they all share the essential details (in bold):

> *Any* person or organisation that is *actively* involved in a project, *or* whose *interests* may be positively or negatively *affected* by execution or completion of the project.[12]

> (emphasis added)

While this chapter will consistently refer to 'stakeholders', other terms will be encountered elsewhere, two examples being 'actors' and 'knowledge workers'. This diversity may initially be a little confusing, but regard it as part of the elaborate knowledge needed as the starting point in developing your stakeholder management expertise. Also, if a focus on individuals, groups or organisations whose interests are affected by your project is retained, then it generally does not matter a great deal if they are referred to as stakeholders or actors (unless they are in some way offended by a specific term); they can only be 'labelled' after they have been identified.

The action of identifying stakeholders requires an understanding of the first word emboldened in the previous definition: 'any'. It is important to be aware that stakeholders are not confined within the boundary of what a project manager may traditionally regard as the project (Figure 3.1). Stakeholders may be well outside of those traditional boundaries and, to illustrate this point, a little more of the stakeholder management language needs to be considered: internal environments and external environments. These are useful concepts when considering the identification of stakeholders, in that they encourage the project manager to think outside the traditional (for many) approach to the project boundary. The consideration of internal and external environments is particularly relevant when dealing with global projects, in that these projects go considerably beyond the boundaries of any physical construction site. In terms of environments, there may be an inclination to regard the construction site as the internal project environment, and everything taking place elsewhere (the off-site supply chain for example) as being the external project environment. While this is a reasonable starting point, it is not the whole story and therefore not fully accurate.

A project manager has a range of responsibilities to deal with and these generally relate to issues of control, either explicitly or implicitly – a client brings in a project manager because they want, among other things, their project to be completed on time, within budget and at the required level of quality. In other words they want someone, particularly if they are adopting the traditional perspective on the client–project manager relationship, to exert a desired level of control over the project, and it is this control aspect that allows the concept of project environments to be developed a little further. It would seem reasonable to expect the project manager to have control over the on-site activities that could traditionally be regarded as forming the internal project environment. However, the global project manager also has to have some control (albeit perhaps through the actions of a selected champion) over activities that take place beyond the site, such as procurement of labour, materials, etc. (the supply chain). On this basis, the project internal environment could be expanded to include off-site activities, although doing so can blur the boundary between the project 'proper' and the supply chain. By regarding the boundaries between the project and the supply chain, etc. as having varying levels of porosity/permeability, the project manager can ensure that problems within the supply chain that have no impact on the project remain within the supply chain, whilst allowing communication concerning relevant problems (or opportunities) to flow in a controlled manner. This may well relate to the salience (see Section 3.5) of the various stakeholders, and also the players within the supply chain.

The project manager does not have responsibility for, or control over, everything project-related that happens off-site. There could be a considerable number of stakeholders off-site that the project manager has no responsibility for or control over. Such stakeholders operate within the project external environment (see Figure 3.1) and therefore typically present the greatest challenge in terms of effective stakeholder management. Project managers need a means of classifying stakeholders so that the priority ones can first be identified and then engaged with.

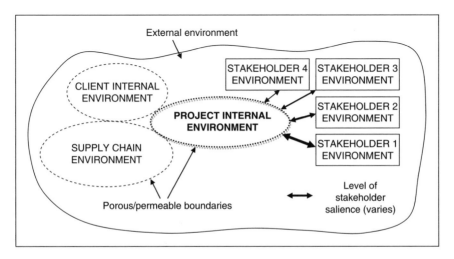

Figure 3.1 Stakeholder environments external to the project

3.5 Classification of global stakeholders and the construction process

In terms of effective use of resources, particularly the project manager's time, it is not sufficient to simply identify all stakeholders relevant to a specific project. While it may not seem an equitable approach, project managers have to find an answer to the question of which stakeholders actually require their attention and which do not. The stakeholder management term applied to this situation is 'salience'; recognition that different stakeholders will compete for the project manager's attention but will do so with differing degrees of success. Typically, some stakeholders will achieve complete success, others will achieve varying degrees of partial success, and some may achieve no success at all. This differing success is evidenced by the project manager's decision to prioritise stakeholders and thereby not give all stakeholders equal access/attention.[13] Underpinning this decision there may be factors such as the previously mentioned Pareto Criterion, along with the requirement within the technique referred to as *principled negotiation* that if a stakeholder can achieve a result that is above their fall-back position they will regard the outcome as being at least a partial 'win' as opposed to a 'loss'. Any project manager considering the use of principled negotiation as a means of seeking to distribute resources amongst stakeholders on an equitable basis must have a pretty good idea where each stakeholder's fall-back position lies. The negotiation can then proceed on the basis that the project manager will, as far as possible, seek not to cause a particular stakeholder to enter into an agreement that drops them below their fall-back position (Figure 3.2).

Project managers may apply a wide range of tools and techniques that facilitate the classification activity, and may take into account a range of factors when selecting a particular tool. One factor that applies in the context of complex projects (such as global construction projects) particularly is the need to prioritise stakeholders in terms of the amount of attention they should be given by a project manager. This is simply recognition that in such projects there may be hundreds of stakeholders seeking the project manager's attention and that it is simply not possible in such circumstances to give each stakeholder an equal amount of attention. While there are several tools that will help in doing this, a relatively straightforward one is the Stakeholder Matrix (in some instances

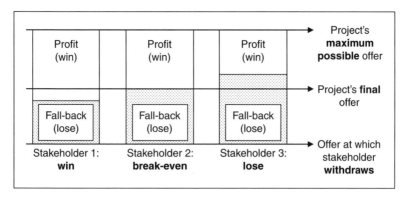

Figure 3.2 Stakeholder objectives, fall-back positions and outcomes

also referred to as the Power–Impact grid/matrix) as used by the UK Office of Government Commerce (OGC). The simplicity of the matrix is one reason for selecting this particular tool; a further reason is that the OGC lists common causes of project failure, and number three on the list is lack of effective engagement with stakeholders.[14] It is also useful to be aware of a suggested link between stakeholder management and risk management (failure number four on the OGC list); the two are, to some extent, intertwined and should not be regarded or implemented as stand-alone disciplines (a subtle hint to also consider the material in Chapter 10, covering risk and uncertainty).

The stakeholder matrix's simplicity comes from considering only two 'measures': the impact of change on the stakeholder and the importance of the stakeholder to achieving project success (hence the power–impact label sometimes applied). Note that the first measure has a focus of the impact of the project on the stakeholder, rather than the impact of the stakeholder on the project (this perspective is applied in the second measure – power). All projects are about achieving one or more intended changes (in the FISH Norway programme, for example, the intended change is to bring about a stronger market presence of Danish suppliers, architects, research institutions and advisors; these could, under other circumstances be regarded as four different stakeholder groups, but in the context of FISH may be regarded as a single stakeholder 'community' in which all stakeholders have equal saliency). Generally, the changes brought about by a project are regarded as being positives (otherwise, why would a project be implemented?) but not all stakeholders will consistently take that perspective. It is vital that the project manager is clear about this possibility, without regarding a negative stakeholder perspective as being 'wrong'; such an attitude simply interferes with the effectiveness of that key activity within stakeholder management – identifying common ground through engagement.

The assessments of a stakeholder's importance and perception of project impact allow an initial classification of a stakeholder as being one of the following four possibilities:

- *Low impact/low importance*: monitor these people but do not engage in excessive communication.
- *Low impact/high importance*: these people should be kept informed; they can often be helpful with detail issues.
- *High impact/low importance*: engage with these people just enough to keep them interested but not so much that they become bored.
- *High impact/high importance*: the people with whom you should fully engage.[15,16]

The latter of the four classifications is the one that will generally be taken as guidance that any such stakeholder should have the highest priority for the project manager's attention. However, it does not perhaps give sufficient understanding of the individual stakeholder's stance with regard to the project – a relatively simple two-measure 'tool' has its benefits in terms of quickly identifying key stakeholders but may make use of relatively little of the knowledge that could be gathered about each stakeholder. In the context of a global construction project in a developing country, the simplicity of the stakeholder matrix may result in subtle but important cultural values being missed; if a local is held in respect by his or her community then that person may reasonably (from their perspective) expect to be 'consulted' and thus suffer from a loss of face if the expected consultation does not occur. In such circumstances it is essential that additional understanding be achieved through creating more elaborate knowledge around the initial two measures so as to provide the following five classifications:

- **Advocates:** the only group driving the change or project; 'internal' champions and sponsorship.
- **Opponents:** have high understanding but low agreement to the project; will potentially 'lose out' in some way from the activity.
- **Indifferent:** individual or groups yet to take a definitive position on the project; have a medium understanding and medium agreement.
- **Blockers:** shows resistance to the project or its aims, principally due to having a low understanding and low agreement, which can be driven by:
  - a lack of communication;
  - a (perceived or actual) loss from project.
- **Followers:** have a low understanding of the project aims and objectives; support the project and tend to 'go with the flow'.[16]

This gradual increasing of detail within classifications can result in some highly detailed and complex approaches to classification, two examples of which will be covered more fully in the next section. The important point to note here is that, as with all analysis, detail should never be added for its own sake. The more detailed the analysis, the longer it will take to complete and the greater the resources it will require; the project manager has to be aware that the clock is always ticking and the meter is always running – time and money are not infinitely available resources in the context of projects. In the context of global projects the clock may well be ticking 24 hours a day – it is quite possible for such projects to have stakeholders in several, if not all, of the world's time zones; just as the project manager is finally managing to get some well-earned sleep, a stakeholder on the other side of the planet may be just starting their working day.

Box 3.1   Vignette – a global project in Afghanistan

Global projects tend to be regarded as large and complex but the 'large' aspect is not always the case. If a global project is characterised in terms of many separate technical and financial actions combined to achieve a high-level objective, then the management of a global project may become largely concerned with 'combining' many small projects, each of which may be regarded as regional or even local in nature. The 'global' nature of the management activity flows from the oversight of the many small projects, with the oversight being from a perspective of a uniting high-level objective. A common example of this kind of global project is the aid project.

Aid may be directed at a country or region as a response to a whole variety of factors but typically a disastrous event will be the spark that ignites the altruistic desire to help. However, aid projects can be focused on less altruistic and more political objectives, with Afghanistan being argued by some as being an example of, at the least, a country 'benefitting' from a combination of altruistic and political objectives identified (largely) by external funders.

Afghanistan has a turbulent history and is a complex environment from geographical, social and cultural, ethnic and religious perspectives. In many respects it is a challenging environment in which to 'do business' even when that business is intended to bring benefit to the country. Aid workers, for example, have been the target of armed groups 'resisting' the changes that are being attempted by intimidating aid workers and indigenes either working with them or receiving aid from them. In 2011, over 3000 civilians were killed, with nearly 80 per cent of those deaths as a result of the actions of armed groups. In addition to such challenges, there is also the problem of determining if any given project has been completed and if so, was it a success? In many countries, such questions would be answerable through the collection of statistical data. However, in Afghanistan even the collection of data has required a capacity-building project; the National Institution Building Project's subproject of the Capacity Development Plan 2011, 2014 with its focus on the activities of the Central Statistics Organisation (CSO). This subproject commences with a recognition that efforts in this area over the previous ten years were not successful due to a variety of factors, with the key factors (from a stakeholder management perspective) being:

- attempts to do too much too quick, without focus on longer-term capacity development;
- donor-led agenda, undermining the national strategies;
- failure to develop national stake in developing the statistical service;
- poor coordination of development initiatives and donor activities with inevitable overlaps and duplication of efforts. In most instances CSO has been bypassed.

The CSO now has the responsibility to compile 72 indicators (some of which are acknowledged as overlapping) for monitoring and evaluation of key goals (are projects achieving their stated objectives?). In order to progress towards achievement of this considerable responsibility, CSO has attempted not to repeat some of the mistakes made previously and has carried out a stakeholder analysis focused on those individuals and organisations that have an interest in accurate statistical data within the areas covered by the identified indicators (see Table 3.1).

While the CSO project no doubt has a general value, it is probable that the push to collect more comprehensive and accurate data comes at least in part from the problems experienced with regard to the completion of US-funded construction projects that have been awarded to Afghan contractors. Since 2008, it seems that around $200 million of construction projects awarded by the US Corp of Engineers have experienced problems ranging from work being at a quality below that required, serious delays and complete failure. These problems have resulted in different responses 'on the ground', including the rewriting of

Table 3.1 Stakeholders matrix of Afghanistan statistical system (CSO, 2011[17])

		Interest	
		High	**Low**
Stakeholder Power Level	**High**	**Key players**	**Keep satisfied**
		Technical Committee on Statistics and National Statistics and Census Committee **President** General, Central Statistics Organization (CSO), the apex official statistics generating and coordinating agency	**Parliament ministers** and deputy ministers of 'line ministries', **heads** of departments of CSO and key functionaries, **heads** of provincial offices, **political** leaders, **media**
		Administrative statistics generating and user ministries and organisations (e.g. Ministries of Finance, Economy, Rural Rehabilitation and Development, Public Health, Women Affairs, Commerce and Industry, Labour and Social Affairs, Agriculture, and Education, Da Afghanistan Bank, CMRS, JCMB, etc.)	
		Independent Administrative Reforms and Civil Service Commission (IAR and CSC)	
		UNDP represented by Capacity Development Adviser, Project Manager of National Institution Building Project (NIBP) and Country Director	
		International donors (e.g. World Bank, ADB, DFID, USAID, EU, ILO, UNAMA, UNFPA and other UN agencies and donor countries)	
	Low	**Keep informed**	**Minimum effort**
		Data users (industry and commerce associations, research institutions, NGOs, international organisations, etc.)	**Staff** at lower levels **Service** providers **National** consultants
		Project Implementation & Coordination Team (PICT) of SRF and NRVA Team	
		Authorities and officials of the twinning partner	

contracts by the Afghan government in order to fit with any work actually completed so that the project can then be presented to the external funders as a success. It is interesting to note that CSO's stakeholders matrix identifies all international donors as 'key players' (in stakeholder management terms these are the stakeholders with the highest salience), while the staff gathering the data are classed as having the lowest salience. Given the challenges faced by these 'low-level' staff, it could well be that their enthusiasm (or lack of) could be a key factor in the success of the CSO project, and that they should be moved up the salience scale, possibly even to the top level of 'key players'.

3.6 Tools and techniques for stakeholder analysis

Depending upon a particular project's aims and objectives – the environment in which it is being carried out, the extent of buy-in evident from stakeholders, and a number of other factors – specific tools and techniques may be suggested as being more appropriate than others. However, it is perhaps most appropriate for the expertise level intended here to consider one factor as being most relevant to the selection process: how comfortable an individual project manager feels in using any particular tool or technique. The following sections outline one relatively simple tool and one technique so as to introduce the manner in which stakeholder analysis can be undertaken.

3.6.1 An example tool

Tools can be regarded as the starting point in the development of expertise as far as stakeholder management is concerned; they are not as extensive in terms of their demands concerning an understanding of concepts or the 'elaborateness' of knowledge. A reasonably robust but not overly demanding example of a stakeholder analysis tool can be found on the Mind Tools website (the template for practising the use of stakeholder analysis is also available freely as a download).[18] The tool follows the standard structure in that it runs through several steps of gathering information about the stakeholders. Typically this commences with identifying the stakeholders without any attempt to classify them. It is important to be clear that, as far as possible, stakeholder analysis should not be rushed and the project manager should not (tempting as it may be) 'accelerate' the analysis by simply applying preconceptions (stereotypes) with regard to issues such as stakeholder identity and objectives; this approach will almost always cause more problems than any perceived time-saving is worth.

Commencing with identification is a similar starting point to that of various risk management tools – you have to identify the risk before you can respond to it in any meaningful manner (as per the risk hierarchy). Similarly, if you have not identified your stakeholders (as mentioned previously, identification is rather more involved than the simple application of a title or label to a particular individual or group) then you cannot, in any meaningful manner, manage them. Identification is arguably the first step to engaging with and understanding a stakeholder, and it is therefore worth taking some time over this step of the analysis.

The second step in applying a tool is to then value or prioritise the identified stakeholders. As with the stakeholders matrix (in the previous section) the Mind Tools template considers only two measures, in this case interest and power. However, both are only assessed as being high or low, thus the level of detail required by the tool is not overly demanding (but be aware that there will be more detail available than is required by this particular tool – do not base your subsequent approach to the *management* of a stakeholder solely on the simple knowledge required by this tool). Again, the prioritising is seeking to identify those stakeholders with high levels of both power and interest; these are the ones that are the main priority in terms of focusing your resources on the development of a connection. That connection is usually in the form of 'understanding' and this is typical of the third stage of most analysis tools. The action of 'understanding' is trying to establish the motivation of these stakeholders, and if they are positively or negatively oriented to your project. In the event of a priority stakeholder being negatively oriented, analysis tools will then typically direct the user towards questions intended to determine if there is any way that the stakeholder can have their orientation changed so as to become either less negative or, ideally, positively oriented to the project.

While analysis tools are a useful starting point, particularly in terms of guiding the 'finding' and 'understanding' stages, they are not generally useful in terms of the management of stakeholders. This aspect is somewhat more qualitative in nature – in that cultural and relationship values can be added – than the analysis aspect and can thus be more difficult to deal with for those project managers who lack 'connection' skills. It is at this point that a consideration of the 'political' nature of stakeholder management becomes relevant to the development of expertise and the focus therefore moves from the analysis tools to the management techniques.

3.6.2 An example technique

As the extent of knowledge elaboration increases there is a risk of tools becoming too mechanical (being done repetitively without any thought); simply gathering large quantities of information or data without an understanding of why it is being done does little to develop skill and then expertise. Thus it is appropriate to consider stakeholder analysis tools as being relevant to the developing of initial understanding of what is being undertaken. In order to develop deeper expertise there is then the need to consider elaborate knowledge in the context of a different approach, which can be referred to as the use of technique.

The definition of 'technique' usually refers to proficiency in a particular practical or mechanical skill. In this chapter the term will be taken to refer to proficiency in gathering and processing elaborate knowledge above the level applicable to the use of tools. An illustrative example can be found in the UK National Health Service (NHS) approach to stakeholder management.[19] This is indicative of how increasing elaboration tends to move an activity (such as stakeholder management) away from being based on solely or largely quantitative/objective data/information and towards being more qualitative (NOTE: There is no intention here to suggest that a qualitative measure is any way better or worse than a quantitative measure; there is merit in them being different but any difference in value simply arises from one being more appropriate than the other in any given set of circumstances). A key term in this regard is 'political' – increasing expertise in stakeholder management requires the project manager to become more political. This is something that any normal human being does many times every day when interacting with other humans but, in the context of stakeholder management, it is given greater emphasis as a result of its recognition as a key factor in a successful outcome.

The term 'political', when applied to stakeholder management, is simply a reflection of the social nature of humans and the manner in which we interact when seeking to achieve objectives in a non-confrontational manner. The majority of professional managers typically define politics in terms of alliance building to achieve objectives (sometimes these will be purely personal objectives; these can at times run counter to the success of a project) and the emphasis on this approach is suggested as increasing in the future, largely in recognition of a perception of an ever more complex and interconnected world.[20] Thus a global construction project manager will be faced with an environment in which stakeholders become ever more focused on the need to be political and to identify others with similar aims and objectives who can then be brought into an alliance, thereby increasing the power (and possibly the impact) of those stakeholders. While large alliances of stakeholders can have the benefit of presenting the project manager with an apparently smaller number of stakeholder groups, there is the disbenefit that the resulting alliances may be more resistant to attempts to manage them. It is not unknown for particularly adept stakeholders to form alliances that present a credible basis for them to 'manage' the project manager. Time to start developing your political skills!

Box 3.2   Vignette – creating multi-stakeholder alliances

Construction industries are typically presented as being team-based and people-focused. There is no doubt that national construction industries around the globe are highly dependent upon human resources. It therefore makes sense for industries in the developed nations to look for ways to make more productive use of that resource. The resource aspect of construction also includes non-human resources, and the diversity of this requirement – particularly on complex global projects – can result in a complex supply chain. Effective management of that chain has led to the development of various approaches to building alliances. Such alliances are, however, focused on economic gain of one form or another; the stakeholders within such alliances may actually have little in common beyond the profit motive. Expertise in the management of such alliances is therefore not always transferable to the management of stakeholder alliances that are focused on not-for-profit aims and objectives. Dealing effectively with such stakeholders requires the project manager to look outside of the construction environment for appropriate models. One possible example is the Medicines Transparency Alliance (MeTA). MeTA has the aim of bringing together all stakeholders in the medicines market in such a manner as to improve access to medicines for those denied access (for reasons such as cost and availability). At current estimates, around one-third of the global population fall into MeTA's target group; this is truly large-scale alliance building, the resource requirements of which require the support of organisations such as the World Health Organization (WHO) and Health Action International.[21] In addition, the complexity of the environment that MeTA operates within is such that the organisation recognises the key significance of bringing together the many interest groups (stakeholders):

> A 'neutral space' in which to work is created, a shared understanding of the problems, common ways of working and an agreed agenda among everyone taking part. In MeTA's case, this means establishing a forum for representatives of everyone involved in the medicines supply chain – manufacturers, governments, international organisations, traders, medical workers, academics, the media, and patients.[22]

To help achieve this, MeTA has developed an alliance-building 'toolbox' of components, and one component that is worth further investigation comprises a tool kit intended to facilitate the analysis of multi-stakeholder processes. The tool kit is intended to function without any technical assistance from elsewhere but does require a 'local' independent consultant to carry out the assessment. It is also worth noting that the assessment process is not a quick-fix; 30+ days are indicated as being required to complete one assessment (perhaps not surprising when it is noted that the tool kit comprises 18 different tools).[23] Such is the impact of MeTA in terms of alliance building that the UK's Department for International Development (DFID) includes it as one of the three multi-stakeholder initiatives that it supports. DFID also supports (in conjunction with the World Bank) the Construction Transparency Initiative (CoST).[24] While CoST and MeTA share similar objectives regarding transparency, it is clear that CoST does not have such a well-developed approach to creating multi-stakeholder alliances (it has a narrower focus for its objectives in that it operates only in the context of public-funded construction projects). Neither is there much evidence of crossover learning, on the part of CoST, from MeTA's experience. In terms of building effective multi-stakeholder alliances, MeTA appears to have the greater expertise and potential for success.

The NHS technique outlined previously is worth looking at in further detail as it is, in part, based on work carried out in the context of sociodynamics (study of critical mass causing change within societies or social groups) with its connection to the political nature of the more advanced (complex and interconnected) levels of stakeholder management. A focus on the sociodynamic basis of 'political' stakeholder management allows the possibility of classifying individual stakeholders/actors/knowledge workers within a larger number of more narrowly defined classes. Such an approach at least allows the possibility for the project manager to have a better understanding of the individuals and groups that the project deems to be stakeholders. Ideally, it should at least start to provide answers to questions such as:

- What motivates this particular stakeholder most of all?
- What is their connection to the project? Is it financial or emotional?
- What is their perception of my project? Is it negative or positive?

This in turn presents an opportunity for the project manager to be more effective in connecting with the stakeholders, particularly those with a high level of importance (power) and impact. Guidance on stakeholder management in the NHS approach makes use of work that was applied to the management of so-called sensitive projects and therefore is applicable to other environments beyond health care. The basis of the suggestions for managing sensitive projects was a model of the interaction of eight different classes of stakeholders (you may not initially feel comfortable with this as it may seem a high level of differentiation amongst stakeholders but there is a value to approaching stakeholder management on such a basis) and these are used in NHS stakeholder management:

- Zealots: support the project without question.
- Golden triangles: generally supportive but some criticism.
- Waiverers: can easily move between support and criticism.
- Passives: little support or criticism.
- Moaners: very little support and some criticism.
- Opponents: consistently question the project.
- Mutineers: high levels of criticism and almost no support.
- Schismatics: highly critical and highly supportive.[25]

For those who have not encountered this approach before, a useful exercise is to consider which of these eight classes you regard as being the most valuable and which you would regard as being the most dangerous. Before we tell you the answer, let us just say that most people get this wrong! The most valuable class is the passives, and this is for two reasons: first, they are always the most represented class in any project – usually 40–80 per cent of project participants will fall into this class; second, they are usually the easiest to turn into either supporters or critics of the project, and if the project manager can manage them so that they become supporters then the probability of a successful project increases. The most dangerous class is the mutineers, again for two reasons. First, because they have no interest in the project being a success; second, they rarely respond positively to standard approaches to stakeholder management. For a mutineer it is generally a case of all or nothing; they would rather the project failed completely than lose any aspect of their argument. It is therefore very useful if the mutineers can, at least, be identified.

3.7 Analysing global stakeholder perspectives

The previous section on techniques suggested quite high levels of differentiation (eight classes of stakeholder) and it would be reasonable to expect that a similar, if not greater, extent of differentiation could be identified regarding stakeholder perspectives (at least one perspective per class). However, given that it has been argued that all of the millions of fiction books that have ever been written boil down into no more than seven plot lines[26] – tragedy, comedy, overcoming the monster, voyage and return, quest, rags to riches, rebirth – it possibly seems overanalytical to seek to identify at least one perspective per class of stakeholder. It may well be that effective stakeholder management can be achieved by considering fewer perspectives. After all, whatever analysis tool that has been applied should have provided the basis of understanding the stakeholder, and the management tool should have identified the classification for the stakeholder. Overall, these should give a sound basis for developing the connection (engagement) with a stakeholder and, therefore, perhaps only a small number of perspectives needs to be considered as context for the activity of connecting. By viewing the stakeholder management activity in the context that every actor with an interest in a project is seeking to either maximise or protect that interest, it can be argued that there are, in essence, only three perspectives that need to be considered: separation, ethical and integrated. It is important to note, however, that each of the perspectives is quite broad and therefore there is the potential for some subtle nuances to be lost. This is where the development of expertise in the previously noted political skills is of benefit.[27]

3.7.1 The separation perspective

The basis of this perspective is effectively a recognition of the fact that, as far as projects are concerned, typically a high proportion all of the actors involved are employees and thus have little independence, perhaps particularly so if they are acting in a management role. Thus a project manager is acting on behalf of the projects 'owners', usually referred to as the client, and the separation perspective argues that a manager should therefore act in the best interests of the client only. While such an argument could seem reasonable in the context of commercial projects (where the intention is ultimately to make a profit), in non-commercial (not for profit) projects the situation can sometimes be less clear cut. However, even in commercial projects there is the possibility that managers can apply this perspective too literally. A key component of the separation perspective is that, even though managers are encouraged to act in the client's best interests, they should not do so at the expense of other stakeholder interests (the ideal is that no stakeholder or actor should lose). Again, this is an aspect of stakeholder management where an understanding of the political nature of 'connecting' with a stakeholder is helpful.

Consider, for example, the possibility that responding positively to the interests of a non-client stakeholder in a manner that – to the less politically adept – seems not to be in the interest of the client, may actually give benefits to the client. Such a situation may arise in the case of a project manager responding to a need of the local community by recruiting unskilled labour who are then trained, with the expense of that training covered by the project. Initially this may seem as though the project is being burdened with extra cost (both in terms of the training and the expectation that such newly trained labour will not be as productive as more experienced labour). Nonetheless, such an action may well result in an increase in terms of what is referred to as 'social capital' – social relations that result in productive benefits.[28] An example of social capital in the context of a global construction project would be the changing of a community (locally, nationally or even globally) from a critical stance towards the project to the point where it becomes a positive supporter of the project. Put simply, the project has connected with the community by offering positive benefits (training and jobs) and the community responds reciprocally because they now regard themselves as being valued through being placed in a 'win' situation rather than a 'lose' situation. Thus the client may not reap a benefit that is directly evident in terms of the profit margin (it is always difficult to estimate the cost of delays that might have happened if a stakeholder group had chosen to exhibit antagonism towards the project) but certainly reaps one in terms of marketing and a feel-good factor that can ripple out beyond the project environment.

3.7.2 The ethical perspective

This runs counter to the separation perspective in that it does not place the interests of any single stakeholder above the others. The argument is that any business should go about its activities in a manner that treats all stakeholders fairly; they should be concerned with behaving in an ethical manner. The potential downside of the ethical perspective is that it can impact negatively on the profit-making function of business owners. The most extreme interpretation of the ethical perspective does actually result in a situation that seems to treat owners unfairly; constraining the right of an owner to a financial reward that is in proportion to the risk taken if allowing this would then be counter to what is ethically best for stakeholders who are non-owners. Adopting an ethical perspective may result in a stance that is equally opposite to the separation perspective and, therefore, there is a benefit to finding a perspective that allows for a stance that is closer to being balanced. This can be found in the form of the integrated perspective.

3.7.3 The integrated perspective

Recognition that a business – commercial or otherwise – cannot function sustainably without some level of support from what is referred to as the non-owner stakeholder environment (broadly the same as the previously introduced external project environment in which the project manager has no control over stakeholders) is the basis of the integrated perspective. Such recognition brings with it an operational model that incorporates awareness that any decision made on behalf of owner-stakeholders (clients) without considering the needs of non-owner stakeholders has two possible detrimental outcomes:

1. It may miss, or even increase the severity of, threats to the project and possibly the client organisation(s). If, for example, a payment is delayed to a non-owner stakeholder (supplier) because to do so would be beneficial to the client's cash flow, that supplier may cease trading. Without realising the nature of the suppliers' 'need' (money to pay its own mounting

debts), a decision to delay payment increases the threat to the supplier, and that in turn presents a threat to the project. Can it find another supplier?

2. It may miss entirely, or not appreciate fully, an opportunity that brings benefit to the project and/or the client organisation. Returning to the previous example, it may now be the situation that by paying the supplier on time (or even better, ahead of schedule) the project reaps a benefit of not only retaining a supplier but also the goodwill that would flow from preserving jobs within the supplier's community.

The integrated perspective may initially seem very similar to the separation perspective but there is a subtle difference between treating all stakeholders equally and balancing the needs of differing stakeholders; the latter needs the decision maker to make more complex evaluations of stakeholder needs. For the decision maker to do this effectively, he or she will require to practise the previously mentioned political skills that are so important, particularly with regard to managing the relationship with the client.

3.8 Relationship with the client

The relationship between the project manager and the client can be a difficult component of the project to manage, irrespective of the complexity of the project; clients are people and they bring with them all of the usual human behaviours. When a client is actually a group of people, the dynamics of the relationship become more complex. When the client group is a collaboration of otherwise independent businesses and/or individuals, the level of complexity rises further still. In the context of a global construction project, the level of complexity within the project manager–client relationship could be very high. The project manager, therefore, needs a structure within which to manage the relationship in the context of such complexity.

Mention of a 'structure' may bring to mind something that is regimented and ordered but this need not be the case; it is arguably more appropriate to consider a structure that simply overarches all aspects of the relationship in a manner that allows the project manager to respond to events or changes in a contingent manner, without seeking to impose any specific order on the relationship. Gray and Larson (2006)[29] encapsulated the demands of this situation when noting that 'project managers must (find out what needs to be done and build a collaborative network to deliver it) do so without the requisite authority to expect or demand cooperation. Doing so requires sound communication skills, political savvy, and a broad influence base.' A rigid, traditional monitor and control structure is not appropriate to this new manner of working, particularly given the need to respond to the seemingly ever-increasing level of rights (in legal terms) and complexity of needs in relation to external environment stakeholders. Consider, for example, the complexity of needs and level of rights of external environment stakeholders in any project linked to the FISH Norway programme previously introduced.

> **Box 3.3   Vignette – possible external environment stakeholders in a FISH Norway project**
>
> - The Norwegian government? By 2015 Norwegian building regulations will require all new houses to be of Passiv standard.
> - House buyers? The population of Norway is projected to increase (longer life expectancy, more immigrants, increased birth rate, etc.)
> - Local government? Predicted increased urbanisation will have an impact on local government budgets, demands on services, etc.
> - Construction industry professionals generally but architects and engineers in particular? There is some recognition that Norwegian designs are perhaps not as generally 'pleasing' (more technically than aesthetically oriented) as the Danish equivalent.
> - The Danish building industry? Norwegian companies have demonstrated greater competence in the re-engineering of solutions so as to provide cheaper and more feasible high-volume products than has been the case with Danish companies.
> - Universities? The proposed involvement of several universities in the programme should provide opportunities for knowledge development/transfer.

One approach to responding to ever more complex stakeholder environments, and that is becoming more widely adopted, is the use of capabilities within modern information and communications technologies (ICTs). Two forms of this approach that are relevant to global construction projects are visualisation and building information modelling (BIM) tools. The latter of these commenced as a relatively low-key way of modelling the three dimensions of a building design but has since had its functionality increased by the addition of developments such as conflict detection software. Since commencing with a 2D (flat) version of BIM, the technology has undergone development to provide 3D (length, width, height) and then 4D versions (3D plus time). Currently we are at the point where 5D (4D plus cost) versions are not uncommon and some users are openly talking in terms of up to 11D (5D plus ADA(automated data acquisition), sustainability, maintainability, acoustics, security and heat) versions. The sheer quantity of information that a BIM system can potentially hold makes this approach to managing complex client relationships in global construction projects one of the most promising upcoming developments for connecting with stakeholders.

3.9   Response decisions in changing stakeholder perspectives for effective global project delivery

The key objective with regard to stakeholder management is to achieve a desired outcome for specific clients whilst managing the various stakeholders so as to either retain/improve their existing support or minimise their negative impact on the desired outcome. Thus far, this chapter has dealt with a number of factors relevant to that but has done so without addressing the worst-case scenario: a high-interest and high-power stakeholder who simply does not wish to connect/negotiate with the project manager. In such cases, the project manager has to essentially go back to basics and recognise that he or she is dealing with a stakeholder who is unwilling to change their behaviour with respect to the project. All other stakeholders will give the project manager at least the possibility of connecting and, hopefully, will change their stance regarding the project. So how can the project manager deal with this worst-case scenario once it has been acknowledged that a particular stakeholder has shown that they are unwilling to change? The project manager should, perhaps paradoxically, commence by considering the alternative behviours that they could exhibit in dealing with such a stakeholder. This stage is generally referred to as the response decision, in that the project manager has several different responses available and these can be classed under four headings:

- concession (offence, involvement, concession and adaptation responses);
- compromise (trade-off, collaboration and accommodation responses);
- defence (avoidance and dependency/attachment reduction responses);
- hold (monitor, reaction and dismissal responses).

These responses are presented in order of increasing robustness/toughness, hence the need for the project manager to decide upon the most politically appropriate response. However, there is little evidence that project managers consistently consider the selection of the most appropriate response to a given stakeholder. Whilst this is not established as fact, there is a general impression that most project managers are not even aware of the range of response strategies available to them and typically go immediately to a 'hold' response (the toughest of the responses available), thereby making the situation more difficult to manage. Assuming that you, as a project manager, will not default to the 'hold' response, how can your interaction with the stakeholder be guided so as to increase the probability of a successful (on whatever basis this is to be measured) outcome? One possibility is to consider the Transtheoretical Model of Change (TTM).

Asking a project manager to consider using TTM may become more difficult once it is explained that the model is not based in the project management literature and that it actually comes from the health promotion field. Achieving consideration may become even more difficult when it is explained that TTM was developed to promote positive behaviour change in addicts.[30] However, a good project manager is someone who is always open to new ideas and will evaluate a suggestion before making a decision concerning its implementation. The fact is that TTM has evidenced good results with regard to behaviour change and has started to move outside of the health promotion field and into the organisation behaviour field. On that basis it merits consideration.

Managers operating within organisations that have an objective of achieving specific changes are becoming aware of the possibility of bringing about change at the organisation level through the use of TTM. The presumption is that by focusing on individuals and bringing about change at that level, the behaviour

Table 3.2 TTM comprises five stages

- Pre-contemplation: individual has the problem (whether he/she recognises it or not) and has no intention of changing.
- Contemplation: individual recognises the problem and is seriously thinking about changing.
- Preparation for action: individual recognises the problem and intends to change the behaviour within the next month. Some behaviour change efforts may be reported. However, the identified behaviour change criterion has not been reached.
- Action: individual has enacted consistent behaviour change (i.e. consistent condom usage) for less than six months.
- Maintenance: individual maintains new behaviour for six months or more.[31]

of the organisation can incrementally be changed.[32] However, in the context of stakeholder management, the focus is typically on individuals and therefore there is no need to be overly concerned with change at the organisation level. In terms of actions by project managers, a key action is for them to perhaps start by considering their own behaviour with regard to any expectation of a linear relationship between a forceful expression from them that others need to change and that change actually occurring. Such an approach will simply not work in the context of TTM, nor is it generally of any great value with regard to stakeholder management.

A key action regarding the application of TTM is to analyse the stakeholder involved and accept whatever stage the analysis indicates they are at: pre-contemplation, preparation for action, etc. The project manager should then work to move the stakeholder to the next stage. In the case of the most antagonistic stakeholders this may mean that the project manager has to work at moving a stakeholder through several stages, and this is obviously demanding in terms of resources (political skills and time). It is important, for example, to allocate sufficient resources so as to allow the identification of leverage points and also to minimise the possibility of any change achieved by the stakeholder being 'lost' through regression to old behaviours. There is some indication that any change achieved through the application of TTM can be reinforced by combining TTM with an understanding of behaviours related to decision making. Given the emphasis within TTM on a stakeholder making the decision to change, this combining of the two approaches appears to be an appropriate route to bringing about and then maintaining stakeholder behaviour change.

One possible route to gaining an insight to decision making would be the means–end theory. The value of this theory comes from its focus on identifying the criteria that are important to a stakeholder when making a specific decision, and then determining why those criteria are important. This can be regarded as a more formalised approach to the 'connecting' activity within stakeholder management. The simplest form of the means–end theory comprises three steps: attributes, consequences and values. On more advanced models, the attributes and consequences steps can both be subdivided to provide a greater depth of analysis, if this is required. As mentioned previously, this will add to the demand on resources available to the project manager, and the extent to which either of these theories can be applied will be constrained by factors such as budget and time. Nonetheless, if the assessment of project stakeholders has indicated a particular actor as being a priority for the success of the project, the resources have to be made available (as per the integrated perspective). The only other alternatives are to either reconfigure the project so as to remove a particular stakeholder or to decide not to go ahead with the project.

3.10 Chapter summary

All individuals and organisations have to deal with change. Some of this will be planned and some will be happenstance. In both cases, there will almost always be others involved, either by bringing about the change, or through being either affected by the change or by the impact of the change on someone else. These 'others' can be regarded as stakeholders, the majority of whom will never be directly known to or contacted by the 'changed' individual or organisation. Nonetheless, that individual or organisation may find there is a need to, or a benefit from, considering stakeholders before implementing a response to change. This is particularly the case when considering the change brought about through a global construction project. The manner in which construction resources (both human and non-human) can be moved around the globe is, in itself, a major change to the nature of national construction industries.

For those organisations that value strategic thinking and planning, the extent and composition of stakeholders affected by various strategic options may be a major factor in the selection of a strategic direction for them for up to ten years ahead. A 2011 report on the global construction economy up to 2020 presents a future construction industry wherein companies will have to decide whether to stay within their traditional markets (and fight for a share of a decreasing level of business, relative to emerging markets) or move into new markets that will be providing the majority of any growth.[33] Of the emerging markets, India and China are predicted to provide the largest drive to growth, with China remaining the world's largest construction market up to 2020. Such a prolonged period of growth is bound to be attractive to those companies outside China that have the resources to move into the region and seek a share of the wealth generated. However, such a decision should only be based on a clear understanding of the needs and requirements of key stakeholders, without which things could go badly wrong!

3.11 Discussion questions

1. Outline the two key aspects of 'engagement' with project critics.
2. Identify the common essential details that all definitions of the term 'stakeholder' should contain.

3. Outline the meaning of 'salience' with regard to stakeholder management, and how this can lead to the classification of individual stakeholders.
4. What is the meaning of the term 'political' in the context of stakeholder management?
5. Outline the four response 'options' available to a project manager seeking to alter a stakeholder's perspective.

3.12 References

1. George Macaulay Trevelyan. Quotable Quotes. http://www.goodreads.com/quotes/show/28425 [Accessed June 2011].
2. Moore, D. R. (2002). *Project Management: Designing Effective Organisational Structures in Construction*, Oxford: Blackwell Science (Chinese translation 2006).
3. Goldschein, P. and Bengston, B. (2009). How to Engage Stakeholders on Sustainability. http://www.greenbiz.com/blog/2009/09/17/how-engage-stake holders-sustainability [Accessed July 2011].
4. The contrarian manager. Bloomsberg Businessweek. http://www.businessweek.com/chapter/jenrette.htm [Accessed August 2011].
5. Pareto Criteria. Encyclo. http://www.encyclo.co.uk/define/Pareto%20criteria [Accessed August 2011].
6. Norway. Expatica. http://www.expatica.com/hr/story/norway-20323.html#6 [Accessed July 2011].
7. Goldschein, P., Bengston, B. (2009). How to engage stakeholders on sustainability. http://www.greenbiz.com/blog/2009/09/17/how-engage-stakeholders-sustainability [Accessed August 2011].
8. Goldschein, P. and Bengston, B. (2009). How to Engage Stakeholders on Sustainability. http://www.greenbiz.com/blog/2009/09/17/how-engage-stake holders-sustainability [Accessed June 2011].
9. FISH Norway. http://norway.fishclusters.dk/30376 [Accessed January 2013].
10. Fouche, G. (2009). Norway's sovereign wealth fund: £259bn and growing. http://www.guardian.co.uk/business/2009/sep/20/norway-sovereign-wealth-fund [Accessed June 2011].
11. Kremer, J. (2009). Norway's sovereign wealth fund sold all US mortgage bonds. Bloomberg. http://www.bloomberg.com/news/2011-10-28/norway-s-sovereign-wealth-fund-sold-all-u-s-mortgage-bonds.html [Accessed July 2011].
12. Project Stakeholder Management. http://www.projectstakeholder.com/ [Accessed August 2011].
13. Stakeholder snalysis. http://en.wikipedia.org/wiki/Stakeholder_analysis [Accessed July 2011].
14. Beyond conventional stakeholder management. Moorhouse. http://www.moorhouseconsulting.com/beyond-conventional-stakeholder-management [Accessed January 2013].
15. Stakeholder analysis. World Health Organization. http://www.who.int/hac/techguidance/training/stakeholder%20analysis%20ppt.pdf [Accessed June 2011].
16. Stakeholder planning and management. http://www.stakeholdermap.com/stakeholder-management.html [Accessed January 2013].
17. CSO (2011). Capacity Development Plan (2011–2014). Central Statistics Organisation, Government of the Islamic Republic of Afghanistan, Kabul (June).
18. Thompson, R. Stakeholder analysis. http://www.mindtools.com/pages/article/newPPM_07.htm [Accessed July 2011].

19. Stakeholder analysis. NHS Institute for Innovation and Improvement. http:// www.institute.nhs.uk/quality_and_service_improvement_tools/quality_and_ service_improvement_tools/stakeholder_analysis.html [Accessed July 2011].
20. Hartley, J., Fletcher, C., Wilton, P., Woodman, P. and Ungemach. C. (2007). Leading with political awareness. Chartered Management Institute. https:// www.managers.org.uk/sites/default/files/user35/CMI_-_LWPA_June_2007_-_ Executive_Summary.pdf [Accessed June 2011].
21. Welcome to MeTA. http://www.medicinestransparency.org/ [Accessed July 2011].
22. Multi-stakeholder alliance. http://www.medicinestransparency.org/key-issues/ multi-stakeholder-alliance/ [Accessed August 2011].
23. Institute of Development Studies. Assessing multi-stakeholder pocesses (AMPS): guidance & tools from an access to medicines project. http://www. medicinestransparency.org/fileadmin/uploads/Documents/Resources_MSH/ Toolkits/Assessing_Multi-stakeholder_Processes/Assessing_Multi-Stakeholder_ Processes.pdf [Accessed July 2011].
24. Construction sector transparency initiative. http://www.constructiontransparency. org/ [Accessed July 2011].
25. D'Herbemont, O. and Cesar, B. (1998). *Managing Sensitive Projects: A Lateral Approach.* London: Macmillan Business.
26. Haig, M. (2011) The seven stories that rule the world. http://abagond.tumblr. com/post/9584885600/the-seven-stories-that-rule-the-world [Accessed January 2013].
27. Droege, S.B. Stakeholders. http://www.referenceforbusiness.com/management/ Sc-Str/Stakeholders.html [Accessed August 2011].
28. Claridge, T. (2004). Definitions of social capital. http://www.socialcapitalresearch. com/definition.html [Accessed August 2011].
29. Gray, C. E. and Larson, E. W. (2006). *Project Management: The Managerial Process.* New York: McGraw-Hill.
30. Velicer, W. F, Prochaska, J. O., Fava, J. L., Norman, G. J. and Redding, C. A. (1998). Detailed overview of the transtheoretical model. http://www.uri.edu/ research/cprc/TTM/detailedoverview.htm [Accessed August 2011].
31. Stages of Change. http://www.uri.edu/research/cprc/TTM/StagesOfChange. htm [Accessed January 2013].
32. Wirth, R. A. (2004). Organizational change through influencing individual change: A behavior centric approach to change. http://www.entarga.com/orgchange/ InfluencingIndividualChange.pdf.
33. Global construction perspectives (2013). *Global Construction 2020.* Oxford Economics, http://www.globalconstruction2020.com/.

4

Measuring and Improving Global Project Performance

4.1 Introduction

Performance has long been a key factor in the management of projects. In the days before the concept of a project manager emerged, the approach to performance measurement was relatively rudimentary. Consequently, attempts to improve performance were generally lacking in sophistication. Nonetheless, there were some notable successes with regard to project performance: the Welsh castle-building programme of King Edward I, the Gothic Cathedral of Notre Dame in Paris and the Crystal Palace of the Great Exhibition could all be regarded as examples of 'good' project performance. However, while some of the criteria used to judge performance in each case were consistent across the three projects, other criteria were considerably different – Edward I was not as concerned with producing great architecture as was the case with Notre Dame, he was more concerned with exhibiting the authority of the Crown.

At the simplest level, the measurement of project performance appears deceptively uncomplicated – the use of time, cost and quality criteria allow for the most simple of measures. Any overspend on the budget and the project may be considered a failure and the performance as poor. In the absence of technologies or techniques that allow for a more robust monitoring and assessment of performance, a simple budget-based assessment can certainly be used to determine the success or failure of a project. The main problem with such an approach is that it tends to be reactive in that the project manager only knows how much of the budget remains once he has all the bills for the materials purchased and the labour hired. In the context of a global construction project, such a basic approach is simply not acceptable; the project manager has to adopt a robust approach to the management of performance in order to achieve project success. Managing in today's environment provides many challenges and global construction organisations (GCOs) are frequently confronted with situations that challenge the traditional ways in which construction projects have been managed. To facilitate an understanding of the nature of global construction challenges, this chapter will focus on project successes and failures in the industry.

4.2 Learning outcomes

The specific learning outcomes addressed in this chapter are to enable the reader to gain an understanding of:

>> global project success;
>> global project failure causes;
>> global critical success factors and key performance indicators;
>> international measures of project performance.

4.3 Defining global project success

All clients should be focused on the performance of their projects; each project will represent a means of obtaining something of importance or value to the client. The terms 'importance' and 'value' tend to be interpreted in financial terms such as achieving a desired return on investment or return on capital employed, both of which are valid measures of a project's end performance. However, such measures may not be particularly useful in terms of representing in-progress performance (they are essentially final measures applicable at the completion of the project) and they may not capture what it is about a project that is of importance to a client – in addition to the client there are also stakeholders to consider. The stakeholders may have no financial benefit to gain from the success of a project – it may be, as one example, that they would judge success in terms of damage limitation – and may have made no financial contribution to the project but they may, as was noted in Chapter 3, have considerable power in terms of delivering a successful project. An apparently simple question such as 'Was this project a success?' can therefore become quite complex when the project environment contains a myriad of measures used to indicate success (or failure).

A project manager needs to be as clear as possible with regard to how the client, along with any significant stakeholders, will be measuring success; a consistent definition of success is desirable but it has to be recognised that this will not always be the case. Success criteria and definitions used at the start of a project may well change considerably during the unfolding of the project. This can be a particular issue in the context of large and complex projects involving many stakeholders and clients (both as individuals and as groups) and having long durations. One example of how success criteria may change can be found by examining the Sydney Opera House project (Box 4.1).

Box 4.1   Vignette – Sydney Opera House project

The Sydney Opera House is a building that is in danger of becoming famous for illustrating how a project that, by either of the traditional success criteria of time and cost, should be classed as an abysmal failure – but which can actually be reclassified as a success simply by applying new criteria. In terms of the success criterion of cost, the project went approximately 16 times over the defined budget (the final cost in 1973 was

$AU102,000,000) – not the greatest achievement in terms of financial performance. When the success of the project duration is considered, the outcome is not particularly successful either – the project actually took four times longer to complete than was planned at the outset. And yet the completed building has become iconic to the extent that the 'failure' (in terms of budget and schedule) of the project has largely ceased to be of any significance (other than, perhaps, those having an interest in the study of project management). The words 'marvel' and 'wonder' are commonly used when describing the building that has become as representative of Australia as the Pyramids are of Egypt, and therein lies the measure of success that has become applied to the building: its impact on those who see it. The Sydney Opera House is of relevance when considering a definition of success in that it is a good example of how the product of a project can be judged a success (in this case almost entirely by stakeholders as opposed to the client – at several points during the project the client considered pulling the plug but the completed building is used by around 2 million people per year), while the project that delivered the product can be judged a failure. The project manager has to be as clear as possible with regard to how success is to be defined – is it to be on the basis of delivering good project performance (typically using quantitative measures) or is it to be on the basis of delivering a good final product (typically using qualitative measures)? In other words, is it the project or the product that is most important?

Source[1]

The irony that a poor project can deliver a great product, while a great project can deliver a poor product, is not lost on the majority of project managers (just ask the project managers involved in the Millennium Dome in London. Initially dismissed as a government vanity project, the completed building (the product) never made money during its planned use for millennial exhibitions (the company running the venue at the time ran out of money within four weeks of its opening) and was said to have been disposed of by the government for £1. As a project, however, the Dome was a success in that it was completed on time and within budget (although this was revised on a number of occasions) and was the largest dome in the world at that time). It is entirely understandable that a project manager focuses their effort on delivering good project management processes and procedures 'by the book'. As the 'custodian' of the client's interests, they rightly regard it as a matter of professional pride that they should focus their efforts on delivering a successful project. However, if the project manager has failed to appreciate that the client (and stakeholders, where appropriate) is actually more focused on the product as their determinant of success, then it is unlikely that project success will win the project manager any plaudits (from anyone outside the project management community) if the product is deemed a failure.

Determining the relevant success criteria over the duration of a project can be extremely difficult, but this can be eased by concentrating effort on agreeing the means by which everyone involved in the project can contribute to the determination of the success criteria to be applied. Rather than seeking to rigidly define a number of success criteria that will be enforced throughout the duration of the project, it can be more effective to adopt a stakeholder management approach that allows for the development (rather than imposition) of success criteria shared by all involved. This may initially seem to represent a duplication of effort – many projects (particularly those in the public sector) will be believed to have carried out appropriate stakeholder analysis prior to determining project scope and objectives. Nonetheless, it is important to be aware that any stakeholder analysis carried out prior to the implementation phase will rarely have focused explicitly on determining success criteria. There may well have been a focus on identifying 'desirable' and 'undesirable' outcomes for stakeholders but, as was shown in Chapter 3, covering stakeholder management, this is typically carried out in the context of a negotiation. In such circumstances it is usual for stakeholders to inflate criteria and thus the stated criteria will not always be reliable indicators of where the tipping-point between success and failure lies. What is required is a change of emphasis; rather than focusing on identifying stakeholder needs, the emphasis moves to involving stakeholders in expressing the factors moulding their *perception* of project success. Put simply:

> Project success criteria can also arise from the expectations, sometimes unstated, of individual stakeholders about what the project will accomplish and how it will accomplish it. Since these expectations may not be shared by – or even known by – the entire project team they can be sources of miscommunication, misdirection, and confusion later in the project. [2]

Even within the project team it is entirely possible that there is not a consistent clear understanding of the success criteria being applied. By focusing effort on fully defining applicable success criteria at the earliest stage of the project, any differences of perspective (it is highly probable that the majority of project team members will have similar but not exactly the same perspectives) can be identified and then resolved, thereby increasing the probability of the completed project being deemed to be a success. Along with working to define what would constitute project success, an understanding of project failure is also useful.

4.4 Understanding global project failure

Many traditional project managers approach their role on the basis of seeking to avoid failure. In many respects this is obviously a responsible approach to the delivery of what may be considered successful projects, but it does have one aspect that can be problematic: a fixation on risk avoidance. Whilst a robust approach to risk management is entirely appropriate for any scale of project, the pressure of being responsible for delivering a large and complex project successfully can result in an overly robust approach, in which the simple perception of some aspect of uncertainty becomes regarded as a risk that must be avoided. The problematic aspect of this is that it can result in opportunities being missed; uncertainty may represent an opportunity rather than a risk, but acting to avoid it (rather than analyse it) will mean that the project manager will never know if an opportunity has been missed. Ultimately, the very approach that was intended to 'protect' the project could lead to its failure. It is therefore important to both understand the nature of project failure and appreciate the contribute.on that such understanding can make to achievement of a successful project.

4.4.1 Global project failure causes

The need to understand the causes of project failure has resulted in various structured approaches to the development of a project proposal or business case. In each instance, the intention is to focus the project manager's attention on factors that typically result in project failure if not addressed. While these structured approaches tend to have a context that is specific to the industry in which a project is being carried out, there is a benefit in looking outside of the construction industry. As mentioned in Chapter 3, global construction projects can involve many different industries and stakeholders, so an awareness of how project management is approached in other industries can be beneficial. A relatively straightforward example can be found in the work of the UK's National Audit Office and Office of Government Commerce in identifying common causes of project failure. Eight common causes are identified, as presented in Table 4.1, along with a sample question (from the range available) to be asked in conjunction with each possible cause.

Several of the eight causes of failure relate to poorly handled stakeholder management, but others could be seen as reflecting the problem of assuming that traditional success factors (such as cost) will be applied, when in fact the success criteria may be non-traditional, such as long-term value for money (which is not quite the same thing).

4.5 Strategy-modelling failure

The strategic perspective (as presented in the first of the listed causes) is worthy of further consideration with regard to global construction projects, as such large and complex projects frequently have a 'visionary' value that it is possible to lose when project management processes and procedures – with their focus on operational issues – are developed and applied. In the context of 'normal' projects, success factors typically focus on issues such as having sufficient resources, competent team members, good communications and imposing relevant control mechanisms. In the context of global projects, culture-related success factors become more visible and the focus typically moves to consideration of managing diversity, technology differentials between regions, creation of worldwide strategic alliances and developing a wider sensitivity to the implications of project management decisions. It is therefore appropriate for the manager of a global project to be more focused on strategic issues than would be the case for a manager of a national or regional project. When viewed from this perspective it seems correct for global project managers to be concerned with success factors at the strategic level; failure to appreciate this shift of emphasis should itself be regarded as a possible cause of failure for global construction projects.

Table 4.1 Eight common causes of global project failure

Lack of clear links between the project and the organisation's key strategic priorities, including agreed measures of success.	Have we defined the critical success factors (CSFs) for the project? Have the CSFs been agreed with the key stakeholders?
Lack of clear senior management and ministerial ownership and leadership.	If the project traverses organisational boundaries, are there clear governance arrangements to ensure sustainable alignment with the business objectives of all organisations involved?
Lack of effective engagement	Have we secured a common understanding and agreement of stakeholders' requirements?
Lack of skills and proven approach to project management and risk management	Is there a skilled and experienced project team with clearly defined roles and responsibilities? If not, is there access to expertise, which can benefit those fulfilling the requisite roles?
Too little attention to breaking development and implementation into manageable steps.	Has sufficient time been built in to allow for planning applications in property and construction projects?
Evaluation of proposals driven by initial price rather than long-term value for money (especially securing delivery of business benefits).	Is the evaluation based on whole-life VFM, (value for money) taking account of capital, maintenance and service costs?
Lack of understanding of, and contact with, the supply industry at senior levels in the organisation.	Have we tested that the supply industry understands our approach and agrees that it is achievable?
Lack of effective project team integration between clients, the supplier team and the supply chain.	Have arrangements for sharing efficiency gains throughout the supply team been established?

Source[3]

4.5.1 Law of Requisite Variety and control

One possible cause of failure that is not explicit within the causes identified in Table 4.1 is complexity. While the theory of complexity can seem overwhelmingly complicated, it is not the intention at this point to enter the debate on the definition applicable to the construction environment. The key aspect in terms of the consideration of possible causes of failure in global construction projects relates to variety, and in particular what is commonly referred to as Ashby's Law of Requisite Variety (LRV). LRV is relevant in that it is a possible vehicle to bridge the apparent gap between the traditional project management focus on control at the operational level and the project management focus required by global projects: effective strategic alliances. Looking at the control focus first, the LRV perspective argues that a control mechanism or system can only be effective if it has sufficient *internal* variety to model the *external* system that it is seeking to control.[4] Unravelling that statement a little further leads to awareness that if an external system comprises two possibilities: P1 (building a wall using bricks) and P2 (building a wall using blocks), the control system required by the project manager must be able to model both of these possibilities. In other words, if the control system is only able to model P1 it will not be capable of controlling the wall-building operation if a decision is taken to use blocks (P2). The bottom line is that the control system simply does not possess sufficient internal variety. On this basis, Ashby's law can be used to indicate the maximum level of *internal* variety that any given control system can achieve, and this then represents the level of *external* variety (the project environment) that the system can realistically control. Figure 4.1 represents this relationship in more detail.

Figure 4.1 posits a project strategic aim and objectives, resulting in a project environment required variety level that necessitates eight different factors in order to measure the project's performance in a satisfactory manner. The LRV suggests that an eight-point control system would then be required in order to equal the variety that the project environment model could achieve; any less than eight points in the control system would result in a control shortfall, thereby leading to some degree of project failure. It seems reasonable to posit that the greater the control system shortfall, the greater the degree of failure (e.g. slight failure, partial failure, complete failure). If this relationship is a valid one, it

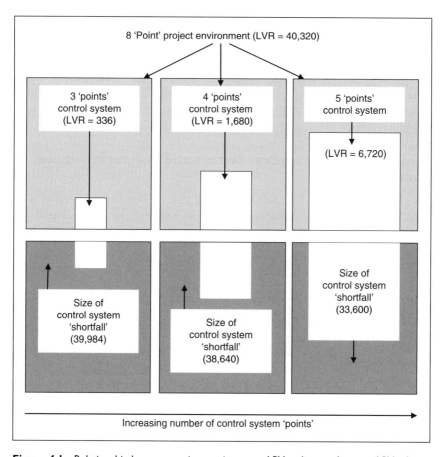

Figure 4.1   Relationship between project environment LRV and control system LRV values

would be of value to the project manager to know how much shortfall, if any, the project control system may have.

Working on the basis that a project's internal variety (LRV 'value') is a result of permutations (rather than combinations; permutations respect the required order of interactions whereas combinations do not) of performance factors/resources, then an eight-point project environment and an eight-point control system can model around 40,000 permutations. However, if the control system drops to containing only four points, it is capable of modelling only 1680 permutations. A two-point control system results in a modelling capability of only 56 permutations, which represents a considerable shortfall in terms of the variety possible within the project environment.

The application of LRV to control system modelling is a novel approach that may initially seem overly detailed. Indeed, until relatively recently the information demands of such an approach to managing a project were greater than the information-gathering capability of the ICTs available. However, ICTs are improving, as evidenced by the rising number of dimensions within commercially available BIM packages, and there is no reason to believe that this trend will not continue for the foreseeable future. In terms of measuring the performance of a global construction project, the traditional two-point (time and cost) approach is simply not up to the job. A complex project taking place over a long duration within a global environment has to be managed on a predictive basis; a reactive approach (seeking to control after an event has taken place, such as being overbudget) is too slow. Also, given that the control of such projects has been suggested as requiring to respond to the strategic needs (e.g. alliance forming) that will be unique to each project, it may be that a 'one-size-fits-all' approach to identifying the points to be measured (relevant to performance) is not particularly helpful.

4.6 Global critical success factors and key performance indicators (KPIs)

KPIs (Key Performance Indicators) are used in several industries, one of which is the UK construction industry. In the UK context, KPIs developed from the work published in *Rethinking Construction* (1998), and were initiated in 1999 as a means of measuring construction industry performance and progress towards achieving the targets set out in that publication:

1. Traditional processes of selection should be radically changed because they do not lead to best value;
2. An integrated team, which includes the client, should be formed before design and maintained throughout delivery;
3. Contracts should lead to mutual benefit for all parties and be based on a target and whole life cost approach;
4. Suppliers should be selected by best value and not by lowest price: this can be achieved within EU and central government procurement guidelines;
5. Performance measurement should be used to underpin continuous improvement within a collaborative working process;
6. Culture and processes should be changed so that collaborative rather than confrontational working is achieved.[5]

Of the six targets listed, the last two (targets 5 and 6) are most explicitly connected to project performance. Target 6 can also be considered as relating strongly to the previous section's suggestion that the performance of global construction projects is dependent upon the creation of strategic alliances. However, target 2 can be regarded as at least implying the significance of alliances with regard to project performance levels. Over time, the strategic aspect of KPIs has developed and the manner of their use by the industry has moved away from the original intention that they would result in detailed performance data for individual companies, which could be used by clients in the selection of contractors. The 'evolved' use has proven to be less demanding in terms of producing detailed statistical data, in that KPIs are more typically used by construction companies to benchmark their performance against their competitors or 'target' companies.[6] This has resulted in less focus on the statistical data gathered and, in recent years, KPIs have arguably become less useful in terms of achieving performance in terms of reaching *Rethinking Construction* targets. The less robust nature of data collection can also be argued to result in a less useful resource for the measurement of individual company performance. KPIs may not, on this basis, be a particularly useful tool in terms of identifying partners/contractors to potentially ally with on your next global project. A further aspect of the manner in which KPIs tend to be used, and that could be of concern, is the mantra that is usually not far from any mention of KPIs: 'What gets measured, gets managed'. All well and good if the particular KPIs chosen by a company are relevant to the performance of your global project. However, given that it is possible to access over 5000 KPIs (across all industries, not just the construction industry) it may well be that any given potential partner has selected KPIs that, while relevant to the company benchmarking its performance against others, are not particularly helpful in indicating their performance when you have selected a different set of KPIs. As far as the UK construction industry is concerned, the problem of incongruence (between 'your' KPIs and 'their' KPIs) may not be a particular issue if:

- your project is wholly or largely focused on the UK (so no need to consider e.g. Bulgarian, Argentine, perspectives on appropriate KPIs);
- your project, while not UK focused, makes use of a wholly or largely UK contractor workforce.

Even when the project is not global in nature, there remains a considerable diversity of KPIs available for selection; the BRE Group claim to have over 200 different measures available as the basis of building a set of KPIs.[7] With regard to construction industry performance, the range of KPIs narrows considerably – the Industry Performance Report 2010 focused on the following:

- client satisfaction – product, service and value for money;
- contractor satisfaction – overall performance, provision of information and payment;
- defects – impact at handover;
- predictability cost – construction;
- predictability time – project design, construction;
- construction – cost, time;

- profitability – return on sales (ROS);
- productivity – value added per human (employee) (VAPH current values) (VAPH constant 2000 values);
- safety – industry, contractors all companies;
- respect for people KPIs – all construction;
- environment KPIs – all construction;
- economic KPIs – all housing;
- economic KPIs – all non-housing;
- construction consultants' KPIs – year on year comparisons;
- mechanical and electrical contractors' KPIs – year on year comparisons.[7]

Such an approach, if applied consistently globally, would provide confidence in identifying, for example, any national construction industries that either wholly or in part scored below the global average on each KPI. Information of this kind could be particularly useful when considering an alliance with one or more contractors from a nation of which you have no previous experience. The ability to at least have an indication that country X appears to have problems in achieving average scores on KPIs – such as safety and respect for the environment – would be valuable. However, are KPIs approached consistently across the construction industries of all countries?

4.7 International KPIs/contractor comparisons

The KPI 'market' is one that covers players across a variety of industries in many countries. Nonetheless, there remains no clear consistent application of measures relevant to the range of KPIs typically focused on the UK market. Seeking to find one set of KPIs that will allow a like-for-like comparison between construction industries, construction companies and the completion of construction projects across a global spectrum is therefore not a viable approach. The tendency may then be to 'default' to more standard measures that are seemingly recognised globally. The concept of profit, for example, is one that is recognised in all countries irrespective of the cultural perspective on the making of profit; not all cultural groups adopt the 'pure' economic theory perspective that profit is good and should be the primary objective of all economic activity. The Saudi government, for example, have supported the development of the Saudi Responsible Competitiveness Index (SARCI) that is part of the government's desire to support the business climate by building responsible competitiveness through actions such as: 'through smart philanthropy; from their policies to attract, develop and retain a talented and diverse workforce; by going beyond legal requirements to comply with best practice business standards; through managing their supply chains to support local businesses and better environmental and social conditions'.[8]

The structure and methodology of SARCI is somewhat similar to that of the KPI structure and methodology used in the UK but the measures used are different; profitability, productivity, certainty of time, are not explicit within the measures used for SARCI. In addition, there is a sector within all industries that is essentially operating on the basis of being not-for-profit; there is recognition that income needs to at least match expenditure, but the concept of rewarding shareholders, for example, is not part of the culture of the not-for-profit organisations. Nonetheless, this sector of the global construction industry can be involved in some significant projects. Mawhinney noted that there are (as an acknowledged simplification of a complex environment) essentially four sources of funding for construction projects: public sector, private sector, DBFO (design, build, finance and operate) and aid.[9] The relationship between these sources varies over time (for example, as a country's economy rises and falls) and between regions. Aid funding is a significant component of construction activity in underdeveloped countries/emerging economies, as evidenced by the activities of organisations such as UNESCO, WHO and the Aga Khan Foundation. The latter organisation, even as an essentially private organisation (presenting itself as a not-for-profit, non-denominational development agency) funds aid activities (typically in conjunction with other development agencies) in countries ranging from Pakistan to Portugal and covering health, education, urban development, environment and other areas.[10] The foundation's development network (AKDN) employs in the region of 80,000 people globally, had a budget of US$625 million in 2010 and its projects generated revenues of $2.3 billion (reinvested into future projects, etc.) and is therefore a significant global 'client'.[11] As a simple comparison, a country with a current account balance of $2.3 billion would, in 2010, have ranked as the 42nd wealthiest sovereign state in the world (displacing Angola at $2.089 billion), while a current account balance of $625 million would result in 50th position (above Ukraine).

The sums of money classed as 'aid' can seem huge – since 1970 it is estimated that $3.04 trillion has been donated – but remain under half of the money that was promised by rich nations (the shortfall is approximately $4.17 trillion.[12] However, Mawhinney suggests that public sector funding is roughly equal to aid funding as a country develops, with the typical model being one of aid funding reducing over time and public sector funding 'making up' the reduction as national wealth increases. Ultimately, a country develops to the point where public sector funding starts to decrease and then the private sector takes over the responsibility of making up the reduction. Overall, there is a slight upward trend, over time, in terms of the value of work available to construction companies, and this can be attractive in terms of global companies moving into new markets. However, such companies need to consider if they are applying performance measures relevant to the overall distribution of funding types within a nation's economy at any given point in its development. If, for example, a country is classed as 'emerging' it may be relying heavily on aid funding to pay for infrastructure projects, etc. Funders in this sector will typically have different success criteria than those from the private sector.

In 2010 there were 192 'recognised' countries in the world. Of these, 131 had negative current account balances; the natural assumption would be that such countries would largely be the developing/emerging countries and be heavily reliant upon aid funding. However, that assumption would not be entirely accurate; while the first country listed as slipping 'into the red' is Guinea-Bissau (number 62 on the list) there are also a number of developed, supposedly rich nations, listed as being in deficit. Iceland (−US$0.042 billion, number 69), Belgium (−$1.126 billion, number 139), New Zealand (−$4.504 billion, number 169), Canada (−$52.60 billion, number 184), UK (−$66.60 billion, number 188) and, by far the most indebted country, the

US (−$599.90 billion, number 182). As a global construction organisation, which country would you rather do business with − China ($305.40 billion, number 1) or the US? In terms of an explicit measure of national performance (current account balance) then China would seem preferable simply because it appears to have more cash available. Unfortunately, when dealing with global issues things are rarely as simple as they may seem. It is worth considering, for example, that the largest holder of US debt outside of the US is China (with around $1.2 trillion of US debt), with Japan owning the second largest amount (£912 billion) and the UK owning the third largest amount ($347 billion).[13] Furthermore, in 2010 China overtook the US as the country spending the most on construction projects. As mentioned previously, global players need to be very careful when selecting the measures of performance that they use when considering both the markets to enter into and the organisations to form alliances with.

4.8 International measures of project performance

The international market for construction is difficult to quantify accurately in terms of money spent − inevitably there will always be an indeterminable amount spent on projects that are either too small to be recorded or too 'informal' to be recognised as a structured project. Those projects that are captured in official statistics do, however, add up to a considerable amount of money. In 2010 global construction spending was estimated as $4.4 trillion (a decrease for the third year consecutively) split between five trading blocks: Africa, Asia Pacific, Europe and the Middle East, the Americas, and Australasia.[14] Generally, predictions are that construction projects will be increasingly based in developing countries and that the currently developed countries will drop below 50 per cent of the market sometime around 2020. It is inevitable that the established global construction players will increasingly have to undertake work in developing countries, with China and India being predicted as the two main markets. However, is this of concern to construction companies globally?

One perspective on the changing make-up of the 'international' construction market is that it is really of little concern to the majority of construction companies. Girmscheid and Brockmann proposed a market structure based on six different scales of organisation:[15] regional, national, international, multinational, global and transnational. At the regional and national levels, the typical distribution of companies (based on size: number of employees, turnover, etc.) is around 95 per cent of companies at the SME (small to medium enterprise) level, leaving only 5 per cent of companies in the 'large' category. The market strategy of SMEs is based on increasing numbers of local networks as the company increases in size, and consequently these companies do not usually concern themselves with work at the international level and above, and so these markets are the domain of the large construction players. Girmscheid and Brockmann argue that it is difficult, given this kind of distribution, to regard the various national construction markets as being a global market. Nonetheless, there are a relatively small number of construction players that operate at the international market level and above; such players should be regarded as national players operating in a series of regional and national markets, both of which are composed of local networks.

Regarding organisations such as Skanska, Hochtieff and Bouygues not as global construction organisations, but as national organisations operating across a number of discrete national markets structured around local networks, is a significant change of perspective. However, in the context of seeking to identify international measures of project performance, such a perspective is a valuable one in that it allows the project manager on a 'global' project the freedom to think locally and measure performance on the basis of relationships between each level of market (local to transnational). In such a scenario a global project manager may have to deal with relationships between regional, national, international, multinational, global and transnational market strategies. Given that the global construction company (as an organisation) should be focused on implementing its transnational development strategy, the company's success criteria should, as discussed previously, be congruent with those strategic aims and objectives. However, given that there will inevitably be a high proportion of local networks being relied upon in the delivery of any global project (mega projects included), the selected performance measures should be in tune with the dominant culture within each of those localities.

4.8.1 Linking local and the transnational performance measures

Construction companies, as with any other commercial organisation, have to make a profit in order to survive. Admittedly, any cash reserves or credit facilities available could be used to 'smooth' any cash-flow difficulties but, over the longer term, the requirement for sufficient profit to survive remains. In this regard global construction players have both an advantage and a disadvantage in that they are not locked into a single regional or national market (this both distributes risk and exposes them to risk depending upon regional changes). While the consideration of effectively managing risk is a key activity for all construction players, it seems that global construction players are potentially exposed to a particular form of risk: achieving a profitable market share. There is some evidence (Mawhinney[9]) that market share can be used as an indicator of performance achievable with regard to profitability. A rule of thumb proposed by Kotler[16] is that profitability increases with rising market share up to a ceiling of around 40 per cent of the market. Applying this heuristic in a quite arbitrary manner could result in a strategy by a global construction player to take 40 per cent of the global construction market. Based on the 2010 figure of global construction spending as $4.4 trillion, our ambitious global player would be aiming for in the region of $1.8 trillion of business very year.

To put this in context, in 2002 the top international contractor (Skanska) had total revenue of $13.951 billion – a few zeros would have to be added before even Skanska approached the 40 per cent market share ceiling for profitability. Given the previous assertion that global construction players are essentially national companies operating transnationally in national markets structured on local networks, it may be assumed that such companies are major players in their home market, but this is not always the case: Skanska's home market accounted for around 17 per cent of its total income in 2002. In the same year, the fifth largest multinational construction company (France's Technip-Coflexip) had almost no home market share at all (based on a contribution to total revenue of around 0.007 per cent). Such a company could be regarded as having no significant concerns regarding the economic situation or the performance levels of construction projects within what would be considered its 'home' market. It would, however, be very interested in the same factors within the regional and national (local) markets falling within its transnational strategy.

At the transnational level the company focus would realistically be expected to be on strategic issues. Bouygues Construction (a global player in the industry), for example, states that its strategy is 'to increasingly operate within the framework of end-to-end contracts (public–private partnerships or concessions) which allow it to deliver high added value'. As with all strategies, the intention is to outwit the companies that are being competed against, and one standard approach is to apply or present a means of clearly differentiating 'your' company from all its competitors. In the case of Bouygues this differentiation is on the basis of sustainable construction being embedded in all of its projects.

Bouygues Construction is therefore in the situation of having a strategic-level performance criterion of increasing the proportion of end-to-end projects within its portfolio, whilst also having a 'local' performance measure of evidencing support for sustainable development. If these two performance measures become disconnected then the company could find itself internally conflicted; transnational managers may be solely focused on obtaining end-to-end projects (of any focus), while local project managers are focused on delivering projects that support sustainable development. Such a situation is less likely to occur if there are clear performance measures across all objectives at all levels within the company. Unfortunately, global and transnational construction players cannot apply the standard tactics as employed by equivalent players in other industries (of various forms of manufacturing/producing for inventory) and they therefore have to approach the development of performance measures from a different perspective.

Box 4.2   Vignette – global branding of a construction player

Global branding essentially relies upon a standardised 'product' or 'service' that can be recognised within, and moved between, an increasing number of local markets as the brand expands. The construction 'product' is dissimilar to most non-construction products in that few construction players have a product that can actually be moved to new markets. A construction product is typically 'localised' in that it requires a physical site and is built using largely local networks of suppliers. There are a small number of exceptions to the latter constraint; companies such as Huff Haus (http://www.huf-haus.com/en/home.html) market their 'brand' within six different market sectors (single-family houses through to commercial buildings) and across 16 different 'local' markets (Norway, UK, Canada, Italy, etc.). The Huff Haus 'brand' is manifested in high-quality, energy-efficient, tailor-made, prefabricated (kit) buildings, typically made predominantly of wood and glass, all of which are made in a single factory.

In strategic management terms, Huf Haus would be regarded as building their brand through marketing to a specific niche of buyer. Such a combination of niche marketing within a global market environment is highly unusual for a construction player, and a key factor in the company's success is the realisation that each 'product' (building) has to be site-specific (responding to the nature of the site and local regulations, building codes, etc.), yet recognisable as a Huf Haus even though each house is tailor-made and of a consistently high quality across all markets in which it is sold.[17] For construction players trying to build a global brand without the kind of specificity formulated by Huf Haus, the process becomes more difficult.

A player such as Bouygues Construction, for example, has no 'product' – it builds whatever its clients require it to build and each product will be unique. It is therefore difficult for Bouygues to achieve economies of scale in the manner that Huff Haus can (with its focus predominantly on wood and glass materials within its products). Nor can it achieve efficiency of production by manufacturing for inventory (effectively placing completed buildings into storage until someone comes along to buy one) in the manner that global brands such as BMW and Ford can. Bouygues Construction, as with other global construction players such as Skanska, has to accept that by focusing on construction projects it is actually disadvantaged if it is seeking to build a global brand. There is evidence that Bouygues Construction has indeed accepted this constraint and are not focused on the strategy of building a global construction brand: 'Bouygues Construction operates on international markets on a long-term basis through local subsidiaries.'[18] The key component of this statement is that the company is seeking to operate on a long-term basis; their local subsidiaries therefore have to perform well in their own context or it will be difficult to obtain repeat work. Bouygues also appear to have recognised this within their use of sustainable development, as a factor differentiating their 'brand' from those of other global construction players.

The company's outward-facing vision of sustainable development involves activities such as participating in the economic and social life of individual regions, respecting human rights, and exhibiting patronage through the activities of its charitable foundation (Terre Plurielle). However, a more internally focused vision of sustainable development is its value in terms of 'creating' future projects. By starting with an individual project, such as an office block, the company aims to gradually broaden its approach to the scale of the neighbourhood, and then to the city. If this approach is to be successful (and thereby contribute to delivering financial performance at the global/transnational level) it has to perform sufficiently well at the local level to be seen as a credible 'partner' in delivering value-added projects to neighbourhoods and then cities. Being successful at the local level is highly dependent upon understanding locally relevant and culturally acceptable measures of performance.

A requirement to achieve sustainable long-term involvement in a local market can only be met by fully understanding that market and the networks that operate within it. By collaborating, rather than competing with, and adding value to such networks, the probability of long-term success increases. Performance measures applicable at the local level should focus on determining if the appropriate level of value has in fact been added to the functioning of the existing local networks, whether this is through 'traditional' approaches such as upskilling of local labour and providing management training, or through some other 'non-traditional' activity.

4.9 Collaboration and project performance measurement

Earlier in this chapter it was noted that aid funding is a key component in the level of global spending on construction projects, and that many of the resulting projects have a 'social' aspect with regard to the judgement of them being either a success or a failure. The social aspect of such projects also contributes to one of the typical causes of failure: a lack of clear understanding as to what would be considered relevant (and therefore of value if facilitated by a particular project) by the local 'consumers'. An example of this problem is the Lake Turkana fish-processing plant, a $22 million project that 'failed' for a few simple reasons related to the nature of the local market, which will be outlined below.

4.9.1 Collaboration failures

Lake Turkana is in an otherwise arid region of Kenya and supports around 300,000 people, with the single largest group being the Turkana people. In 1971 the Norwegian government, in collaboration with the Kenyan government, initiated a project to provide the Turkana people (a largely pastoral group) with employment opportunities and enhance their social status amongst other local groups (which had suffered due to a variety of factors, but largely because they were perceived as being poor and unable to live up to the requirements of a nomadic way of living). This led to problems such as the Turkana being perceived as unsuitable marriage partners. The proposal was to address such issues through building a fish-processing plant, with the project being funded by NORAD, the Norwegian aid agency. This proposal built on earlier work by the Kenyan fisheries department, which had sought to encourage the Turkana away from the pastoral way of life (with its pressure on arid land for grazing, etc.) and towards a less vulnerable way of life based on fishing and fish processing.

From a technical perspective the project seems to have been entirely reasonable in that all of the required resources were identified and supplied.[19] A particular resource required was an ice-making plant (for refrigerating and storing of the processed fish for export) and this seems to have absorbed around half of the project budget. However, ice-making plants demand large amounts of energy and water. The latter resource began to be less available around 1983 and, along with making it more difficult to justify the manufacture of ice, resulted in increasing levels of salinity in the lake (which is fed by only one river that flows all year). While the link between salinity levels and fish population has not been fully established in the context of Lake Turkana, fish stocks began to fall as salinity levels increased. Overall, the fish-processing plant project was starting to be squeezed by a number of critical factors. One further factor that is cited as being the crunch factor is that the Turkana people have no history of fishing or eating fish as a consistent component of their diet – the assumption being essentially that they did not manage to make the transition from a largely pastoral way of life to one based largely on fish: 'The factory proved to be an unsustainable business due to its geographical remoteness, the nomadic culture

Table 4.2   Typical weaknesses leading to failure of aid projects

Weakness number	Form of weakness
1	One- or two-dimensional definition/understanding of poverty or socio-economic status.
2	No consultation or needs assessment with the active participation of the community itself before implementing projects.
3	Disregard of the large political and economical context and history/genealogy of development issues and sectors.
4	Poor implementation due to a lack of understanding of on-the-ground economic and cultural context, societal norms, gender dynamics, etc.
5	Sectoral/fragmented approach to development initiatives – ignoring the complexity and overlaps in situations.
6	The mentality that the project's business goals and human rights or any social mission are mutually exclusive – the two can be integrated within a 'for social profit' scheme.
7	The mentality that installing hardware or technologies can solve complex issues without adequate integration of socio-economic and political factors.

Source[21]

of the workers needed to keep it up and running, and the cultural perspective on fishing in general in a society where owning cattle is a sign of wealth. The factory is now largely unused and has not contributed to the growth or development of the region as intended.'[20] Overall, the project was one of good intentions, manifest in collaboration between two countries and a number of government and NGO groups; it was adequately funded and appears to have set realistic targets for scheduling of work etc. Unfortunately, it is classed as a failure because its good intentions did not take account of several of the typical causes of failure in aid projects (as listed in Table 4.2).

A particularly ironic aspect of the Lake Turkana project is that, since the mid 1980s, the impact of changing climate conditions have been added to by another, much larger project that has a more global span in terms of the players involved: the Gibe III dam in Ethiopia. This project is near to completion and is projected to meet not only all of Ethiopia's electricity demands but should also produce a surplus for export. One of the countries identified as an export market is Kenya. The Gibe dam is also built on the only river that has a year-round flow into Lake Turkana, and this has raised concerns that water levels in the lake will drop further as the Ethiopian government slows the flow of the river in order to fill the dam.

A final point for consideration is that, while the Ethiopian government claims that the cost of the dam (estimated at around €1.6 billion) will be met entirely from government funds, there is a suggestion that China's Dongfang Electric Machinery Corporation has signed a memorandum of understanding to provide machinery for the project and that this is backed by a loan (to the Ethiopian government) from the Industrial and Commercial Bank of China for $500 million. Other countries are also suggested as being involved: Italy to the extent of $250 million and the African Development Bank ($250 million may yet be made available). The World Bank is suggested as having decided against any involvement in the project, due to concerns of a lack of competitive bidding prior to the engaging of the main contractor (Salini Costruttori of Italy). Salini was also the main contractor for Gibe II (which hit technical problems shortly after completion and was unable to generate electricity for 11 months) and has been engaged as the main contractor for a new dam project in Ethiopia with an estimated budget of $4.8 billion.[21] Only time will tell if Gibe III is classed as a failure (as a result of its environmental and societal impacts) or as a success (as a result of its energy and revenue generation impacts).

4.9.2 Collaboration successes

Khabarovsk is a Far Eastern Russian state that borders on the north of China and traditionally relies on forestry and fishing as its main areas of economic activity. The Nanaiski Raion, or region, covers around 22,000 square kilometres and has a population of roughly 20,000 people. In recent years, the post-communist reforms have resulted in this region suffering from overaggressive timber extraction and fishing activity. Combined with escalating unemployment (at one point the average was around 60 per cent and in some villages would be 100 per cent) Nanaiski suffered badly during the 1980s and 1990s.[22]

The problems of Nanaiski were recognised internationally when the Canadian Model Forest Network (CMFN) offered its expertise in sustainable forest management (in the form of model forests) to the region. CMFN defines a model forest in terms of a framework composed of six 'philosophies': partnership, landscape, commitment to sustainability, programme of activities (relevant to CMFN's vision and local stakeholder needs, etc.) and knowledge sharing, capacity building and networking.[23] This framework was the basis of the proposed Gassinski Model Forest, the first such forest in Russia and, from a project management perspective, more importantly the first international aid development project to be attempted in Nanaiski: there were no previous attempts that the project could 'learn' from. Based on CMFN's framework, the first step was to undertake partnership-building activities by discussing needs, desires and requirements with all who relied upon the existing forest. Commencing in 1994, three years of discussing, planning and working towards the identification of objectives agreeable to all stakeholders culminated in the Natural Resources Based Economic Development Project. The objectives of this project were as follows:

- Development of value-added wood-processing capacity.
- Development of non-wood forest products enterprises.
- Development of a regional tourism strategy, infrastructure and enterprises.
- Development of commercial projects between Canadian and Russian indigenous people.
- Creation of specially protected areas.

In comparison to the Lake Turkana aid project, the development project structured around Gassinski Model Forest seems significantly more ambitious. Under such circumstances it would be reasonable for the more risk-averse project managers to recommend caution and perhaps a scaling-down of the project (particularly given that such an ambitious project had relatively little funding – $2.7 million over four years from the Canadian partners[24]). However, the Lake Turkana project failed, in large part because it commenced with a relatively simplistic perception of the problem being addressed: it connected an apparently impoverished community with a resource that, from a Western perspective, had considerable financial value. Bring the two together and the problem is solved! The Gassinski Model Forest project took a much broader perspective on the problem and, while not wholly successful, proved to be more successful than the Lake Turkana project. The areas where the Gassinski project was overall successful, and also where it enjoyed little success, related to the degree of multidisciplinarity applied. An example of where a multidisciplinary and multicultural philosophy was not effectively applied relates to the first objective: development of value-added wood-processing capacity:

> ...to reinvent the thinking of local producers was much more consuming of time and effort. Convincing enterprise managers to invest in new equipment was not as successful as the project planners would have liked it to be – the managers had never used the kind of equipment the Canadians showed them and did not believe that their old processes for handling wood needed to be transformed so dramatically.[25]

The project managers retrospectively concluded that:

> ...the slow integration of new kinds of equipment and business ideas could have been more successful than the attempt made to change all local production methods within the first year of the economic development project. The Russian project participants would have been more accepting of the foreign ideas if their own traditions, even if economically inefficient or unsound, were not immediately labelled as such and rejected.[25]

An example of where a multidisciplinary and multicultural philosophy was effectively applied relates to the development of non-wood forest products:

> ...the Russian attitude towards this part of the economic development project was more positive from the start also because the initiative showed locals that there is brand new economic potential in their communities and thus allowed Russians ownership of the products from beginning to end.[25]

The Canadian partners focused the business training on the creation of new businesses and entrepreneurial activity in appreciation (arising from their experience regarding the first of the project objectives) that focusing on making old businesses 'successful', and thereby labelling them as being 'failures', would not gain the trust of the Russians. The result of this revised approach was a more positive investment in the 'new' business models, thereby contributing to success in achieving the second objective of the project. Arguably the success achieved by the project flowed from the multidisciplinary project team partners creating a stable environment in which the objectives of the project could be worked on without the distractions of the turbulent Russian economic environment.[26]

4.10 Project performance measurement and improvement

Weaknesses that lead to the failure of aid projects (see Table 4.2) indicate problems that may also affect the performance of non-aid projects. Obviously many of the environmental factors will differ between the aid and non-aid projects, but where those projects are both 'construction' and fall within the context of a strategy such as that of Bouygues (to start with a single building, then move to the neighbourhood and finally to the city level of involvement in a local market), then the non-aid projects would benefit from considering the causes of failure in aid projects. Indeed, there is a case to be argued, based on the societal aspect of projects when considered in terms of 'branding' a global construction player within a local market, for non-aid projects to seriously consider the use of performance measurement criteria based around aid-project weaknesses. Developing such measures in the context of an awareness of the positive use of multidisciplinary and multicultural approaches, as a factor in the success of aid projects, would add to their value both in terms of measuring current performance but also in identifying areas that require further work if future projects are to perform better. Whilst a simple measure such as profit achieved may be accepted as an adequate measure of performance, it does not have the same value as a multifactor approach to performance measurement when the objective is to improve project performance over time.

Table 4.3 presents suggested objectives for performance measures and local 'interpretation' of each. The suggested interpretations are not intended to be definitive – as previously discussed, the application of a 'one-size-fits-all' approach to the management of global construction projects is not desirable, even though its apparent simplicity may suggest it to be a relatively easy one to adopt. A global organisation seeking to apply a monocultural approach to measuring project performance and success is understandable in terms of its

Table 4.3 Objectives for performance measures

Form of weakness (as Table 4.2)	Possible performance measure
One- or two-dimensional definition or understanding of poverty or socio-economic status.	Develop a multidisciplinary approach to understanding of key issues that is relevant to a specific local market.
No consultation or needs assessment with the active participation of the community itself before implementing projects.	Identify and establish regular communication with key communities or community leaders
Disregard of the larger political and economical context and history/genealogy of development issues and sectors.	Establish the economic, political and social perspective taken locally.
Poor implementation due to a lack of understanding of on-the-ground economic and cultural context, societal norms, gender dynamics, etc.	Identify cultural 'green lights' and 'red lights' relevant to the project's interaction with the local market. Monitor over time.
Sectoral/fragmented approach to development initiatives – ignoring the complexity and overlaps in situations.	Actively seek to identify and collaborate with others undertaking projects with overlapping objectives.

The mentality that business goals and human rights or any social mission are mutually exclusive – the two can be integrated within a 'for social profit' scheme.	Develop measures for the creation of 'social' profit relevant to specific local markets (one size does NOT fit all).
The mentality that installing hardware or technologies can solve complex issues without adequate integration of socio-economic and political factors.	Assess the skill level of local employees and users, along with any cultural barriers to training.
Beneficial multidisciplinary and multicultural objectives	
International development projects need to be based in the communities that they hope to help in order to achieve their goals.	Avoid 'remote' decision-making procedures and management structures.
Explicit recognition of the value in achieving general long-term sustainability of the community and its traditions.	Community (local market)-centred activities that are achievable within current conditions and within any short–medium term training/economic developments.

need to convince the financial markets to invest in the company. When the degree to which a company is attractive to investors is assessed against measures such as return on capital employed (ROCE) and internal rate of return (IRR), there will inevitably be a tendency for the company to apply the same or similar measures to its individual projects, irrespective of the project being in Berlin or Bishkek. Whilst such an approach may possibly be applied by an organisation with a clear brand identity and 'branded' products, global construction players do not fall within these parameters. They therefore have to satisfy their global investors while building a business base in potentially different local markets, by collaborating with a range of other organisations and identifying and connecting with relevant local networks. Buying a sewage-treatment plant is not the same kind of purchase as buying a pair of trainers, and project performance is not an 'instant' measure in the manner of counting the number of T-shirts sold today.

4.11 Chapter summary

Measures of project performance, whilst responsive to the need to determine profitability over time, should not, in the context of global construction projects, be applied in an indiscriminate manner. There is a real need for global project managers to consider the typical factors that contribute to aid projects failing to deliver their intended objectives, irrespective of whether the market in which the project is carried out is in a developed, developing or underdeveloped country. Project-level performance can be measured on a variety of bases and, unless there are specific local requirements to collect and present data and information in a particular manner, global construction project managers should regard this as an opportunity to be creative.

4.12 Discussion questions

1. Outline the suggested nature of the link between LRV (Law of Requisite Variety) and project failure.
2. Discuss the significance of aid funding to global construction companies and how this type of funding can introduce different success criteria that such companies may not be familiar with.
3. Outline the form of uncertainty that can occur when contractors implement different strategic objectives at differing levels (such as international and regional) within their operations.
4. Collaboration is a common strategy adopted by global contractors. Using the Lake Turkana project as an example, discuss possible areas of uncertainty that can lead to the failure of such collaborative ventures.
5. Discuss the problem of uncertainty arising from 'branding' in the context of measuring project performance as opposed to company performance.

4.13 References

1. Sydney Opera house. Australia. http://australia.gov.au/about-australia/australian-story/sydney-opera-house [Accessed June 2011].
2. Honeymoon hangover: the happy beginnings of a creeping scope. USA. http://www.federalpm.com/articles/startup_p3.php [Accessed May 2011].
3. Common causes of project failure and their remedies. Chapter 9 (pp. 169–175), Scottish Capital Investment Manual, Scottish Government Health Directorate. Scotland. http://www.scim.scot.nhs.uk/PDFs/Manuals/BC/BC_Guide_6.pdf [Accessed June 2011].
4. Heylighen, F. and Joslyn, C (2001). The law of requisite variety. *Principa Cybernetica Web* (Principa Cybernetica Brussels). http://pespmc1.vub.ac.be/REQVAR.html [Accessed August 2011].
5. Bishop, P. (2005). What is rethinking construction? Local Government Task Force, UK. http://www.constructingexcellence.org.uk/pdf/lgtf/lgtf_newsletter_jan_05.pdf [Accessed June 2011].
6. MBS (2008). Review of construction key performance indicators. Manchester Business School, University of Manchester (April).
7. Hutchinnson, V. (2012). Key performance indicators (KPIs) for the construction industry. BRE Group, UK. http://www.bre.co.uk/page.jsp?id=1478 [Accessed June 2011].
8. Saudi Arabia responsible competitiveness initiative (2008). *The Saudi Responsible Competitiveness Index*. Saudi Arabia. http://www.accountability.org/images/content/0/8/080/The%20Saudi%20Responsible%20Competitiveness%20Index_January.pdf [Accessed July 2011].
9. Mawhinney, M. (2001). *International Construction*. London: Blackwell Sciences.
10. Aga Khan Foundation (2010). *Current List of Projects*. Aga Khan Development Network, Switzerland. http://www.akdn.org/akf_projects.asp [Accessed August 2011].
11. About the Aga Khan Development Network: Aga Khan Development Network. http://www.akdn.org/about.asp [Accessed July 2011].

12. Shah, A., (2012). Official Global Foreign Aid Shortfall: $4 Trillion. *Global Issues*. http://www.globalissues.org/article/593/official-global-foreign-aid-short fall-over-4-trillion [Accessed June 2011].

13. Murse, T. (2011). How much US debt does China really own? US Government Info, US. http://usgovinfo.about.com/od/moneymatters/ss/How-Much-US-Debt-Does-China-Own.htm [Accessed August 2011].

14. Crosthwaite, D. (2011). *World Construction 2011*. (review/outlook) Davis Langdon, UK.

15. Girmscheid, G. and Brockmann, C. (2006). *Global players in the world's construction market*. In Proceedings of the Joint 2006 CIB W065/W055/W086 International Symposium, Rome and Naples. Swiss Federal Institute of Technology Zurich, Institute for Construction Engineering and Management.

16. Kotler, P. (1997). *Marketing Management: Analysis, Planning, Implementation and Control*. 9th edn. Hemel Hempstead: Prentice Hall International.

17. *Inspired Living*. Huff Haus, Germany. http://www.huf-haus.com/en/home.html [Accessed January 2013].

18. Strategy. Bouyges Construction, France. www.bouygues-construction.com/96i/group/strategy.html [Accessed July 2011].

19. NORAD (1985). *Fisheries Development*. Oslo: NORAD.

20. Keene, C. (2007). Case studies of a few 'failed' projects. Globalhood, US. http://www.globalhood.org/casestudies.shtml [Accessed August 2011].

21. Muchira, J. (2011). Ethiopia awards engineering contract for contested Nile hydro project. *Engineering News*, South Africa. http://www.engineeringnews.co.za/article/ethiopia-2011-05-13 [Accessed August 2011].

22. Globalhood (2007). Annual Report: http://www.globalhood.org/news.shtml [Accessed July 2011].

23. *About Model Forests*. Canadian Model Forest Network, Canada. http://www.modelforest.net/index.php?option=com_k2&view=item&layout=item&id=4946&Itemid=142&lang=en [Accessed July 2011].

24. http://www.mcgregor.bc.ca/russian_activities/ [Accessed August 2011].

25. Korchumova, S. (2007). Development projects that work: Multidisciplinarity in action. Globalhood, US. http://www.globalhood.org/articles/briefingnotes/Development_Projects_that_work.pdf [Accessed July 2011].

26. Heathman, R. About McGregor's Russian activities McGregor, Canada. http://www.resourcesnorth.org/russian_activities/about_ra.htm [Accessed January 2013].

5

Relevance of Global Project Structure and Organisation

5.1 Introduction

Within the context of managing global construction projects there is a need to consider the strategic nature of what is being attempted by the project directors or project managers; marshalling the resources required for the completion of such demanding projects may realistically be likened to engagement in a battle focused on a single location (the battle-field) but making use of resources (allies) from many different countries elsewhere in the world. Given the geographically distributed nature of the allies involved in conflicts since the turn of the century (e.g. Iraq, Afghanistan), this aspect of 'managing' the manner in which multiple allies work together in the battle-field engagement has attracted increased research and development (Northrop Grumman, ND). Likewise, civilian (rather than military) projects of a global scale have increasingly focused on managing 'engagement' through application of theory relevant to issues such as collaboration, communication and supply chain management. For example, Lamber and Cooper (2000) noted that supply chain management on a global scale required a broadening of the traditional definition to something more reflective of the current environment:[1]

> Supply Chain Management is the integration of key business processes from end user through original suppliers that provides products, services, and information that add value for customers and other stakeholders.

Such a broad perspective on obtaining resources (both human and non-human) requires leadership of the overall strategy, along with the implementation of an appropriate structure (or range of structures) for the project itself and also the organisation(s) involved in the delivery of the project. Ultimately, a project manager needs organisational methods that facilitate teamwork, maximise the use of limited resources, and drive efficiency and quality. In this way, the gradual completion of the project achieves the goals identified. Those goals should flow from an overall strategy deemed appropriate to the stated objective(s) of a specific project, and their achievement depends upon several factors. One factor is the selection of a project structure and form of organisation that is relevant to that particular strategy. However, managers of global projects should not succumb to the temptation to impose a particular structure simply because they are familiar with it and feel comfortable when using it; global projects are typically more demanding of the project manager and he or she must be aware of the need to be both flexible and creative when responding to the demands of such projects. Thus, this chapter will consider the relevance of strategy, but this is not the sole focus; relationship building (hierarchy), the nature of structure (traditional or transformational), and the nature of value chains are all valuable components of managing a global project. Prior to dealing with these components, it is worth noting that each is supported by a considerable body of literature that cannot realistically be covered in a single chapter; the intention here is to introduce the key items relevant to each component in a manner that provides a framework for their integration in the context of managing a global construction project. Through consideration of the resultant framework, the would-be global project manager will be better able to determine an appropriate structure or structures (a long-duration project may need to change structure more than once) for a specific project. The order in which the components of the framework will be considered reflects the suggested order in which the project manager should consider and apply them for the majority of global projects (although there will always be at least one exception to any management 'rule'!):

1. Strategy.
2. Relationship-building hierarchy.
3. Possible traditional (transactional) structures.
4. Value chains and possible new forms of structure.
5. A possible non-traditional (transformational) structure.

5.2 Learning outcomes

> The specific learning outcomes from this chapter are to enable the reader to gain an understanding of:
>
> \>\> modern global project management;
> \>\> structures for creating a global project and client/parent organisation relationship;
> \>\> awareness of global organisational value chains.

5.3 The origins of strategy

Modern-day strategic management can trace its roots back to the early days of the ancient Greek civilisation. The word strategy has its origins in the Greek 'strategos', meaning 'army leader'. Indeed, the term has been retained by the Greek military and is currently applied to the highest officer rank (equivalent to that of a general). Thus strategic management is essentially concerned with outwitting (or in a commercial context, outcompeting) the enemy (or competing organisations). In other words, it is the identification or leveraging of some advantage that will bring about victory through beating (outcompeting) those who are not seen as neutral, friends or allies. Such a perspective requires the project manager to exhibit a number of skills, with the primary skill usually being seen as good leadership. Strategic management can be defined as:

> The systemic analysis of the factors associated with customers and competitors (the external environment) and the organisation itself (the internal environment) to provide the basis for maintaining optimum management practices.[2]

While strategic management has evolved into a commercial 'tool' since around 1950, it has not entirely lost its military origins and from time to time researchers have reminded the management fraternity of this, in some cases by explicitly linking military strategic principles and construction project management. Pheng and Chuvessiriporn (1997),[3] as one example, presented research on the connection between ancient Thai battlefield strategic principles and the leadership qualities of construction project managers. They concluded that 'the military commander and project manager share numerous similarities in the discharge of their duties'. In both cases, the implementation of administration and control were deemed to be high on the list of duties. For the military general, this would be focused on managing, organising and controlling resources such as troops, weapons, logistics and provisions. For the construction project manager, the focus would be on resources such as manpower, finance, plant, equipment and construction materials. In the context of this chapter, the more relevant aspects of the 'strategy' are:

- The selection of one or more leadership behaviours (essentially either transactional or transformational) that will increase the probability of achieving project objectives.
- The implementation and keeping of policies and records.

Both of these ultimately lead to a consideration of project structure and organisation. However, prior to considering those areas, a useful next step is to consider the environment in which the 'leadership' of a global construction project will be undertaken. In other words, what is the nature of modern project management?

5.4 Administer and control strategies in modern global project management

Project management theorists can be placed in a number of different camps and, as with their military cousins, these camps sometimes enter into alliances and at other times enter into conflict. A relatively safe approach to differentiate the various camps is to ask if an individual or group is broadly traditional (also known as transactional) or non-traditional (transformational) in the beliefs and leadership style that they apply to the management of projects. Each of these two groups (transactional and transformational) will adopt quite different stances with regard to the implementation of administration and control.

Typically, the transactional project manager will administer and control on the basis of believing all relationships within the project environment (and also the environment external to the project) to be essentially linear in nature.

A relationship that is deemed to be linear is seen to be fully understandable, consistently repeatable, and able to be controlled (thereby reducing risk in the project). The simplest manner to present a linear relationship is: $X + Y = Z$ (in all cases); take two specific resources and bring them together in a specific manner and you will always get the same result.[4] Such a perspective is valid in the context of relationships such as chemical reactions (cement powder and water, for example) but is increasingly being questioned in the context of the sort of personal relationships that projects (particularly large complex projects) demand for their completion. The nature of relationship building in different cultures was introduced in Chapter 3 and will be returned to in Chapter 10, but in both cases the underpinning material is covered in this chapter: the so-called relationship-building hierarchy, which is more transformational (non-linear) than transactional (linear) in nature. In response to such developments, modern project management is increasingly moving to a more transformational approach.

A transformational project manager remains focused on administration and control but does not underpin his or her approach with a belief in linear relationships. For the transformational project manager, the underpinning belief is that relationships are actually, in the main, non-linear ($X + Y \neq Z$) in nature; a realisation and acceptance that the world is a place of complex interactions and turbulence. Modern projects are becoming more and more complex (over the past 100 years or so this has certainly been the trend[5]) and the rate of change (turbulence) that such projects have to accommodate, in both their internal and external environments, has also increased. Organisations have been formed in recognition of the need to respond to this situation by support-ing project managers in developing appropriate abilities and skills, with the International Centre for Complex Project Management (ICCPM) being one example.[6]

In addition to the growth of special-interest organisations such as ICCPM, there has been an increased willingness by some researchers to contemplate more radical directions for project management as it becomes a 'transformed' function. For example, Mikheev and Pells (2005) posited that there were many alternative futures/development scenarios and opportunities for project man-agement used in a transformational manner on a global scale.[7] This would involve the transformation of individuals, organisations and the global envi-ronment itself. Such step-changes could be regarded as an example of what Weaver (2007) referred to as 'waves of innovation', with the development of modern project management commencing around 1960.[8] It can there-fore be argued that while the origins of strategic management can be traced back thousands of years, modern project management had its origins only a matter of decades ago and is continuing to develop rapidly. Thus the mod-ern project manager is faced with a considerable body of 'advice', even if he or she considers only the formalised bodies of knowledge produced by the various professional organisations. However, the situation is even more demanding in that there is a further mass of material that does not fall within the structures or bodies of knowledge of the professional organisations, and the project manager has to decide what, if any, of this wide range of mate-rial he or she is going to make use of when developing their expertise as a global project manager. In a sense, this is akin to the issue of determining stakeholder salience (determining which stakeholders get most attention) as discussed in Chapter 3. However, in the context of determining the most appropriate form of structure and organisation for a global project, the focus is on identifying key project management knowledge, skills and competences. Global projects typically suffer when the project manager cannot effectively implement what can be referred to as the hierarchy of relationship building: interact/initiate, communicate/connect and collaborate/care. The complexity of a global project can, understandably, push a project manager into imposing a particular structure on the project as, essentially, a means of providing confidence that he or she is in control. If, however, this is done without any consideration of the relationship-building hierarchy, the selected structure may well impede the success of the project by imposing barriers instead of building relationships. Ideally, it is better to build relationships and then determine the most suitable structure to support them throughout their required duration.

5.5 A relationship-building hierarchy

The cultural basis of relationship building tends to lead individuals, particularly in developed countries, to perceive work-based relationships as being different from home-based relationships. The increasingly post-industrial nature of many developed countries seems to be reinforcing the perception that work-based relationships are essentially just a means of supporting career development. In contrast, the relatively large proportion of time spent in the work environment has arguably contributed to an ongoing devaluing of home-based relationships. Overall, in developed countries the approaches to building relationships in the work and home environments seem to be regarded by individuals as being somehow different. In fact, the two types of relationship are built in the same manner, and this fact is regarded as being the norm in many developing countries. Developed countries seem unwilling to accept that effective work-based relationships require a bond of trust between all parties and this trust (and therefore the desired relationship) takes time to be 'built'. Developing countries however, seem to more readily accept this fact, with one result being that both 'home' and 'work' relationships are familial in nature. It is therefore appropriate to consider relationshipbuilding in the context of global projects to be essentially no different to building a relationship with a family member or loved one outside of the work environment, thereby allowing the use of the relationship-building hierarchy that commences with the initiation of interaction between two or more individuals.

5.5.1 Interact/initiate

Humans are generally regarded as being inherently social animals. While our social activity will frequently involve forms of communication, and possibly be seeking to achieve collaboration, it is primarily a response to a need to interact with others; few humans voluntarily withdraw from all interaction with others. The initial interaction (initiate) can be likened to testing the temperature of a swimming pool prior to deciding whether or not to dive in. For the majority of individuals, daily life is composed of a series of initiations:

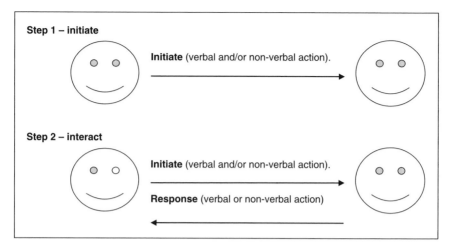

Figure 5.1 Initiation and interaction

making an initial contact with another individual, group or organisation and then awaiting a response. In this sense, initiating is a one-way action, while interaction moves up a level to become a two-way (or more) action (see Figure 5.1).

Deliberately choosing not to initiate any potential interactions is typically, within a Western culture, regarded as being indicative of an extreme need to escape from failure to interact appropriately with more 'normal' people. However, in other cultures, deliberately reducing initiations/interactions can be seen as worthy of respect. In the case of Josef Stawinoga, both situations combined as he withdrew to live in a tent on the central reservation of a Wolverhampton road for 40 years, during which time he became revered as a saint by local Hindu and Sikh communities.[17]

Interaction in the context of projects can be regarded as having at least a partial basis in what is referred to as the psychological contract: '... the perceptions of the two parties, employee and employer, of what their mutual obligations are towards each other'.[18] The psychological contract can be a powerful component in the interaction between the project manager and players within the project environment, particularly with regard to the 'rules' within which any interactions will be seen as being reasonable. The nature of the mutual obligations, for example, may be a key focus for initiating interaction in that they would be regarded as valid reasons for interacting. In the context of negotiation between parties, these mutual obligations should represent zones of accordance (items that all parties agree on) and thus an opportunity for building trust within the relationship.

The project manager may need to consider any 'rules' of interaction when considering the appropriate form of structure and organisation for a project. A key consideration is the rules to be applied to the opening (initiation) and closing (response) sequences of the interaction; these can vary between cultures according to factors such as gender, religion, perceived hierarchical position and age. If these rules are broken, it is possible to cause significant offence and thereby undermine any possibility of developing the required level of trust for business to proceed. In Thailand, for example, the customary greeting is a single word, irrespective of the time of day, accompanied by a placing of the palms in a prayer-like gesture called a *wai*; within the rules of interaction, this could be regarded as representing an acceptable opening sequence (initiation). However, the rules applicable to this particular form of initiation result in an expectation that the younger, or lower social status, of the two persons interacting should initiate the *wai*. Also, in terms of the rules governing the response (closing sequence) there is no need to return a *wai* initiated by a child or by waiters and such like (to do so would actually cause them to be embarrassed).[19]

In a multicultural global project, interaction rules may vary considerably across the players involved, and any construction project management director or project manager who is seeking to adopt a transformational approach will need at least to consider them when structuring and organising his or her project. In doing so, the step to the next stage of the hierarchy (communication) will be made more easily in that potential 'rule-breakers' (as perceived by some players) within the project structure will not exist to impede the building of trust. By engaging in effective communication and reinforcing the connection, the nascent trust created by 'proper' interaction can begin to blossom.

5.5.2 Communicate/connect

The theoretical act of communication is essentially a simple one: I send a message and you then respond. Unfortunately, the actual act of communication invariably proves much more difficult to carry out effectively. Perhaps the most basic reason for not communicating effectively is that the individual or group with whom communication is being attempted must be paying attention; if they do not realise that communication is being attempted, they will not respond. This lack of attention may be difficult to accept when the sender believes their message to be explicit and clear, such as in the case of spoken or verbal communication. If the project structure clearly identifies individual responsibilities for different aspects of the project, then this aids the act of communication in that the 'target' for the message is explicit; it is clear who the communication should be sent to. However, communication comes in different forms, and if an inappropriate form is selected then it is entirely possible that the intended recipient may not be aware that you are attempting to communicate. This can become particularly difficult with non-verbal forms of communication such as body language.

Failed projects throughout history are evidence of the results that can arise from poor and ineffective communication. A 2007 survey for the website Projects at Work identified poor communication as the number one cause of project failure.[11] Within the context of information systems (IS) projects undertaken within the EU in 2004, failure resulted in the loss of €142 billion.[12] In that year, around 24 per cent of the projects surveyed were cancelled and the majority overran budget and/or time objectives. While the causal factors identified were labelled as 'business process', the majority of the problems encountered could be addressed through achieving more effective communication.

Box 5.1   Vignette – Scottish Parliament

There were many laudable reasons for the desire to build a new home for the Scottish Parliament, with one of the key reasons being the need for a communication of an independent Scottish identity. However, the project overran the budget by a factor of ten and was delivered three years behind schedule. Shortly before its opening, the Auditor General for Scotland produced a report in which it was stated that £160 million was added to the cost of the project through factors such as disruption, delays, design changes and inflation. He also noted: 'There was not a clear, single point of leadership and control for this very complex and challenging project. This has been a weakness in the system.'[13] Even at that point, the project would have been considered a failure (in terms of failing to meet the budget and schedule; the issue of success in terms of creating a perception of independence is not for consideration here) but the problems continued. Some of these gained significant publicity, with perhaps the most publicised being the 'collapse' of part of the building.[14] Overall, a key lesson from this project is that attempting to structure a project without considering how its leadership is to be incorporated into the structure allows far too many possibilities for ineffective communication, and that this significantly adds to budget and schedule overruns.

The manner in which the project is structured and organised, particularly when the project itself is complex (in terms of both the objectives and the extent of interconnections), can have a significant impact (positively or negatively) on the level of effectiveness for communication. As shown in Figure 5.2, this requirement is not easily addressed when there is a lack of clarity regarding what constitutes effective communication; there are many models that claim to represent the 'ideal'. Such models range from the simplistic (Figure 5.2) to the complex (Figure 5.3):

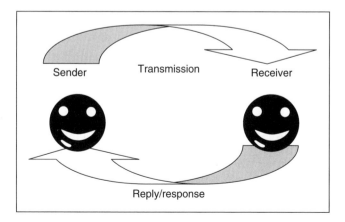

Figure 5.2   Simplistic communication model

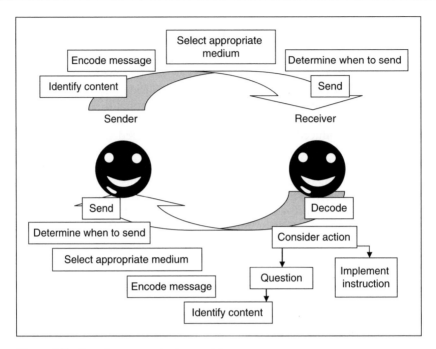

Figure 5.3 Complex communication model

In all cases, the project structure and organisation must not introduce 'noise' (anything that diminishes the effectiveness of communication between sender and receiver) into the communication process. With regard to global construction projects, the most problematic form of noise is semantic noise: words that are received correctly but not understood in the correct context. This can be particularly problematic when jargon, specialist terminology and 'community' (especially a community of practice; a wholly or largely invisible (to anyone not a member) body of individuals who form a community focused on a single issue of 'practice' of relevance and interest to all) phraseology are used, but can also cause confusion when the sender and receiver are not from the same cultural background. A few examples of semantic noise (for those with English as their first language) are as follows:

- Sign in Norwegian cocktail lounge: '*Ladies are requested not to have children in the bar.*' For many English speakers this would be decoded as 'Ladies should not give birth in the bar'.
- Detour sign in Japan: '*Stop. Drive Sideways.*' For many English speakers this would be difficult to decode into a sensible message as it is not generally regarded as being possible to drive 'sideways' (as opposed to the intended message of 'turn left or right').
- Elevator in Germany: '*Do not enter the lift backwards, and only when lit up.*'[16] There are so many different possible forms of decoding for this message that it is difficult to know where to start!

Effective communication, even when using explicit mediums (such as verbal communication), can be difficult to achieve, and this in turn has a negative impact on the development of trust, but the manner in which a project is structured can support the development of trust between parties, even if only through not imposing (knowingly or not) barriers that introduce semantic noise. However, in the case of non-verbal forms of communication, the situation can become even more difficult to resolve. Given that estimates of the amount of communication that takes place in a non-verbal manner reach as high as 80 per cent, the project manager of a global project really needs to be aware of the potential for non-verbal 'errors', which could undermine the possibility of any relationship building that has survived the initiate/interact stage and, subsequently, its progress to the final stage of collaborate/care.

5.5.3 Collaborate/care

The action of collaborating can be summed up in one word: trust. Without an appropriate level of trust, there will be no effective collaboration. There may be interaction and communication, but if these have not resulted in a sufficient level of trust being established then true collaboration will simply not happen. This is simply a reflection of the fact that projects are essentially about people – the non-human resources required to complete a project are obviously important, but if the right people are involved it can be truly amazing what can be achieved, even when there is an absence of the 'right' materials and equipment. To return to the military analogy, unbelievable victories have been achieved when the right people have been involved.

Box 5.2   Vignette – the Battle of Narva

The Battle of Narva, while not the most widely known of battles, is an example of what can be achieved by having the right people and the right collaboration. Between 1700 and 1721 the Great Northern War involved the Swedish Army in a conflict with a number of other nations. The Battle of Narva had a Swedish force of around 8500 pitted against a Russian force of up to 37,000. The two forces had very different leaders; while the Russian leader dismissed the threat of an attack from the much smaller Swedish force, particularly given that a blizzard was raging, and invited his officers to dinner, the Swedes prepared for battle. They were ready to take the initiative (in modern project management parlance they exhibited high-level synergy), when the wind shifted direction and blew snow directly towards the Russian force, effectively blinding them. The Swedes were able to get within 50 yards of the Russians before they were spotted and therefore had a significant competitive advantage with the result that, by the following morning, they had killed around 10,000 Russian soldiers and captured 20,000 or so whilst losing only 667 of their own soldiers.[9]

Collaboration is typically defined in terms of achieving multiple interactions between multiple players over time, and is therefore a more challenging proposition than the (relatively) simple – but nonetheless challenging in the context of a global construction project – actions of interaction and communication.

The challenging nature of collaboration is suggested as being difficult to capture in the context of a standard form of contract (the traditional, transactional approach to allocating responsibility and risk in projects), largely because contracts tend to focus on performance (e.g. completion on time, within budget) and allocation of risk and reward. There is also the problem of collaboration relying on trust; if you have enough trust to truly collaborate – and you truly care about being fair to others involved in the project – why do you need a contract? To some extent, any attempt to combine the act of collaboration (particularly when defined in the context of the relationship-building hierarchy) and a form of contract seems counterintuitive. Nonetheless, various standard forms have attempted to encapsulate the nature of collaboration. The NEC form is one example that has a number of clauses/options that could be seen as attempts to 'contract' in a collaborative manner. Two options that are worth specific consideration are Option X12 and Option X20. Option X12 is the NEX Partnering option and has several objectives for use:

- partnering for any number of projects (i.e. single project or multi-project), internationally;
- projects of any technical composition;
- apply as far down the supply chain as required.[10]

In addition, the Option contains specifics that are clearly intended to aspire towards a collaborative environment, such as item 3(8): 'The Partners should give advice and assistance when asked, and in addition whenever they identify something that would be helpful to another Partner.' However, there is the problem that the Option (Option 12) also seeks to (as identified previously) incentivise performance, as with item 12.4(1): 'If one partner lets the others down for a particular target by poor performance, then all lose their bonus for that target.' From the 'true' collaboration perspective, this may well be interpreted as threatening parties to the contract with punishment if they do not, in effect, police each other.

Option X20 is noted as not to be used along with Option X12. The focus of Option X20 is, again, to 'incentivise' contractors to achieve and, where possible, exceed performance objectives set against identified KPIs. A requirement of this Option is that the contractor should submit regular reviews of their own performance to the project manager and, if it appears probable that the benchmark figures are not going to be met, identify proposals as to how performance will be improved. As with Option X12, the tone could be interpreted as being one of compelling (rather than truly collaborating). Overall, the focus returns to that which can be so difficult to achieve in any project: trust. Without this, there has to be a legal contract as a means of ensuring specific aspects of project performance, but if there is truly complete trust between all parties, is a contract (even one that seeks to encourage collaboration) required at all? It is conceivable that such a contract could be of value in providing a form of structure for the global project (through legally defined relationships and responsibilities) but if the required level of trust is not achieved, that structure is likely to become little more than a series of barriers as players mitigate the perceived lack of trust by working strictly to their contractual requirements. The traditional organisation structures can be problematic in this regard and their use should be approached with this in mind.

5.6 Traditional structures for creating a global project and client/parent organisation relationship

This section deals with structure in terms of creating or formalising a relationship between an organisation (the client/parent organisation) and projects carried out either by or on behalf of that organisation. In such circumstances, the client/parent organisation may choose to impose a specific structure (usually the one that it uses in its own day-to-day operation) on the project – they may have their own preferred organisation structure that they use on all of their projects and will expect that this is adopted by the project manager. Construction project management directors and project managers need to be aware that just because a particular organisation structure works well in one environment, it is not a given that it will work well in all environments. There can be a tendency to take the easy route and simply replicate, at the project level, whatever structure is used by the client organisation(s), but there are some good reasons for not taking the easy option with regard to project organisation structure. Maylor (1996) identified reasons for project managers to define organisation structures at the project level (rather than simply replicating the parent organisation's structure):[20]

- Structure defines the 'distribution' of authority and responsibility.
- Reporting arrangements (as in team members to manager) should be clearer.
- Management overheads (salaries, etc.) can be more clearly determined.
- Structure should represent the organisational culture for the project.
- Project activity stakeholders are explicitly determined.

Typically, such replicated structures will be one of the standard structures: functional (as in structuring around a functional specialism), pure project, or some form of matrix. These will be described within this section. Thus all of the structures presented are used on projects, but there are other forms of structure that can be regarded as being more effective for complex projects such as global construction projects. These will be presented later in the chapter.

5.6.1 The nature of structure

For most project managers, structure is so prevalent in their work environment that it becomes almost invisible, in a similar manner to the gradual 'disappearance' of features and landmarks within a frequently used commuting route as the driver slips into autopilot; there is a subconscious awareness but the conscious mind is busy processing other information. In terms of structure, project managers frequently know who it is within both their company and project structures that they need to deal with; an organisation chart may have been helpful in initially finding that person but once an interaction is initiated and a relationship begins to build, the chart tends not to be consciously considered. They also are usually aware of the relationships between individuals within

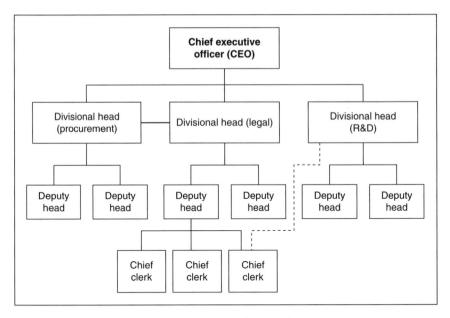

Figure 5.4 An example formal project structure (functional)

the project and company environments, and also of relationships with others who are not formally part of any stated structure but nonetheless may be of value to the project. In combining such strands of knowledge and information, project managers are effectively creating a structure, albeit of a highly informal nature, for a specific project. This *informal* structure may, however, be quite different to the structure that is *formally* presented for that same project: a situation that presents particular risks with regard to maintaining control of a large and complex global project. Figure 5.4 gives an example of the type of relationships that are typically included in a formally presented structure (in this case a functional structure); essentially these are presenting broad functional relationships in a manner that clarifies the chain of command in terms of authority and responsibility. The use of varying thicknesses and types of line connecting each box in the structure allows some communication about the nature of the intended relationships between those individuals who are populating it.

It is important to remember that each box effectively represents at least one *person* (typically that one person is the point of contact but there may well be a considerable number of other individuals acting as support for that person) and not just a specific *function*; simply regarding the individual person as a function creates difficulty with regard to relationship building in that it is more difficult to collaborate with a function than with a person representing that function. At the level of a single relationship, a reduction in the effectiveness of collaboration may not have a significant impact on a project, but if the majority of relationships experience a similar reduction then there is a cumulative impact that can severely hamper complex projects. Project managers must not forget that they are dealing with *people* and not just functions within the structure of the project.

5.6.2 Functional project organisation structure

A functional organisation structure follows the spirit of functional specialism very closely by assuming that the parent organisation structure is based on functional divisions (see Figure 5.4). This then imposes on the project manager the requirement to place the project within the division whose function represents the optimum home for it. This is usually on the basis of a further assumption: a functional division that recognises a project as focusing wholly or largely on goals and objectives that fall within its function (is it a 'legal' project or an 'HRM' project?) will want the project to succeed and will work to achieve that. However, even if this is actually the case, in situations where a project is highly differentiated (many different functions, as is typical of a global construction project) identifying a 'home' for it can be difficult. The different functions within a project may be of equal significance (no single clear focus), and none of the possible divisions may express sufficient enthusiasm at the prospect of being responsible for such a project (based on the need to work as partners with other specialisms that are frequently interacted with – and their perceived, stereotypical behaviours).

Allocating each individual with a function – a functional specialism – is a concept that has been around for centuries. The various infrastructure projects (e.g. castles, temples) carried out around the globe for several thousands of years of recorded history are strong evidence that someone was capable of organising large numbers of people and resources over long periods. The functional specialism-based project structure has been until recently the only structure used in the Western European context from the time of the Industrial Revolution, when point production moved away from low-volume, craft-level production to full-scale high-volume, production based in a factory environment; the production line had arrived. This required individuals to specialise in only one part of the production process, and this in turn required that these specialisms within the production process be coordinated so that an output from one specialist became the input for the next, as each added their little area of expertise to the production of the completed item. Each of these new, narrower stages of the process allowed individuals to become more expert and thus productivity increased. The proven effectiveness of this new model gradually moved it from use in factory-based production to site-based production in the form of construction projects, to the point when it became the standard basis of any industrial organisation structure. Consequently, the concept of functional specialism is one that anyone who has spent a year or two in any traditional industrial environment will be aware of (even if they have not applied the appropriate title to it).

Such a functional specialism approach to structuring is characterised by a focus on avoiding failure through emphasising analysis and planning as major activities in the control of the project or organisation, through the implementation of transactions (I will give you something that you need in exchange for something that I need) between players within the value chain. On this basis, the traditional forms of structure are sometimes also referred to as transactional structures. In either case, such structures are therefore essentially about control and are based on assumptions that an environment is highly linear and therefore open to being controlled. A functional specialism that is essentially a transactional (linear or $X + Y = Z$ in all cases) approach to management has been proven to work well in production-line environments (particularly those exhibiting a slow rate of change regarding technologies), but it is less effective in non-linear ($X + Y \neq Z$ in all cases) and/or rapidly changing environments (as may be found in a complex global project). Any assumption that a complex global project's environment is as linear as that of a car production line is inevitably going to lead to a false impression of 'controllability'. Additionally in this approach, production staff may be, in essence, seconded temporarily to a project, and therefore they do not always have any depth of project management expertise.

In the case of functional structures, it may well be appropriate for a project to structure itself on the basis of functional specialisms but it is unlikely that the specialisms represented will be entirely the same as those of the parent organisation. In essence, the project structure may be regarded as a snapshot of part of the parent's structure (e.g. R&D, Legal). Overall, functional structures are not generally regarded as ideal for the management of projects, except in situations where the parent organisation involves itself in only a few projects, perhaps at irregular intervals, which are characterised by either a need to meet a specific (and rapid) change programme, or are focused on the in-depth application of a single, specific technology. In the event that a global project falls within those boundaries, then a functional structure may be appropriate. However, these are tightly defined boundaries and it is a rare global project that falls within them. Nonetheless, it is worth noting the claimed advantages and disadvantages for this approach to structuring. Advantages for the functional organisation structure are:

- staff can be used with maximum flexibility;
- many projects can make use of an individual expert;
- knowledge and experience can be shared by grouping specialists;
- division allows technological continuity when an expert leaves;
- division allows a clear career path.

Disadvantages for the functional organisation structure are:

- the focus of concern moves away from the client;
- divisions do not deal with problem solving in the manner needed by a project;
- lack of a clear 'home' and individual responsibility for project success results in disorder;
- projects tend to be suboptimised due to lack of enthusiasm **throughout** any division to which the project is allocated (rather than the division having initiated the project);
- enthusiasm of **individual** staff assigned to the project by a division not having overall responsibility for the project tends to be low;
- does not achieve a truly holistic approach to managing the project.[21]

The recognised disadvantages of the functional structure when applied in the context of projects rather than production environments led to the development of an alternative structure that more specifically reflects the nature of projects: the pure project organisation structure.

5.6.3 Pure project structure

Working on the basis of there being a spectrum of structures relevant to the organisation of projects, the functional structure would be at one end (in system terms this would be the 'closed' end), while the pure project structure would be towards the opposite end (again, in system terms this would the 'open' end). This is due to the pure project structure providing self-contained 'homes' (free of any rigid connection to a specific functional division) for each project within the parent/client organisation environment. Such an approach does not usually cause the integration problems typically arising from attempts to integrate projects into functional divisions that are more focused on production than projects.

Pure project structures seek to provide a project home that has the minimum possible ties to the parent organisation. Most, or all, of the project staff are not sourced from a (functional) division elsewhere within the parent organisation; they are, in functional specialism terms, project management specialists who move between projects. This differs from the situation in the functional structure, where the project staff may be seconded temporarily to a project, and therefore do not have the depth of project management expertise that can be developed by staff working within a pure project structure and environment.

When considering the minimal level of tie achievable between the project and the parent/client organisation, this is usually in the form of the parent/client only being concerned about final accountability; if the project objectives are achieved, there is no concern with regard to *how* they are achieved. This can be contrasted with the maximum level of tie that can be imposed on the project in other forms of structure, typically involving the use of procedures prescribed by the client/parent organisation for administration, financial, personnel and control activities. Figure 5.5 illustrates a typical pure project structure.

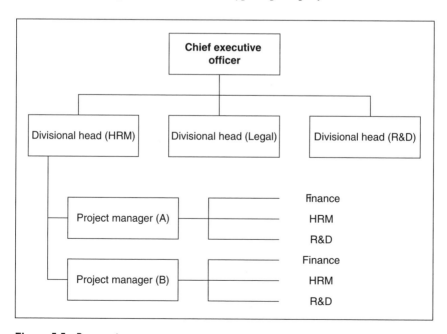

Figure 5.5 Pure project structure

Advantages typically cited for the pure project organisational structure are:

- the project manager has a clear and full line of authority, with one, and only one, boss;
- shortened lines of communication as a result of not being in a functional division;
- a near-permanent group of expert project managers can be developed;
- team members are encouraged to be highly motivated and committed;
- it is easy to understand and implement;
- it supports a flexible and holistic approach to the project.

Typical disadvantages are:

- duplication of staffing and effort across many projects;
- project managers may operate on the basis of stockpiling resources not currently required;
- development/maintenance of expertise may be negatively affected as experts are removed from their functional 'home';
- possible inconsistencies between projects with regard to application of procedures and policies;
- *projectitis* (focusing only on the project and ignoring its external environment) may occur as projects start to develop a life separate from the parent organisation;
- uncertainty of re-employment prospects after the project is completed.[21]

Pure project structures are generally most suitable for the following two different situations:

- For organisations that are involved in a large number of projects with similar characteristics. This allows for the development of project management expertise both within the project-focused 'environment' of the client/parent organisation and within the client/parent organisation 'environment'.
- For one-off projects of a highly specific nature (that do not fall within the 'norm' for the client/parent organisation). These are difficult to link to a specific division and they also need careful control.

The functional and pure structures both have disadvantages that constrain their use to specific situations. In order to deal with those situations, where neither the functional nor pure structures would be considered appropriate, a third possible approach to structuring the relationship between a client/parent organisation and a project was developed: the matrix structure.

5.6.4 Matrix structure

In creating a third form of structure, the reasoning seems to have been that combining the advantages of the functional and the pure organisation structures would give the benefits of both in a single structure. However, the matrix structure is essentially a compromise that is achieved by taking the parent organisation's relevant functional divisions (the 'vertical' component of the structure)

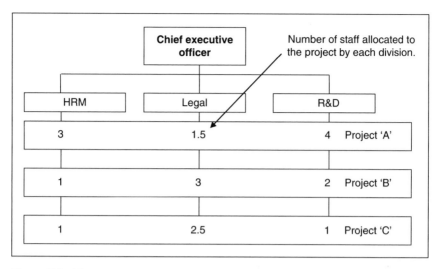

Figure 5.6 Matrix organisation structure

and then imposing a pure project structure (the 'horizontal' component of the structure) over each functional 'silo' (see Figure 5.6).

The compromise of layering a pure project structure over a functional project structure results in a what is generally seen as being the biggest problem of the matrix structure: the so-called 'two boss' problem in which the project manager controls what tasks individuals and groups within a specific project do, and when they will do it, but the functional manager has the decision as to which individuals (or specific technologies) from their division will be allocated to each project. In effect, the project manager is reliant upon the functional division manager in terms of the staff available. The project manager is also reliant upon the functional division manager not seeking to manage those staff allocated to the project from their division. It is also known for staff to refer back to their functional division manager for guidance and advice rather than asking the project manager – the result being that problems may be 'solved' without the project manager ever having been aware of them. Because of the compromised nature of the matrix structure, it does require a certain type of individual in order to achieve a successful project outcome, and these individuals will usually have been trained at some point in how to work within a matrix environment. This is particularly relevant given the role of ambiguity that results from conflicting priorities between the project and the division within this structure – a situation that cultures in developing countries typically do not encounter and therefore may experience considerable difficulty in dealing with. The nature of the structure also means that particular attention must be given to the quality of key support systems, such as administration and career support. In addition, the following advantages and disadvantages can be identified.

Advantages:

- Individuals can take responsibility for managing a specific project.
- Reasonable access to all technological and expertise resources within each division.

- Produces less duplication than pure project when multiple projects are being carried out simultaneously.
- Response to demands or opportunities, both externally and internally (two or more environments), is rapid.

Disadvantages:

- Can appear that as soon as something seems to be working it is changed.
- Group approval is usually required before individuals will make decisions.
- Time taken in achieving group consensus means the group becomes a creativity barrier.
- Can result in high overheads.

Achieving a successful outcome with a matrix structure depends, as with any structure, upon it being used for a project having compatible success criteria. Matrix structures work best when dealing with success criteria that are typical of projects having goals and objectives relevant to achieving high levels of quality (typically so-called 'high-tech' projects). They also work well in conjunction with what are referred to as 'crash' projects: attempts to gain the most productivity at the least cost in the shortest time, usually through reducing activity durations by increasing the resources available.

The matrix can result in a 'flat' structure that adds to its responsiveness through possessing fewer communication barriers; more people are cooperating more closely, with the result that information flows more freely (both vertically and horizontally). This also adds to the level of democracy within decision making (along with the relatively flat structure bringing previously upper management staff closer to the decisions relevant to the operational level), which in turn can improve the level of motivation within the structure. While these advantages may prove to be especially relevant to certain projects, the disadvantages of the matrix structure lead to it being viewed as generally expensive to maintain due to greater reporting requirements; uncertainty regarding the chain of command, resulting in decision-making anxiety and high staff turnover; the effort required to address perceptions of role ambiguity. As with any structure, the matrix is more appropriate to some organisations and projects than others; projects staffed largely by professionals and semi-professionals are usually the ones that result in the best outcome when using this structure.

A frequently cited application of a form of matrix that may be attractive to transnational construction companies carrying out global projects features what is essentially an attempt to integrate three axes rather than the more normal two. Along one axis lies the project/production environment. The second axis represents the domain of the client/customer/parent company, and the third axis represents the domain of the overseas operations/division(s) of an international firm. In such a three-dimensional matrix there is, potentially, a fragmentation of management between the project/product/market environment, the various functional divisions within the parent company/client environment, and the cultural and legal environment of the 'host' country (for the project). In terms of structure, the typical arrangement for a construction company would be comprised of a three-way relationship between a worldwide project/product manager, a worldwide functional divisional manager, and a project manager specific to the country in which the project is being carried out. This kind of structure is, however, highly variable as individual organisations seek to place their commercial activity in what they regard as the most favourable version of the three-axes matrix structure.

Box 5.3   Case study – global organisation and the matrix structure

Skanska is a construction organisation that has its headquarters in Solna, Sweden but operates in many countries throughout the world. As it has grown in size it has also developed an approach to organisation that it regards as being decentralised but integrated. The decentralisation per-spective results from the classification of projects as profit centres. Given that Skanska claims to be involved in thousands of projects on a daily basis, the number of profit centres being operated by the company will be considerable. These project profit centres are split between national boundaries across the globe but a key feature of the company's culture is that it regards these different counties/regions as being 'home' markets for the company, even though the local business units involved are claimed to be firmly established in their individual market environment. An example of the locally specific nature of these business units is the claim by Skanska, within their code of conduct that they will:

> maintain organisational structures, management systems, procedures and training plans that as a minimum ensure compliance within all relevant laws, regulations and standards.

Given that Skanska operates across 'home' markets of Sweden, Finland, Norway, Poland, Czech Republic, UK, USA and Latin America, there will be a considerable variation in those all-important 'relevant laws, regulations and standards', and thus the country-specific axis of a three-dimensional matrix would come into being as the organisation addresses the attempt to operate within the confines of its values and beliefs in a decentralised manner. A second axis would be the Skanska 'brand' – the values and beliefs of the parent organisation that 'integrate' the var-ious home markets as the organisation expands globally. The third axis would, in a product-focused organisation, typically be the functional divi-sional manager, but Skanska is a project-focused organisation and therefore would not have any such managers. However, it operates on the basis of a synergy between its construction operations and its project development operations, with one 'supplying' the other. Thus it is reasonable to regard the third axis as being the construction/project development 'function', with the resulting structure as shown in Figure 5.7. While the structure may initially seem somewhat chaotic, it is actually a simplified version of what the 'full' structure could look like, in that not all of the national business units are included and no more than three projects are shown for each country. By removing the detail for each of the three axes, the boundary of a 3D organisation structure can be presented as shown in Figure 5.8. A key consideration to be aware of regarding construction projects involving global organisations is that the boundary of the

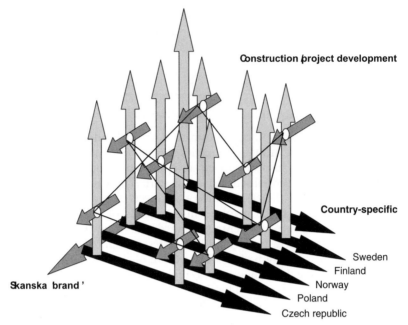

Figure 5.7 3D Functional organisation structure

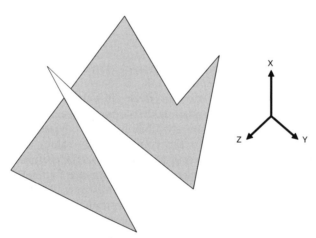

Figure 5.8 3D matrix organisation structure boundary

structure can be highly fluid. This results from the potentially large number of projects that are started and completed across a number of countries, Thus the 'structure' rarely remains static for any length of time; a situation that can cause considerable stress to those not used to this form of organising a global business.[22]

The three-dimensional nature of the matrix can also be considered purely in the context of the management roles. As previously mentioned, a project or product manager (the first dimension) will usually be focused on project/product-specific issues that are common across regional or national boundaries. The second dimension will be the functional manager who will usually have one of two different focuses: international issues (such as global HR), or alternatively, local/regional issues (such as national legal requirements). The third management dimension is typically a country manager with responsibility for project/product and function activities. This dimension of the management matrix usually becomes more embedded in a specific country and develops a deep understanding of marketing and project/product requirements in their particular country. While such a three-dimensional management structure does have its advantages, it is almost always regarded as being 'at risk' due to the extent of role ambiguity between managers across the different dimensions and different countries (a tendency to 'empire build' can result in undue competition between national managers). This ambiguity requires considerable expertise on the part of upper management if the 3D matrix structure is to deliver its benefits without becoming overly bureaucratic, inefficient and expensive to operate. However, consideration of the matrix approach to structuring an organisation's activities leads to two factors that are particularly relevant in the context of global construction projects: how people relate, and the management of organisational value chains.

5.6.5 How people relate in global projects

A problem that can afflict large global organisations such as Skanska and its French equivalent Bouygues is that the upper management become so distanced from the day-to-day operations of the organisations that they forget the functional division manager or the national project manager is actually a person, as opposed to being a point on an organogram (a drawing or chart identifying all key members of an organisation, their responsibilities and the connections between them). While representations such as organograms have many advantages, they do have one potentially significant weakness: they only represent the formal, planned relationships within an organisation. They rarely capture the informal and unplanned relationships that people, by their very nature, crave and put considerable energy into creating and maintaining. Managers should not lose sight of the fact that they are relying on people to deliver projects successfully, and that people can be quite inconsistent and unpredictable, irrespective of the structure that may be 'imposed' upon them.

An example, albeit quite an extreme one, of how relationships between people can develop, initially in a near-invisible manner but then, when given the appropriate circumstances appear to burst seemingly from nowhere, can be found in the use of online social networking to build a 'community' with a single but powerful focus: the removal of a hated government.

Box 5.4   Vignette – social networking and organisational change

A key factor in the success of the Arab Spring's achievements lies at the very heart of the construct that is the Internet; porous boundaries that allow for open communication. The concept of porous boundaries is not a new one; Miller and Rice (1967), for example, regarded porosity of system boundaries as an important factor in the success or failure of a system. However, since that time 'systems' have undergone a significant change in the context of the environment; it has increasingly become a digital, online environment and this brings significant benefits in terms of the ability to organise structures. Digital tools can be used to bring together individuals who would otherwise be geographically separated and bring together groups that would otherwise not have been aware of each other (at least in the short term). It is also possible to bypass the 'official' structure that, amongst other activities, would effectively control communication and the spread of information. The Internet brings the possibility of making the official structures more porous through the use of social media to connect and then organise individuals and groups. In terms of imposing control on 'revolutionary' groups, the traditional structures would ideally take the apparently simple step of turning off the Internet. Perhaps fortunately, turning off the Internet is not as simple as turning off your TV, particularly when the users are knowledgeable enough to 'hide' both their identity and location through the use of proxy techniques; the best that can be achieved is to slow, rather than stop, the flow of information.

As a consequence of the Internet, unrest and agitation can move beyond national boundaries to find supporters and sympathisers elsewhere. While there may no longer be the impermeable national boundaries that dictators of old were able to enforce, this does not mean that there are no boundaries. The boundaries that remain are arguably cultural rather than political or technical; social media seems to support political change when its use is congruent with 'local' norms for acceptable social behaviour. If these norms are observed then social media can have significant impact simply by showing protesters that they do in fact have supporters elsewhere – they are not alone; and the online community can be shown that protesters are not lone and isolated voices.[23]

Social and organisational change can also build on the porosity within official structures by seeking to make the structure even more porous but in a planned manner. The UK government, for example, may not explicitly refer to 'porosity', but in adopting a formal policy of encouraging the formation of communities around specific problem areas it is, in effect, making the boundary between the political and public bodies significantly more porous (and some would argue more democratic). Such communities can, for example, develop around initiatives such as the Communities of Practice for Public Service website that has the stated interest of providing:

> ... a community platform supporting collaborative networks for those involved in local delivery: central departments, local authorities, other public bodies, frontline staff, health staff, people working in charities and the private sector who are in some way delivering for the public.[24]

The 'planned porosity' that results from the provision of both an environment in which a community can be supported and flourish, and a point of interest common to sufficient individuals to form a viable community, can be regarded as an attempt to add a component of agility to an otherwise traditional (in the sense of governments and bureaucracy) and slow-to-respond organisation structure. In one sense, it is a similar approach to the attempts to produce a 'better' matrix structure; do not fundamentally change the core structure but try to encourage it to work in a different manner by adding an additional component. In the case of the 'communities of practice' approach, the desired component is essentially agility: a collection of minds (not all of whom need to consider the original structure as their 'home' – each could be based within different organisations) all focused on a common problem that is regarded as being worth solving. Such an approach could be seen to be very much concerned with adding value to a pre-existing environment within a formalised structure by connecting to it additional resources (that frequently are 'free', in that they are individuals or groups sufficiently interested in solving a problem that they do not charge for their efforts), which have combined in an informal structure. Such a possibility is just one reason for the manager of a global construction project to give serious consideration to identifying the 'true' value chain for their project.

5.7 Management of global value chains

The concept of value has tended to be considered in terms of an organisation's explicit supply chain, as presented within the traditional forms of organisation and project structure. However, as other less formal forms of organising working emerge, there needs to be recognition of value that can be achieved through the 'informal' interaction of individuals and (more frequently) groups within the project's internal and external environments. Such recognition may involve the use of perspectives that the project manager may not be familiar or comfortable with. However, some comfort may come through the application of a traditional perspective on management – if you cannot identify it, you cannot manage it. On this basis, recognition (identification) of the true, rather than assumed, value chain has to be the starting point for managers of complex global projects seeking to increase the probability of achieving a successful project. One example of a perspective that may not be familiar to many project managers is that of the psychological contract's (as introduced earlier in this chapter) reference to 'mutual obligations', not all of which relate to items such as pay and holiday entitlements.

5.7.1 The value of play

When applied in an organisation context, particularly when considering the traditional emphasis on structure of organisations, the psychological contract can be argued to become a key factor in the development of 'organisations' such as communities of practice. While such communities are essentially virtual and, therefore, the usual 'rules of engagement' applicable to face-to-face contact become less significant, they nonetheless require some rules (in the same manner as social media being most effective when established social 'rules' are observed) if the project is to realise the potential value of such a resource. An emerging problem (particularly for the more traditional project managers) is the recognition of the value of 'play' within the work environment. There are probably very few project managers who would feel comfortable with the Google approach to managing the social aspect of work within the so-called Googleplex; essentially a blurring of the traditional boundaries between the 'work' environment and the 'play' or 'home' environment. Employees within the Googleplex can get free haircuts, benefit from on-site medical staff, use the laundry facilities, go swimming (albeit in quite small pools where the water moves and the swimmer stays in place), and a number of other benefits. The benefit to Google is that, if employees can get most of what they need on-site, they are less likely to go home – and productivity will increase. The so-called Google Play-Time is probably the most radical part of the overall employee package.[25] This allows each employee to take 20 per cent of their time (equivalent to one day per week) to work on projects that are unrelated to their normal workload but are of interest to them personally. This has had the claimed benefit of operating as a form of feeder route by which some 'personal' projects become 'work' projects to be developed in Google laboratories.

Google, however, is not the only organisation to add play time to its structure. 3M have long recognised a value to both problem solving and new product development from making space to play; in 1968 the Post-it note was developed in one 3M employee's 'free time'. The initial idea came to him while singing in his church choir, but his free time allowed him the space to develop the idea through to a commercial reality. However, even with the free time available to him, it initially seemed unlikely that the Post-it note would have come to market, largely because of the type of glue it relies on – something that is not especially sticky! Fortunately, another 3M employee had developed just such a type of glue some time previously but had not found a product that it was suited to.[26] Bringing the two together was the real synergistic breakthrough for both ideas. Back in the 1960s the bringing together of such ideas was more difficult to do than in a contemporary environment with the wide range of ICT (information communications technology) available. There has to be recognition, however, that factors such as ICT, increased porosity, and communities of practice inevitably result in a need to reconsider concepts of organisation structure. The organisation structure of 3M allows employees 15–20 per cent of their working week as 'free time'. The payback to 3M is that if any of the ideas developed in this free time become marketable, the employee(s) involved gain an equity share in a spin-off company organised around 'their' product; in psychological contract terms, very much a win–win outcome of recognising and adding value. This raises the possibility of a different perspective on organisation structure; the recognition of value-related communities within the project environment suggests the possibility of a structure that comprises organisational value chains.

5.7.2   Value of supply chain management

Management theory when applied to project environments places considerable emphasis on various aspects of building and maintaining efficient supply chains. Such chains are of obvious importance in the context of large and complex global construction projects such as the Yangtze River Three Gorges Project. In 1993 the cost estimate for this project was 90.09 billion yuan (US$8.3 billion), with an acceptance that inflation would add to this over the projected 17-year project duration; the projected final figure was 203.9 billion yuan (roughly $19.8 billion) but the actual final figure ranges from the official $23 billion, through the general consensus of most experts outside China of around $46 billion, and to the highest estimate of $88 billion.[27] For this project, significant quantities of resources had to be obtained, shipped and used over the long project duration. Within such large projects it is understandable that the emphasis has been on control and that this results in a reliance on techniques such as supply chain management in order to create added value. In this case, the added value is a projected internal rate of return on investment of around 15 per cent.[28] In the context of contemporary global construction projects there is a valid argument for a different approach to adding value.

The changing market environment (e.g. the ever increasing rate of changes) has resulted in an evolutionary 'push' being exerted on all areas of management. Supply chain management, as one example, has now spawned extreme supply chain management in response to a realisation amongst supply chain professionals that the environment has reached a level of volatility that traditional supply chain management simply cannot deal with.[29] Extreme supply chain management is intended to be a means of dealing with conditions such as systemic volatility with few, if any, rest periods between constant and variable oscillations in the environment, all of which combine to render the traditional reactive and just-in-time approaches to supply chain management ineffective. By moving the focus to achieving collective risk management (rather than the traditional focus of sequential risk management), extreme supply chain management claims to provide an anticipatory approach allowing for collaboration on a larger scale than is possible with traditional methods.

There is no doubt that market conditions globally are changing at a faster rate than many senior project managers are accustomed to. The act of accepting this to be the case is an example of where knowledge of a situation can result in a change of perspective. Thus the development of extreme supply chain management is one example of how knowledge itself changes the established order of doing things. Why not then take this a step further and focus on knowledge explicitly (rather than implicitly through knowledge 'embedded' in resources and products)? The value then becomes the knowledge rather than the product. To some extent this is happening through the increasing application of knowledge management.

5.7.3 Value of knowledge

Knowledge management (KM) is a relatively new addition to the project management 'tool kit'. Its origins can be traced back to the 1970s but serious development only began in the 1980s and KM was recognised as a formal discipline in 1991. While it has been in mainstream use for a relatively short period, this discipline has developed rapidly, facilitated by the close connection between KM and the development of IT tools and processes enabling the rapid capturing and interrogation of large amounts of information and data. Such a relationship enables KM to be potentially both highly agile (useful in the volatile environments encountered in some global projects) and versatile (able to support capture, storage, creation and dissemination of information), particularly when systemised so as to result in a structured knowledge management system (KMS). A key point to consider in this context is the nature of the relationship between a formalised KMS and the stated project organisation structure.

A KMS will typically be structured around a software package that is usually database-oriented. In essence, a KMS is about creating connections between previously disconnected data and information so as to create new knowledge (ideally) of value to the organisation. By viewing this 'connectivity' in a project management perspective, there is a case for regarding the resultant data/information connections in a similar manner to the connections between individuals and functions within a matrix structure; the data/information connections become the structure for the organisation. Such an approach has the benefit of explicitly recognising that value flows from the creation of knowledge and learning by the organisation (if only to reduce the risk and wasted resource of the so-called Rembrandts-in-the-attic problem; organisations relearning knowledge that has been 'forgotten'). Rather than structure the project around traditional functions or even more contemporary forms of supply chain, why not structure it around the connections required to handle information and create (valuable) knowledge?

The two barriers that have previously been claimed to prevent such a development are the technology and the people. The technology aspect is becoming less problematic – the rise of building information modelling (BIM) systems, for example, means that there are packages (such as Innovaya) available that can routinely process four dimensions of information. Near-future predictions are for BIM systems to be able to handle upwards of nine dimensions of information; at present, 5D packages are becoming available (e.g. Excel, Innovaya, Timberline). The technology 'barrier' is therefore becoming more porous as computers become ever more powerful at ever decreasing (relative) cost, as evidenced by the continued operation of the so-called Moore's Law (the power on a computer chip will double every 18–24 months). This then moves the focus to the 'people' barrier.

The traditional perspective on people as a barrier to the development of a structure based on the value of knowledge results from questioning that they are capable of functioning within an organisation structure that potentially changes several times an hour, as volatility in the project environment provokes the 'closing' of an existing knowledge connection and the 'opening' of a new knowledge connection. Such a question is reasonable but suggests that the questioner has retained a traditional perspective on the control of projects, where a visible organisation structure of formalised titles and functions must be followed or the project will spiral out of control. In many aspects of modern life in developed countries, the reliance on computers is ubiquitous to the point of being largely invisible; in many aspects of daily life the general population seems to be unconcerned that there is no visible structure or organisation, nor does there seem to be any undue concern with regard to controlling the decision making carried out by these 'invisible' computers. Arguably, the value relates to the continued operation of day-to-day systems without any conscious involvement in their management by the recipients of the systems' outputs. In essence, why should there be any need to understand something as long as it continues to work satisfactorily? This perspective is, however, disregarding what may be a greater problem, but could also be a factor in structuring future project organisations: claims that the mode of operation for social media and computer games, for example, is 'rewiring' the human brain and shortening attention spans.[30]

The suggestion that attention span is decreasing can understandably be of concern (will we all become more 'stupid' if we are not paying much attention to information?) but it may also be a benefit in terms of processing large amounts of 'shallow' information very quickly. If this were to be the case, then a comprehensive and rigid organisation structure would actually be a disadvantage in that it slows the flow of information. By basing the structure solely on the connections between information it may be possible that the software maintains the overview (all the connections within the strategy for the project) while the project manager functions on the basis of focusing on real-time changes that are flagged up as being either possible opportunities or possible threats. Such a proposal would understandably concern senior project managers with a background in more traditional approaches to managing (controlling) their projects, but if a recognised discipline such as supply chain management can evolve in a relatively short period to result in extreme supply chain management, perhaps project organisation structuring can evolve to something akin to extreme project organisation structuring? There becomes a need to develop tools and techniques appropriate to developing and managing contemporary and future project structures. One suggested approach to this is to make use of the fractal web concept.

5.8 A transformational structure: fractal web technique for managing project structures

McMillan (ND)[32] proposed the use of a fractal web-based approach to structuring organisations operating in twenty-first-century environments. The proposal was based in part on the principles arising from a complexity-science perspective on environments. In adopting such a perspective it is argued that a non-linear approach (essentially similar to a transformational approach) to organisational design must be adopted. In doing so, it becomes apparent that it then becomes extremely difficult to produce a structure similar to that produced from a traditional linear approach; hierarchical, multi-layered, function-led and tending towards the bureaucratic. Put simply, a non-linear approach does not 'naturally' produce an organisation structure such as, for example, the functional structure. Instead of imposing a specific structure on the project or organisation, a non-linear approach largely leaves the project or organisation to organise itself. The basis of self-organisation in the context of projects is: 'spontaneous formation of interest groups and coalitions around specific issues, communication about those issues, co-operation and the formation of consensus'.[31] Because the act of 'spontaneous formation' could occur at any level within a project, any technique used for the management of the result-ing interest groups (such as communities of practice) will need to be agile (which suggests an essential simplicity) and versatile (so as to achieve com-plexity). A concept that fits both of these requirements is that of the fractal: a complex geometric pattern exhibiting self-similarity at all scales. Fractals typically are connected with non-linear and chaotic systems and are widely used in computing for the modelling of patterns and structures occurring in nature.

The key aspect of McMillan's proposed fractal structure was that the self-similarity requirement would allow the ethos (values and beliefs) of the organisation to be constant at all levels and locations within the environ-ment. This should provide the benefit of a consistent basis for 'control' since all participants are striving for the same outcome, namely a successful project – the nature of which they have agreed on – without the traditional hierarchical approach being imposed. According to McMillan: 'All participate in learning; speculate on the future; take risks with ideas and experimen-tation; work on projects.'[32] At any level within the project, the structure is the same and yet it has not been formally designed or imposed: it has self-organised.

The concept of a self-organising project might seem rather too much like sci-ence fiction, yet in 2002 the prospect of self-organising projects was regarded as possibly being imminent. Software capable of limited self-organisation was available, and a few web pages had basic self-organising capability. It was envi-sioned that the project manager of the near future would, when the project environment became too complex for traditional techniques to retain con-trol, simply select the appropriate self-organising applications (SOAs) and let them get on with interacting and forming the functionality required to address the complexity without any formal structure whatsoever.[31] An example of SOA development can be found in the Pastry Project: a collaboration between Microsoft, Rice University, Purdue University and University of Washington during the early part of this century to produce a generic, scalable and efficient substrate allowing nodes to form a decentralised, self-organising and fault-tolerant network within the Internet. While very few people will have been aware of the Pastry Project, there will be a larger number of people aware of a product resulting from the kind of research that Pastry carried out: Cloud computing.

The essence of Cloud computing is the making use of 'spare' computer stor-age capacity outsidethe computer that an individual is using. In order to work effectively, this environment makes use of what, in extreme supply chain man-agement terms, would be classed as an 'anticipatory approach' to collaboration between agents forming the Cloud. The Cloud itself does not know what the level of demand from users for storage will actually be, nor does it know at all times where all of the capacity amongst service providers is located. For the com-position of the Cloud to be achieved in a manner that satisfies demand almost instantly, increasing use is being made of various self-organising agents: the sys-tem manages its own 'composition' without human intervention and does so in response to a dynamic and volatile environment that a human manager simply could not deal effectively with. On this basis, the self-organising project may be closer than you think!

5.9 Chapter summary

Projects of a global scale have increasingly focused on managing 'engagement' through application of theory relevant to issues such as collaboration, communication and supply chain management. However, there is also a need to consider the environment in which the 'leadership' of a global construction project will be undertaken: what is the nature of modern project management? The answer to that question will depend on an individual or group being either broadly traditional (also known as transactional) or broadly transformational in the beliefs and leadership style that they apply to the management of projects. Holders of these two belief systems (transactional and transformational) will adopt quite different stances with regard to the management of global construction projects.

A particular area where the transactionalists and transformationalists differ is the perception of the extent of alternative futures/development scenarios and opportunities for project management: transformationalists perceive there to be many such alternatives on a global scale. These alternatives would involve the transformation of individuals, organisations and the global environment itself. A less grandiose perspective is to argue that modern project management can be viably condensed to three factors: collaboration, communication and interaction. Collaboration is typically defined in terms of achieving multiple interactions between multiple players over time, and is therefore a more challenging proposition than the (relatively) simple actions of interaction and communication. However, while communication is essentially a simple activity, it must be recognised that the manner in which the project is structured and organised, particularly when the project itself is complex (in terms of both the objectives and the extent of interconnections), can have significant impact (positively or negatively) on the level of effectiveness for communication.

With regard to interaction in the context of projects, the project manager may need to consider any 'rules' of interaction when considering the appropriate form of structure and organisation for the project. In terms of structure, project managers frequently know who it is within both their company and project structures that they need to deal with; an organisation chart may have been helpful in initially finding that person, but once contact is made and a relationship begins, the chart disappears. This informal structure may, however, be quite different to the structure that is formally presented for that same project, a situation that presents particular risks with regard to maintaining control of a large complex global project. Irrespective of the applied structure being formal or informal, construction project management directors and project managers need to be aware that just because a particular organisation structure works well in one environment it is not a given that it will work well in all environments. The typical arrangement for a construction company would be comprised of a three-way relationship between a worldwide project/product manager, a worldwide functional divisional manager, and a project manager specific to the country in which the project is being carried out. This kind of structure is, however, highly variable as individual organisations seek to place their commercial activity in what they regard as the most favourable version of the basic structure.

Managers should not lose sight of the fact that they are relying on people to deliver projects successfully and that people can be quite inconsistent and unpredictable, irrespective of the structure that may be 'imposed' upon them. When project team members begin to feel that they are not being considered, there can be a negative impact on their willingness to be proactive in creating value. It seems that the majority of project managers tend towards recognising a form of value that can be achieved through the 'informal' interaction of individuals and (more frequently) groups. There has to be recognition, however, that factors such as ICT, increased organisation boundary porosity, and the existence of communities of practice inevitably combine so as to result in a need to reconsider the more 'traditional' concepts of organisation structure, and particularly that structure related to the project supply chain.

Management theory when applied to project environments places considerable emphasis on various aspects of building and maintaining efficient supply chains. Such chains are of obvious importance in the context of large and complex global construction projects such as the Yangtze River Three Gorges Project. There is no doubt that market conditions globally are changing at a faster rate than many senior project managers are accustomed to, and acceptance of this is an example of knowledge with regard to a situation, resulting in a different perspective. Thus the development of extreme supply chain management is one example of how knowledge itself changes the established order of doing things. There is also a need to consider the sheer amount of information (some of which may become knowledge) that a global construction project manager will need to process.

By basing the structure solely on the connections between bits of information it may be possible that a software package can maintain the overview (all the connections within the strategy for the project) while the project manager focuses on real-time changes that are flagged up as being either possible opportunities or possible threats. This possibility leads to the suggestion that instead of imposing a specific structure on the project or organisation, the project or organisation is left to organise itself. While such a suggestion may seem implausible, it should be noted that there are already commercially available self-organising applications (SOAs) available. These may well be relatively rudimentary at present, but a future in which the project manager simply selects the appropriate SOAs − and lets them get on with interacting, so as to form the functionality required to deal with a given project's complexity, all without any formal structure whatsoever − may not be too far away.

5.10 Discussion questions

1. Outline the origins of the term 'strategy'.
2. Identify and briefly discuss the nature of the key requirement for effective collaboration.
3. Identify the key advantages and disadvantages for a matrix structure and discuss the requirement for individuals to be trained in working within this structure.
4. Outline the importance of the opening and closing sequences of communication in the context of a multicultural project.
5. In the context of information processing and management, discuss the potential concerns and benefits of the suggestion that the human attention span is decreasing.

5.11 References

1. Lambert, D. M. and Cooper, M. C. (2000). Issues in supply chain management, *Industrial Marketing Management*, **29**, pp. 65–83.
2. Business Directory(n.d.). Strategic management definition. http://www.business dictionary.com/definition/strategic-management.html [Accessed June 2011].
3. Pheng, L. S. and Chuvessiriporn, C. (1997). Ancient Thai battlefield strategic principles: Lessons for leadership qualities in construction project management, *International Journal of Project Management*, **15** (3), pp.133–140.
4. Moore, D. R. (2002). *Project Management: Designing Effective Organisational Structures in Construction*. Oxford: Blackwell Science (Chinese translation 2006).
5. Environmental Turbulence. http://planningskills.com/glossary/30.php [Accessed January 2013].
6. International Centre for Complex Project Management. http://www.iccpm.com/ [Accessed January 2013].
7. Mikheev, V. and Pells, D. L. (2005). The third wave: A new paradigm for project and programme management: How modern project management can transform your business, organisation and people. http://www.pmiforum.org [Accessed June 2011].
8. Weaver, P. (2007). The origins of modern project management. In *Proceedings of the Fourth Annual PMI College of Scheduling Conference*, 15–18 April, Vancouver: http://www.mosaicprojects.com.au/PDF_Papers/P050_Origins_of_Modern_ PM.pdf [Accessed July 2012].
9. Hickman, K. Great Northern War: The Battle of Narva. http://militaryhistory. about.com/od/battleswars16011800/p/narva1700.htm [Accessed August 2011].
10. Patterson, R.(2011). NEC3: Contracts for partnering. NEC. http://www. neccontract.com/news/article.asp?NEWS_ID=782 [Accessed January 2013].
11. 2007 *Poor Communication Top Cause of Project Failure*. http://www.projects atwork.com/content/articles/235492.cfm [Accessed June 2011].
12. McManus, J. and Wood-Harper, T. (2010). A study in project failure. Chartered Institute for IT. http://www.bcs.org/content/ConWebDoc/19584 [Accessed July 2011].
13. Scott, K. (2004). Holyrood plan flaws cost extra £160m. http://www.guardian.co. uk/politics/2004/jun/30/scotland.devolution [Accessed June 2011].

14. Cramb, A. (2006). Dangling roof beam clears Holyrood debating chamber. http://www.telegraph.co.uk/news/uknews/1511950/Dangling-roof-beam-clears-Holyrood-debating-chamber.html [Accessed August 2011].
15. Education for Skills. www.education4skills.com [Accessed August 2011].
16. Madden, D. (2007) *Semantic noise and interpersonal communications.* http://www.helium.com/items/558037-semantic-noise-and-interpersonal-communications [Accessed January 2013].
17. Stawinoga, J. http://www.telegraph.co.uk/news/obituaries/1569599/Josef-Stawinoga.html [Accessed August 2011].
18. CIPD, 2011: What is the psychological contract? Factsheet. http://www.cipd.co.uk/hr-resources/factsheets/psychological-contract.aspx [Accessed August 2011].
19. Wai Properly. http://www.1stopchiangmai.com/how_to/wai/ [Accessed January 2013].
20. Maylor, H. (1996). *Project Management.* London: Pitman Publishing.
21. Meredit, J. R. and Mantel, S. J. (1995). *Project Management. A Managerial Approach.* New York: John Wiley.
22. Our Code of Conduct Skanska. http://www.skanska.co.uk/about-skanska/our-code-of-conduct/ [Accessed January 2013].
23. Williamson, A. *Social Media and the New Arab Spring* Hansard Society. http://hansardsociety.org.uk/blogs/edemocracy/archive/2011/04/19/social-media-and-the-new-arab-spring.aspx [Accessed June 2011].
24. Communities of practice for public service. http://www.communities.idea.gov.uk/welcome.do) [Accessed August 2011].
25. 2011. Did you say work? Nah! Its playtime for the staff at Google's ne Pittsburgh offices. http://tommytoy.typepad.com/tommy-toy-pbt-consultin/2011/02/googles-new-steel-city-satellite-goes-easy-on-the-google-ness-google-is-famous-for-forcing-perfectly-respectable-people.html [Accessed June 2011].
26. Linderman, M. (2009). How playtime is responsible for Post-It Notes, Lasik, and more. http://37signals.com/svn/posts/1804-how-playtime-is-responsible-for-post-it-notes-lasik-and-more [Accessed July 2011].
27. Hays, J. (2008). Three Gorges Dam Project. http://factsanddetails.com/china.php?itemid=1046&catid=13&subcatid=85 [Accessed August 2011].
28. CYTGP development corporation (1999). *Three Gorges Project.* CTGPC, Hubei.
29. Hochfelder, B. (2011). Extreme supply chain management. http://www.sdcexec.com/article/10326695/executive-memo-supply-chain-volatility [Accessed July 2011].
30. Macrae, F. (2010). Facebook and internet 'can rewire your brain and shorten attention span'. *Daily Mail.* http://www.dailymail.co.uk/sciencetech/article-1312119/Facebook-internet-wire-brain-shorten-attention-span.html [Accessed August 2011].
31. Stacey, R. D. (1996): *Strategic Management and Organisational Dynamics.* 2nd edn. London: Pitman.
32. McMillan, E. (date). Considering organisation structure and design from a complexity paradigm perspective. http://www2.ifm.eng.cam.ac.uk/mcn/pdf_files/part5_5.pdf [Accessed January 2013].

6

Global Project Management Process

6.1 Introduction

Traditional construction project management processes and ways of managing construction projects are becoming obsolete as rigid project management processes can no longer cope with the demands of the new economy. This situation has become a real challenge, since most of what has been assumed as standard practice in the past no longer befits current reality. To explore adequately construction project management process, key issues have so far been examined in respect to contemporary project initiation, closure, monitoring, process portfolio management and value chain. The objective of this chapter is to provide advice and methodologies appropriate for managing the project management process in a global environment. It is divided into the following sections:

- project management process in global projects;
- project initiation in global construction projects;
- project closure in global construction projects;
- project monitoring and control in global construction projects;
- process portfolio management in global construction projects;
- programme managing the value chain;
- integrating the process in global construction projects;
- leading strategic transformation of global project management process;
- techniques for a global project management process.

6.2 Learning outcomes

The specific learning outcomes of this chapter are to enable the reader to gain an understanding of:

>> whole life cycle and project phases of a global construction project;
>> project deliverables and their role in sanctioning subsequent phases and managing of a global construction project;
>> strategic transformation and process portfolio in global construction projects.

6.3 Project phases

Traditionally, certain sectors (e.g. construction) have perceived the project to be separate from the organisation responsible for delivering it. This has resulted in poor communication, little integration and too much confrontation. However, recent years have seen increased recognition of the importance of the project life cycle, and hence, value management, whole life costs, investment appraisal and sustainability. In viewing the project as a whole, the interfaces between the various phases have to be better integrated. To assist with the management of individual phases of the project life cycle, certain questions can be asked:

- What is the aim of the phase?
- What are the deliverables?
- What information is required to proceed to the next phase?
- What approval is required to proceed to the next phase?
- Who does what?

In doing this, responsibilities for deliverables can be established and wasteful processes eliminated. There is no standard approach to managing phases. However, a framework typically used in large engineering global construction projects is presented in Figure 6.1.

The framework is applicable to most projects. Some variations in a global environment may occur due to:

- the size of the project;
- the project complexity;
- the need to fast-track the project (overlap the phases);
- the procurement route (change of roles and sanctions);
- innovative projects that are more iterative than the 'norm'.

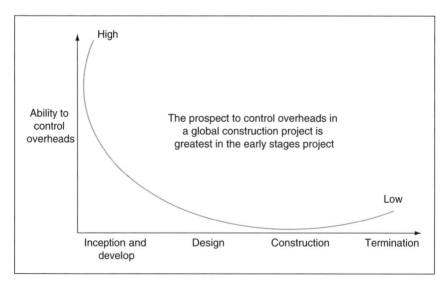

Figure 6.1   Recurrent savings in a global construction project

There are many different approaches to defining the various phases to a project. In the most simple form, the phases are:

- project initiation;
- project monitoring and control;
- project closure.

Each of these project phases is described below in terms of their importance.

6.4 Project initiation in global construction projects

An inherent part of almost all global construction organisations (GCOs) is the requirement from time to time to deliver projects with the objective of meeting the changing global business environment. The efficiency with which GCOs manage and deliver projects is likely to significantly impact upon their overall long-term ability to remain in business. At this juncture, it is worth noting that the initial phase of a major global construction project is usually driven by business profitability. The primary issue is to determine whether, in principle, there is a project proposal that has the prospective to be adequately profitable and fit into the overall business strategy of the organisation.[1] The global economic uncertainty, and the effect it has had on project budgets and resources, has led construction organisations to change the way they select, manage and define projects. It is in this context that we view the initiation phase as that period when schedule, capital and human resources have to be explored with the client. To ensure good utilisation of funds and resources, construction organisations have had to sharpen their initiation, monitoring and closure strategies.

To maximise success in the delivery of large and complex construction projects, clients need to instruct contractors to come up with project plans that minimise risk and waste. Clients should view this is as a prerequisite of their project management model, the successful integration of which involves an effective recurring process. A key element of that model is the organisation of a lean project initiation process. When starting a new global construction project it is important for the client and the appointed project manager to arrange a project planning workshop with contractors, to ensure that all members of the project team agree on how the project objectives will be achieved. The agenda should be: who will do what and when will they do it by? At the workshop planning meeting, the appointed project manager should table the project initiation document and be prepared to address any questions. During the meeting, the main role of the client should be based upon reviewing the project proposals, highlighting how the project will be delineated and controlled, identifying suitable quality standards and making a decision whether or not to formally authorise the project. The workshop planning meeting will:

- ensure that the client obtains the best value;
- encourage well-organised and effective planning of project delivery strategies;
- ensure the clear definition of project objectives;
- encourage project delivery options that are innovative.

Depending on the size and complexity of the global construction project, a large amount of preparation may be essential prior to the project planning workshop. It has conclusively been shown that project results can only be attained with objective and responsible management of the right quality put in place as a matter of deliberate policy at the very earliest phase of the project, and given all necessary decision-making authority on behalf of the client.[2] As illustrated in Figure 6.1, significant recurrent savings can be achieved in the early planning phases of the project through careful deliberation of the need for the project and the most appropriate means of delivery method to be used. The project initiation phase of a global construction project requires definition of:[2]

1. **Risk:** It is vital for the client, project manager and the project team to be fully aware of the type of project and levels of risk and complexity involved. If the right project manager is not selected, substantial unproductive expenditure of capital, resources and time can take place. This is likely to occur at the project initiation phase due to lack of leadership, lack of data and indistinct objectives. Utilisation of the risk management process is essential for effective strategic planning.

2. **Market:** This is one of the key areas that the client and project manager must explore; they must obtain a very high level of confidence in the market demand over a period a number of years. They need to deliberate on all the likely changes to that demand and attempt to propose a project framework in terms of expenditure and design that will give them the greatest suppleness to react to market changes, whilst maintaining the achievability of the project. The client and project manager have to ensure that the utilisation of market research is done throughout the project initiation phase and during the construction phase, so that data on demand and income-generating capability is up-to-date and the design strategy adopted responds to any economic change in the forecast demand. In addition, the client and project manager must also take into account how the project might react to change in demand level over its life. After determining the market status, they should be in a position to progress with the design, operating strategy and policy for the project.

3. **Objectives, policy and budget:** There are two important areas of objectives and policy that the client and project manager have to turn their attention to: those concerned with the management of the venture, and those concerned with the venture itself. From a project management perspective, the main actions that the client and project manager need to take relate to the organisation of objectives and policy for: the form of the project management model to be utilised; selling the notion of project management to the entire project team; definition of roles and responsibilities and, in particular, the role of the project manager; securing the essential resources to support the project initiation work. Failure to discuss these can be a source of much perplexity and disagreement as the project develops. It is very important for the client to ensure that the level of authority and responsibility of the project manager is well defined. At the early stage of the project, the project manager has to ensure that the objectives are as clearly defined as possible and are agreed and sanctioned by the client. The next phase is to delineate the project policy. The policy statement can be used to ensure that work undertaken is in harmony with the client's policies. With the

policy and objectives established, the project manager can propose a budget for the project with a sensible level of accuracy. The project objectives and policies form the foundation of the project and must be communicated to all contractors if their efforts are to be efficiently directed and controlled.

4. **Resources:** Once the project objectives, policy and budget have been established, the project manager has to integrate additional and relatively considerable resources to develop the project. In a global construction project, consultants and construction specialists may be needed in areas such as: structural engineering, legal, taxation and project costing. In the early phase of a project, knowledge management is more important than data collection and analysis capability. In order to utilise resources effectively, the initiation phase should be used for *applying* experience and not for *gaining* experience. Using key components of the project objectives, the project manager can engage both international and regional contractors and request that they present a statement of competence and experience related to the project objectives. Once the statements are received, the project manager should assess the specific individuals who will be undertaking project tasks, so as to ensure they have the skills and knowledge. Once the team has been assembled, the project manager has to develop a detailed programme for all the project objectives. Of particular importance are the release dates for project tasks that must flow between contractors. The project team must be involved in developing the programme.

5. **Value management:** The integration of value management in the project initiation phase of a global construction project will help the project manager to come up with a well-thought-out process into the earliest phases of the project. As an effective project method, the project manager can use value management to enhance best practice throughout the life cycle of the project and also ensure that all tasks are achieved at the lowest total cost consistent with required levels of excellence and performance.

6.5 Project monitoring and control

Subsequently, there is the question of whether a global construction project can be monitored and controlled in a consequential way. One of the challenges faced by senior managers is sustaining an up-to-date monitoring and control scheme. If there is any uncertainty, there will inexorably be doubts about the project's delivery and manageability. Only by controlling and monitoring progress against the agreed project objectives, and by taking the suitable remedial actions, can the project be protected against risks. Success in a global construction environment is not achieved merely by using the right tools and techniques, although these are essential. The implementation of an effective project monitoring and control system is required in order to enable the project manager to adhere to the various project phases. In a global project environment, project phases are very important; one of the indispensable steps of global project management is to delineate these phases. In so doing, project managers can assign appropriate tasks to stakeholders, ensure that deliverables fabricated at the end of each phase meet their objective, and the project team are well equipped for the next phase. To ensure project success in a global environment, the project manager should also:

- identify key performance indicators that determine the status of the project;
- establish an effective responsive monitoring and control model;
- plan for amalgamation at the early phase of the project;
- introduce a corrective, change and communication system;
- know what data concerning the project should be quantified;
- give regular feedback on performance to each project team;
- ensure that design, construction and engineering work is achieved through integration;
- run a team-building session, so that all stakeholders have a clear understanding of how data will be shared.

Conventional and contemporary project control inclines to focus on status schedule and comparisons of actual costs with budgets. Successful project control in a global environment relies on a well-structured cost and performance model. This necessitates careful design, development and implementation to ensure that instant feedback can be attained. It is vital that senior managers compare the schedule, cost and performance actuals aligned with the budgeted schedule, cost and performance figures in the project plans. This contrast must be performed in an integrated way. It is only by frequent monitoring of all the three variables that senior managers can keep a close watch on the big picture and minimise the project scenario of winning the combat but losing the war.[3] The alignment of cost, performance and schedule has always been underlined by clients in construction projects. When performance, cost and schedule are not met, the project manager must apply effective control measures to correct the divergence before the project performance worsens noticeably. Project managers can use control strategies in conjunction with their own technical knowledge to assess and rectify any project performance and variations. However, contemporary project control methods are not as helpful to project managers as they might be.[4] A number of control methods are designed to detect the existence of a performance problem. They are not designed to identify the root source of the divergence, or to propose a remedial action. Effective control of a global construction project entails four discrete actions: assessing, planning, correcting and measuring. As illustrated in Figure 6.2, these four variables need to be evaluated in an integrated way.

The aim of the framework set out in Figure 6.2 is to provide global construction organisations with a base that facilitates an integrative approach across all phases. Before the optimum solution to a proposed concept is established and selected, several alternative solutions should be evaluated. The evaluation, as part of a feasibility study, is an investigation into how a proposed scheme might be implemented, and whether it is capable of being carried out successfully. Within a global project environment, this should be assessed on four standard criteria of technical, social, economic and operational feasibility. The feasibility study becomes a project when the client sanctions an agreed course of action, which has been developed in the form of:

- specification-task and quality;
- sanctioned budget cost (estimated);
- programme of work time.

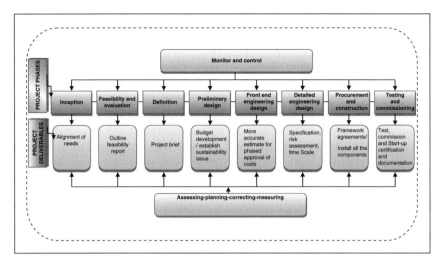

Figure 6.2 Integrative approach in project control and monitoring

The global project manager should use the feasibility and evaluation activities to deliver:

- outline feasibility report;
- detailed feasibility report – this mainly deals with selected prospects when it has been established that those concepts are viable;
- preliminary specification – a statement of requirements.

Feasibility reports are used to seek approval for funds based on assessment of the economic, technical, social and environmental feasibility of options. The feasibility report should contain: summary, basic data, development plans for prospects, recommendations/conclusions, economics and cost estimates. The main definition deliverable is the *project brief*, which outlines what the form of the end product will be. The project brief can be used to obtain sanction to the next phase – preliminary design. A project definition rating index (PDRI) can be used to assess how well-defined the project is at various stages. The preliminary design deliverables can be used to obtain sanction for the next phase (front end engineering design (FEED) or detailed engineering design (DEED)). Additionally, they define the form of the end product and how it will be built through:

- prospect development plan (PDP);
- budget development;
- basis for design;
- pre-project specification;
- project execution plan.

The basis for design should include: summary, objectives, constraints, risks, location, permits, design, technical specification, management strategy, schedule, contractor instructions, pre-project specification or statement of requirements. The project specification develops as the project progresses and defines what has to be built and to what standard. It should:

- record the basis of agreement on the project's technical objectives;
- state the degree of design development;
- identify the codes of practice and standards to be used;
- form the basis of estimates and preliminary programme;
- convey adequate and precise information to the project team.

The project specification should comprise general project information, scope of work, location information, design information, engineering information, procedures for implementation, estimates and cost information. For the project team, accurate estimates are one of the key project parameters. Subject to good preliminary estimates, money can be sanctioned for the front end engineering design (FEED). Early estimates often serve as a baseline for identifying changes as the project progresses and, inevitably, for more accurate estimates to be produced. Estimates and re-estimates should be performed throughout the life of a global construction project and full sanction should be given on the basis of the second estimate. On larger projects, where limited technical definition has been used to prepare the preliminary estimates, phased approval of expenditure may be required. The project manager should use detailed engineering design (DEED) deliverables to procure and construct the project. They include: drawings, specifications, risk assessment, costing and time scale. In a global construction project, the deliverables for the procurement phase of the project will vary depending on the procurement route. Included in this should be details of framework agreements (partnering), contracts and procurement of resources. Construction deliverables should include regular progress reports of the completed works as per the specification and schedule – and on plan. Additionally, the deliverables include details of payments made according to contract. Details of any changes should also be included.

In order for the project to be monitored and controlled in an efficient way, the project manager needs a steady flow of data that is as up-to-date as possible on the status of work being performed, volume of work being completed, and quality of work being executed on each phase of the project. The authors view this as a requirement in a global project environment. Consequently, GCOs ought to invest greatly into the enhancement of their project control and monitoring techniques. To help ensure success in the delivery of large, complex global construction projects, project managers can use project control techniques to identify and monitor deviations but they cannot use control systems to uncover the cause of divergence. A good starting point for project managers is to incorporate project control systems as a formal guideline to their project control models. This entails training project teams and providing essential tools to enable individuals to accumulate and contrast project data.

6.6 Project closure

All global construction projects must reach closure, one way or another. Whereas some may come to a premature end through 'forced' termination, most meet their planned objectives. There are four main reasons for projects to be terminated:[3,5]

1. **Termination by starvation:** This can occur because of political reasons. An organisation may decide to have a project open in their portfolio, even though they do not intend to have it succeed or foresee its completion. This can be because the project is owned by an influential government sponsor who must be appeased.
2. **Termination by integration:** In global construction projects, integration symbolises complexity. In this form of termination, the project team and resources are usually reincorporated within the existing organisation structure following the completion of a project.
3. **Termination by addition:** This form of termination brings a project to a close by normalising it as a formal part of the primary organisation.
4. **Termination by extinction:** This practice will take place when the project is discontinued due to its either successful or unsuccessful completion.

As exemplified in this chapter, global construction projects are designed to produce a specific inimitable outcome. The project manager should use the closing phase to confirm the completion of deliverables across all phases to the satisfaction of the client, and obtain any lessons learnt and best practices to be used for future projects. Applying closure in this manner allows the project manager to commune final project status to the project team and stakeholders. Depending on the global construction environment, and the nature, size and complexity of the venture, the practice of project closeout process can consist of any of the following key components:

- lessons learnt review: the primary purpose should be to assess overall success of the project and ensure transfer of knowledge;
- testing of the end product;
- ensuring the terms and conditions of the contract have been met;
- administrative and operational review, which can be achieved by analysing overall project performance;
- classifying areas of further improvement;
- establishing final project costs and benefits, which should be based on the success criteria identified in the initiating and planning phase of project.

Project closure should be planned at the early phase of the project even though it is often the core process of a project's life. At the initial phase of the project, it is advisable to have a well-defined framework that provides the stakeholders with end-user expectations, requirements, revenue recognition methods, resources, roles and responsibilities. The framework should not be changed once a project is in progress, unless a recognised change management process is used. Without a well-defined project closeout framework, closure cannot be achieved as there may not be any tangible dimension for completion. The need to deal with sociological, environmental, economical, political and technological issues calls for the archiving of information so as to help with the enhancement of future projects. The project manager should gather all data and plan for knowledge transfer where appropriate. Handing over the project to the client can be either an uncomplicated or a complex process, depending on how the project manager handles the terms and conditions of the contract. In a global project environment, the process should involve transfer of knowledge where appropriate, conveying technical designs, sharing long-term sustainability solutions and engineering specifications. The need to deal with political, environmental, legal and economical issues calls for a meticulous handing-over process. The client and project manager may use a project closure and handover checklist to ensure that all activities are completed. If the client is inexperienced in technical designs, sustainability solutions, or engineering specifications, they can obtain support from specialist consultants, general project management consultants, and design and management contractors. Such professionals have extensive knowledge in the overall life cycle of a project.

6.7 Process portfolio management

In recent years, the terms portfolio management, programme management, enterprise project management and multi-project management have been used interchangeably in the literature.[6] Each of these terms seeks to address a particular business incidence and wrestles with the primary problem of coherently controlling, selecting and managing diverse projects. Global construction organisations are trying to find solutions and strategies that will help them manage multiple complex projects and make decisions relating to investments and strategic initiatives in a rapidly changing and increasingly competitive global environment. The application of programme and portfolio management techniques is seen as a possible solution:

> Programme management is the co-ordinated management of related projects, which may include related business-as-usual activities that together achieve a beneficial change of a strategic nature for an organisation. What constitutes a programme will vary across industries and business sectors but there are core programme management processes, whereas, portfolio management is the selection and management of all of an organisation's programmes and related business-as-usual activities taking into account resource constraints. Portfolios can be managed at an organisational, programme or functional level.[7]

Archer and Ghasemzadeh,[8] defined project portfolio as a set of projects that share and compete for scarce resources and are carried out under the sponsorship and management of a particular firm. Project portfolio entails identifying, prioritising, authorising, managing and controlling the constituent projects and programmes and the associated risks, resources and priorities.[9] Project portfolio management is primarily focused on achieving what is needed and ensuring an organisation works on the right projects.[10,11] As illustrated in Figure 6.3, a unique attribute of programmes is that the interrelated projects must be managed together if the desired strategic benefits of a firm are to be achieved. Thus senior managers have to maintain a focus on the delivery of agreed strategic objectives.

Over the last decade there has been a rapid propagation of different types of portfolios, such as product portfolios, resource portfolios, asset portfolios, project portfolios and investment portfolios. Research to date suggests that the concept of project portfolio management has surfaced from two complementary, autonomous drivers, these being the need to make rational investment decisions

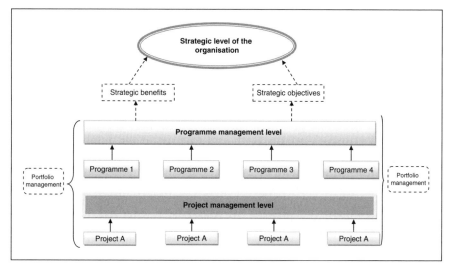

Figure 6.3 The relationship between strategic level, programme level, project level and portfolio management

that result in the delivery of organisational benefits, and the need to optimise the use of resources to guarantee that the delivery of such benefits occurs in an effectual and well-organised manner.[12,13] So far, however, there is a gap between existing tools and the needs of global construction organisations, due to the lack of sufficient management tools to establish priorities among construction projects. Current portfolio management tools and techniques are not enabling organisations to effectively manage the emergent and dynamic nature in which projects are recognised, initiated, managed and withdrawn.[14] This is due to the fact that existing portfolio tools are reliant upon a rational, mechanistic, and linear method to determine the firm's strategy and priorities.[6]

Projects do not operate as single units with an organisation, but instead, there are a number of projects operating simultaneously. In this diverse project context, individual project managers are able to draw on a wide spread of resources that exist within the organisation in order to coordinate their project teams, budget, deliver project objectives and accomplish business results.[6] Each project may be approved depending on its strategic importance to the organisation and, as such, integrated into a portfolio on the basis of providing the best possible balance between risk and returns. If this is performed without consideration for the capacity of the firm's current resource pool, the resource allocation syndrome may transpire.[15] A number of conventional project management techniques and tools used in singular projects are linear, mechanistic in nature, and do not support multiple project decision making or resolve problems of intra-project resource contention that exist in the multi-project framework.[16,17] Moreover, these mechanistic techniques do not permit changes in strategies that may occur due to shifts in the environment or market discontinuities. What is needed in the global construction industry is a set of new tools that will help clients to dynamically identify, select and balance their projects being delivered in the global market. Managing multiple global construction projects requires effective project management tools, techniques, prioritisation and strategies. Albeit multi-project management techniques provide clients with a useful way to understand a complex project environment, these techniques do not address the fundamental resource management and strategic alignment problems faced by global construction organisations.

Innovation plays an important role in the practice of project management in organisations.[18] Growth of an organisation is linked with its ability to constantly deliver new projects and products. Simultaneously, there is economic pressure to minimise the time to market. Both lead to a rise in the number of projects initiated concurrently within global construction organisations and, as a result, to the complexity of managing them. Innovative forms of planning and organising global construction projects have emerged, for example, programme and portfolio management. Project portfolio management is gaining more and more importance in theory and practice. Its objectives can be classified into three broad categories: maximisation of the financial value of the portfolio, linking the portfolio to the organisation's strategy, and balancing the projects within the portfolio in consideration of the organisation's capabilities.[19] The integration of projects within a portfolio delivers benefits beyond the results of independently managed projects. A portfolio has to be balanced along a range of features in order to provide the best value to the establishment. Projects and project portfolios are an important part of strategic management for organisations as they enable a successful strategy implementation. In order to meet the firm's overall objectives, senior managers need to set out the benefits and goals of a portfolio before the selection of projects. It is essential to link global construction projects to organisational strategy.[20] The emergence of portfolio management in construction is correlated with increasing number and complexity of projects. GCOs have been looking into delivering projects more efficiently. To succeed, they will have to integrate effective portfolio management organisational structures. Good project portfolio management is becoming a key competence for GCOs managing projects concurrently. Even though the literature recognises the components that should comprise portfolio success, it still remains difficult to capture the overall management system outcomes. Evidence shows that this might be because project portfolios are dynamic, interdependent systems that constantly change and develop.[21]

In today's global economy, every construction organisation is struggling to do more with less, and performance is a dominant concern, which is why, in order to succeed in lean times, it is important for construction clients to optimise every operation of their programme and portfolios. Given the continued uncertainty of the global economy, cost management and efficiency continues to top the list of priorities for many construction clients. In this environment, it is essential for GCOs to steer resources to the projects with the most impact, ensure visibility into any financial fluctuations and ultimately provide their senior managers with the tools and techniques for managing multiple projects. To help ensure there is alignment between strategic objectives and project objectives, GCOs will have to deploy project portfolio management solutions to identify which project(s) support key business programmes and at what cost. These benchmarking tools will allow construction clients to shift future organisation investments into more productive and cost-efficient tools that can be seamlessly incorporated in the existing organisation infrastructure.

6.8 Leading strategic transformation of global project management process

There is an instantly recognisable trend in the move of GCOs towards implementing project management as a way of working rather than simply a methodology. Project management is also a key enabler with which organisations integrating business improvement models such as Six Sigma or lean manufacturing improve their efficacy and competitiveness. In fact, a number of large global construction organisations have invested significant capital in improving their project management procedures and still do not fully benefit from that investment because the organisation is not fully aligned to supporting project management.[22] It is becoming increasingly difficult to ignore project management/business strategy alignment: organisations must develop and execute innovative project/business strategies in order to stay competitive.[23] The project management/business strategy alignment will help GCOs focus on the right projects, given the objectives of the business strategy. In doing so, projects will be selected as a mode of implementing those strategies. Construction organisations that most effectively apply project management methods have a clear well-communicated strategy and know how each project supports it. It has been conclusively shown that installing effective project management includes putting in place a process to appraise every project for its fit with the strategy, before execution.[24] This needs to be carried out during the project definition phase. In most GCOs, responsibility for the strategic assessment lies with senior management. As each proposed project is evaluated, senior managers need to address the following questions:

- How does this project support our future push for business enhancement?
- How does it fit into our present portfolio and market?
- How is it linked to our potential needs, financial and growth prospects?

If the answers to these questions show that the project is not a good strategic fit, it needs to be terminated. In well-established GCOs, the project management office is charged with decree on the strategic fit of projects. This can only be achieved when the organisation strategy has been well devised and communicated from the strategic level of the company, so that preceding questions can be answered by the project management office. In situations where it is indistinct as to whether or not a project is aligned with the organisation's goals, the project management office needs to verify with senior management before sanctioning it.[24] In today's financial environment, maximum value has to be returned on an organisation's investment (ROI) and it is important that all sections within the company are developed to guarantee that project management is allowed to work efficiently and cross-functionally.

It has been recommended that the executives of organisations should at least have a short executive briefing on project management, if not a longer senior programme to allow them to comprehend how to create the behaviours and support the project management '*way-of-working*'.[22] In addition, senior managers should undertake senior development short courses that educate them on the benefits of utilising good project management practices. Global construction organisations can use these workshops to give senior managers a forum to share existing practices across the organisation and discuss what parts of the project management system are being utilised to their projects, what could be enhanced, and to set proposals to improve it. Senior managers can also set out the organisation policy for project management. The policy should highlight how the organisation will integrate project management concepts. Organisations cannot get maximum value from project management unless senior executives provide:[22,24]

- The vision for project management and the environment, attitudes and behaviours expected in a policy statement.
- Set-up procedures for project management with aligned tools, whilst acknowledging that different sections of the organisation have different requirements.
- Classify and enhance the project management competencies in all sections of the organisation: strategic level, operational level and project level.
- Introduce project management as a central part of the organisation with an identified structured career path.
- Gauge behavioural change and organisation performance early on in the project management advancement process.

GCOs are multifaceted units that operate smoothly only when their project teams work together to produce results. In order to instate strategic transformation successfully in GCOs, senior managers need to ensure that each of the following elements is aligned and integrated into a logical framework for project management: a framework should capture the reality being modelled as closely as practical and include the essential features of the reality whilst being reasonably cheap to construct and easy to use;[25] the use of a framework in a complex situation assists managers in minimising risk, imposing consistency, and provides a common generic and coherent structure in decision making.[26,27] Although each global construction project is unique and has a different set of requirements, common steps are involved in developing a good project framework, as illustrated below:

1. **Objectives:** GCOs need to identify which operational objectives will make a difference in the organisational strategy and install effective methods for keeping these observable to the project teams. Progress against these objectives needs to be monitored, communicated and measured on an ongoing basis.
2. **Operational processes:** within a GCO, the techniques used to congregate, analyse and propagate data must support project-based work.
3. **Human abilities:** effective project management requires the right project team, with the right skills. There is no single best way to configure global construction project teams. When setting up a global construction project team, senior managers need to ensure that they have fully matched the team structure to the projects and business needs. Several substitutes exist for organising individuals for project work. Two common approaches are hierarchical and matrix. Historical and contemporary accounts of projects tend to concentrate on design, aesthetics, materials, technological developments and the impact of projects on the environment, and give little attention to how

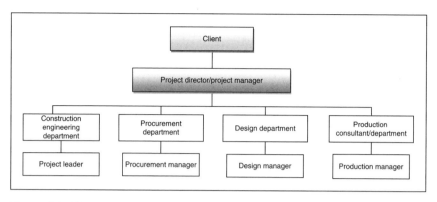

Figure 6.4   Hierarchical structure of organisations of management

people work together and manage their activities.[28] However, what can be observed is that many industries have developed an array of specialisations, each of which must play its part in the realisation of a completed project. Their evolution over the past 200 years or so has resulted in the grouping of tasks according to function (so-called *functional specialism*), the creation of separate professions and the emergence of small- to medium-sized organisations offering a discrete professional service.[28,29] Project organisation of these specialisms (whether consciously or not) has followed classical principles of organisation theory, with a hierarchical structure and clear lines of authority based upon a superior–subordinate relationship as represented in Figure 6.4.

Such hierarchical forms stem from military and church models that served as command and control structures. As such they are not appropriate representation of a range of contributors from a variety of organisations (which themselves may have a hierarchical management structure) brought together to carry out a specific project. Kerzner[29] outlined how management has developed in an attempt to better understand, analyse and improve the performance of such project teams. Van Bertalanffy (cited in Kerzner)[29] considered an alternative way of looking at organisational structures by identifying so-called open systems using anatomy nomenclature. For example, the human body, muscles, skeleton, circulatory system and so on are all described as subsystems of a total system – the human being.

Subsequent research by Boulding, cited in Kerzner,[29] highlighted the communication problems that can occur during systems integration as a result of subsystems specialists having their own language. These concepts are entirely relevant to the global project environment, since it can be described as an open system with specialist subsystems. Similarly, difficulty with coordination is an everyday experience for many, and to have a successful integration take place, all subsystem specialists must talk the same language. The application of these ideas to a business environment, undertaken by Johnson, Kast and Rosenzweig, inspired a management technique that is able to act across many organisational disciplines, whilst still carrying out the functions of management.[29] It has come to be called systems management, or matrix management (see Figure 6.5),

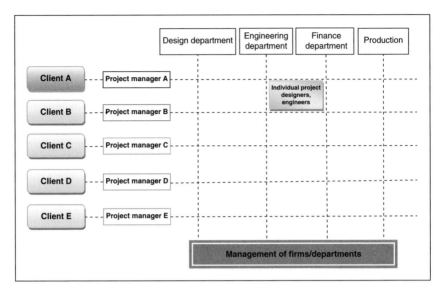

Figure 6.5   Matrix management structure

and in recent years has become increasingly applied in the global construction industry through project management.

The diagrammatic analysis shown in Figure 6.5 arguably offers a more realistic representative of the structure of project management within a GCO. However, as with much analysis, it only helps us to think about projects more constructively; it does not in itself provide any improvement in the performance of project functions. Indeed it only begins to highlight the organisational complexity of projects, the inherent conflict between function and project, and a propensity to fracture along professional lines. Hence, the skills of the project team need to be continuously appraised.

- **Culture and performance system:** the successful transformation of global project management process depends on an organisation's belief that *how* projects are managed is just as significant as *what* they attain. The best way to amalgamate culture, align new systems, techniques and procedures into GCOs is to make them pertinent to the way business is conducted, which means they must prove added value and also be carried out in accordance with the highest project and ethical standards.
- **Implement effective issue resolution systems:** since projects are instigated to meet a one-time client need and recurrently cut across the organisation, existing chain-of-command and escalation procedures may not apply to them.[24] It is therefore important for senior managers to clearly state how decisions will be made within the organisational and project governance structure.
- **External factors:** there are different external factors that affect global construction organisations. In today's global environment, competitive advantage must be extracted from both internal and external factors of the organisation. Senior construction managers need to be familiar with their

external project environments, as this will allow them to discern any threats and opportunities associated with their global construction projects. In addition, government regulations must also be integrated into the organisation's project management process.

In the new global economy, project management entails deliberate planning and action to create the circumstances for success. If strategic transformation of global management process is to be effectively achieved in a global construction environment, construction organisations will have to ensure that their key project decisions are informed by the knowledge and experience of local or indigenous managers. This will require construction project managers to have a better understanding of global strategic planning, global cultural change processes and procedures in the new global economy.

6.9   Programme managing the value chain

The past decade has seen the rapid desertion of traditional organisation forms in the global construction industry. Inflexible functional approaches to management can no longer cope with the demands of the new global economy.[30] Managing global construction projects through project and process-portfolio programmes is gaining popularity. A notable feature to this is that programme management delivers organisational benefits resulting from aligned corporate, business unit, project and operation strategies. It helps project managers to facilitate a portfolio of projects and processes that bring about strategic transformation, innovative continuous enhancement and client satisfaction.[31] Programmes in GCOs can be classified into four levels:

1. Strategic or goal-oriented project-portfolio programmes can be used by senior construction managers to transform strategic decisions. The main advantages of strategic project portfolios are: strategies in construction organisations can be decoded into concrete project actions; emergent changes to strategies during integration can be easily dealt with within the structures of programme management; risk and uncertainty are minimised through iterative programme development; the deliverable of projects is subject to an amalgamation of review and approval based on an assessment of key project indicators.
2. Innovative project portfolio can be utilised to maximise continuous improvements in projects. The major advantages that accumulate from this method are: manifold initiatives are integrated to create actions that will be rational and well organised, whilst short-term project activities can be incorporated into a long-term strategy of the organisation; assessment of derived project advantages, based on key performance indicators, can be proposed with a clear viewpoint; innovative incessant programme can be used to ensure effective coordination and integration of continuous enhancement initiatives across the whole global organisational value chain.[31]
3. Capital expenditure programmes can be employed in mega construction projects. Capital portfolios are prescriptive, and are incorporated around different organisational themes. The main benefits of capital expenditure programmes can be summarised as: better prioritisation of and control

over diverse projects, better allotment and deployment of project resources, suitable recognition and management of dependencies between projects.[32]

4. Contractors can use process-portfolio to augment both internal and external client requirements. Process-portfolio is operational in nature and focuses on enhancing client service, based on senior-manager leadership.[32] The main advantages that can originate from this approach are: delivery of both internal and external client project objectives, which are aligned with corporate strategic goals; organisational effectiveness; efficiency in the project life-cycle chain.

In general terms, global construction organisations can use the four programme configurations to coordinate portfolios of projects in the value chain. The four programme portfolios support each other integratively in the organisational environment and are systematically linked to project operations.[32] In order to realise and enhance efficiency in the value chain, senior managers have to ensure that:

- appropriate resources are available for each project;
- each project is aligned with the strategic intent of the organisation;
- important strategic organisational objectives are realised;
- projects are prioritised in order of importance.

The growing trend in the globalisation of construction projects is giving rise to a need for project-portfolio programmes. For many multinational construction organisations, this will require a highly communicative leadership style at strategic, operational and project levels of the organisation. Senior managers will need to pay attention to long-term strategic organisational objectives. Awareness, recognition and knowledge of long-term strategic issues on global projects would help senior managers to integrate both their strategic planning and project delivery capabilities. Achieving this on global construction projects will need the recognition of thinking and practice relating to the use of strategic business planning, project management, programme management and value engineering. A diagrammatic representation of the above four initiatives is illustrated in Figure 6.6.

6.10 Techniques for global project management process

The essence of project management is to support the implementation of an organisation's competitive strategy to deliver a desired outcome. It is worth highlighting that GCOs are undertaking multiple projects as a growing part of their operations.[34] These rapid changes are forcing GCOs to adopt project information technology (IT) systems that aim to manage the data produced at operational and project levels, and collate it at the strategic level. Many project-based organisations have introduced programme or project management offices, which can have multiple functions but are mostly used to produce information and develop standardised project management practices.[34] Other project-based organisations have implemented portfolio management principles. In all these project tools, the focus is on the management of diverse projects. In the last five years, a number of global construction organisations have moved

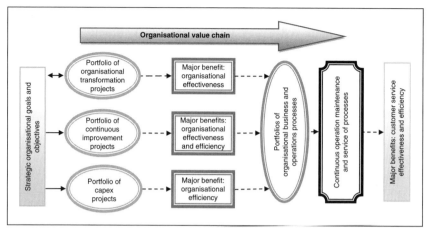

Figure 6.6 A systems perspective of project portfolios and process portfolios in the organisational value chain system
Source: Steyn (2001).[31]

to a more strategic project management model. This shift has generated a greater interest in multicultural stakeholder management; the relatively new discipline of portfolio management and programme management in construction has stemmed from the need to manage benefits from diverse interrelated projects. This has created new challenges for GCOs in the form of technical issues. In order to tackle these issues, GCOs are increasingly integrating new practices and knowledge such as value engineering, multicultural stakeholder management, programme management and portfolio management. As illustrated in Figure 6.7, most project-based organisations have adopted project management office and portfolio management structures that evolved from a traditional pyramidal organisational structure.[33] In this framework, senior construction managers can use the project management office to monitor project work being undertaken, control project performance and develop project management competencies and methodologies. The past five years has seen the rapid development of multiple global project management, but there is currently considerable divergence about how to translate the individual project knowledge to manifold global project management environments. These discrepancies, as well as the general isolation of global project management research to practitioners, seem to ensconce practitioners into easy-to-understand mechanistic project management frameworks. From the reviewed literature, a number of researchers have argued that these mechanistic frameworks are not sufficient to manage complex fast-track projects that are the quintessence of GCOs.[34,35,36]

The need for a more integrated approach in global project delivery could be achieved by a logical project governance approach. A particular issue, which is poorly stated in a global construction environment, is how to generate value through the integration of a project management office, portfolio management and programme management, as well as the double-loop effect of strategy on programmes and projects – and their fragmentary results on strategy. This iterative to-and-from process between executed strategy through projects, and the irreversibility of the result of terminated projects, has yet to be explored. As presented in Figure 6.8 a well-amalgamated GCO would be expected to demonstrate strong interrelationships between its projects and both its operation and corporate approach.

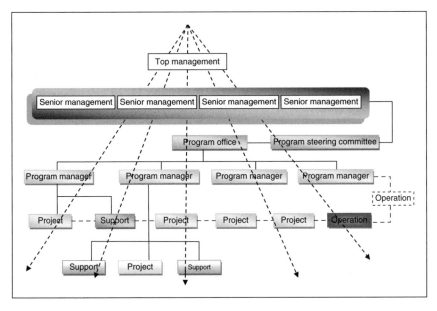

Figure 6.7 A mechanistic framework
Source: Adapted from Thiry and Deguire (2007).[33]

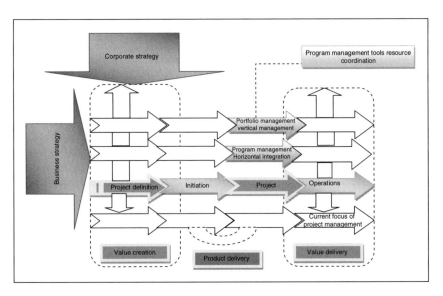

Figure 6.8 Vertical and horizontal integration in a global construction organisation
Source: Adapted from Thiry and Deguire (2007).[33]

In the new global economy, construction organisations will have to incorporate project portfolio management and information technology governance solutions to benchmark and modernise strategic proposals. These solutions will help GCOs to manage multiple projects. In order to build trust and answerability, project management offices must have visibility into every aspect of the global construction project, as well as the tools to manage and monitor project teams. Project portfolio management is an important component in the integration process of projects, and plays an important role in aligning project efforts with corporate strategy and optimisation of resources throughout the organisation. To better manage and prioritise global construction projects, senior managers will have to utilise these capabilities in their projects. As we approach the new uncertain global economy, GCOs will have to utilise holistic, strategic project models, which embrace agile methods and increases responsibility for senior construction managers and project managers. The function of the project, programme and portfolio management has become more significant than ever in supporting organisational initiatives. Success in the new global environment will depend on a combination of tools that will provide visibility into project implementation, best practices and senior management support.

6.11   Integrating the process

GCOs have documented the competitive advantage that management of multiple projects can provide in the fast new global economy. Global economic crisis has led to greater assessment of how construction projects are managed and delivered. This change in focus from clients has led to a more lean and integrative approach of project management practices. Amalgamating project management practices with management practices such as value engineering, simultaneous engineering, multicultural stakeholder management, contract management and cost management will become a coping mechanism for a number of GCOs operating in the new emerging economies. The effective implementation of a multiple project-based strategy and the allied project management practices are seen by practitioners and researchers as having the potential to enhance overall organisational performance, as well as equipping GCOs with the capability to rapidly become accustomed with the changing new global economy. Changing global business environments will require integrating a flexible working culture and diverse organisational forms, which will involve, in essence, an organisation potential in the management of diverse project teams, programmes and portfolios, and its efficiency in implementing project strategies.

One of the central problems that GCOs will have to face is coordination of portfolios. The integration of an overarching framework for portfolio management will assist senior managers to facilitate decisions relative to the priority of projects and the suitability of projects to organisational objectives. Improvements in the aptitude of GCOs to manage and coordinate portfolios will also be essential, along with careful adoption of technologies, and greater attention to the economical, social and political contexts of projects. Recent evidence suggests that when coordinating multiple projects, there is a need to integrate an overarching framework in order to provide sufficient harmonisation to the portfolio of existing and emerging projects.[37,38] Such a framework will assist senior management efforts to balance the requirements of single projects, relative to organisational pressures.

A common problem encountered by GCOs is the difficulty of sustaining control and communication. This occurs due to a number of organisations unable to align and benchmark management of their diverse projects with their organisational objectives. To make a success of the control and communication process, there should be conformity on the requirements, along with the overall strategy to be adopted. Cautious deliberation should be given to the appropriate balance between the various influencing variables, for example risk, schedule, cost and benefit. The key to success with control and communication is for all the stakeholders to understand exactly what is required of them and why it is essential to conform and to be empowered with the right data to do their tasks. From the above, it is apparent that GCOs will have to ensure that systems and processes are in place in order to sufficiently address project challenges and maximise benefits in their organisations. They will have to develop an integrative approach to project delivery, by focusing on the strategic plan of their organisations. In some organisations, this will require the conversion of their business models, for others this will mean creating linkages between existing disparate models and transferring attention to the model as a whole.

6.12 Chapter summary

The authors have shown that GCOs can make significant savings in the early planning phases of their projects, through careful consideration of the requirements for the project. The success of the project initiation phase will depend on how the project manager has delineated the risks, objectives, resources, value management and the market. The implementation of an efficient project monitoring and control system will enable the project manager to manage a global construction through the various phases. In order for the project to be monitored and controlled in an efficient way, the project manager needs a steady flow of data that is as up-to-date as possible on the status of work being performed, volume of work being completed, and quality of work being executed on each phase of the project. The authors view this as a requirement in a global project environment. The project manager should use the closing phase to confirm the completion of deliverables across all phases to the satisfaction of the client, obtain any lessons learnt and best practices to be used for future projects. As shown in this chapter, good project portfolio management is becoming a key competence for GCOs managing projects simultaneously. To help ensure there is alignment between strategic objectives and project objectives, GCOs will have to deploy project portfolio management solutions to identify which projects support key business programmes and at what cost.

6.13 Case study: Dubai World Trade Centre

United Arab Emirates (Dubai) is one of the fastest emerging countries in the Middle East. Dubai continues to be the focal point of construction activities in real estate. A snapshot of key project achievements include: Dubai Metro; the official opening in 2010 of the Burj Khalifa, the world's tallest building; the launch of the first stage of the mammoth Meydan development located on the site of the legendary Nad Al Sheba racecourse, home to the annual Dubai World Cup and Meydan Grandstand, opened to international approval along with the Meydan Hotel overlooking the famous race track. Ongoing work is being carried out at Al Maktoum International Airport, which is situated towards the industrial area of Jebel Ali. The new airport is ten times the size of the present Dubai International Airport.[39] In 2010, both local and international contractors experienced a slowdown in frenetic construction activities across major cities in Dubai as mega construction projects came to completion.

6.13.1 Background to the case study

The existing Dubai World Trade Centre site has undergone a phased US$4.4 billion refurbishment. The project offers approximately 40 high-rise buildings, including residential apartments, hotels, office buildings, shopping malls and car-parking space.

Dubai World Trade Centre has brought in a multi-use outdoor entertainment location called Trade Centre Plaza. The centre offers an ideal networking environment for businesses – bringing together regional and international exhibitor's buyers at their shows every year. The 9000 square metre multi-use location has a capacity of over 12,000 people. There were three companies involved with this project: Mace acted as the project manager and cost manager, WSP Group as structural engineers, and Hopkins Architects as consultant.

Figure 6.9 The Dubai skyline

Mace was appointed as the project and cost manager for the first three phases in January 2006.[40,41] All the three organisations are UK-based construction organisations. Hopkins Architects' design for the project featured a balance of public and private creating a unique mix of urban quarter offices, apartments, hotels and retail use. Hopkins designed a series of internal and external public spaces with tree-lined streets, light-filled atria and rooftop gardens for the community to live, work and enjoy life.[42] The project provides a unique human-scaled urban environment with over 1 million square metres (10 million square feet) of developable area. The office buildings have been designed to the highest environmental criterion to achieve leadership in energy design (LEED) platinum status. The client's design brief required the project to be environmentally responsible, addressing sustainability in general, and in particular directed at achieving a green rating for its building. Based on the design brief, the project team adopted sustainable strategies. Green technologies for the project were assigned to WSP Middle East Limited to provide professional services for LEED assessment, and recommendation on viability of LEED certification. The LEED assessment report focused on energy and atmosphere, water efficiency, indoor environmental quality, materials and resources, sustainable site and innovation. To achieve this, a workshop was conducted by Green Technologies with client representation and project team, which provided the contractors with information on LEED as a design tool to accomplish the project objectives. During the workshop, both regional and international contractors were advised to incorporate design attributes that could minimise ecological footprint of buildings. From the workshop, it was noted that the project could be the largest green building development on the globe, with offices having potential to achieve the highest possible LEED rating.[42] In recent years, both commercial and residential developers in Dubai have increasingly used green building concepts to conserve energy and reduce environmental impact. Thus, the importance of whole life cycle costing and project management has come under the spotlight. The Dubai Trade Centre District is a redevelopment of a premium site in the centre of the Dubai business district, consisting of Dubai World Trade Centre convention and exhibition halls, apartment buildings and the landmark Trade Centre Tower building. The redevelopment has been divided into three stages:

- Phase one is made of a dense urban streetscape environment of high-rise luxury hotels, concierge apartments and a resort club, alongside an additional upscale hotel and a budget three-star hotel. The central zones of stage one are made up of low- to medium-rise buildings and high-quality retail outlets at ground-floor level. The entire project sits upon a four-level subterranean car park.
- Stage two of the project is the nearby Emirates Towers. A signature building has been put up in phase two, which consists of a luxury hotel and a service apartment building.
- Phase three is attached to Sheik Zayed Road.

In 2009, construction work on the first phase of the project was delayed because the developer focused on expanding exhibition space at the Dubai International Convention and Exhibition Centre.[43] The delay meant the delivery of the first office buildings at business hub – which comprised of seven office buildings, two hotels and retail outlets – was held-up by nine months. In a statement, a spokesman for Dubai World Trade Centre noted that 'their immediate focus is on the construction of new exhibition halls in order to meet regional MICE industry demands'.[44] This led to the project team pushing back the build programme and completion of the first office buildings in 2011. Even though there was a delay, all phases of the project were completed successfully. In addition, Foss Network Infrastructure Solutions installed an intelligent cabling system throughout the 25,000 metre square of exhibition halls in less than four weeks. The most innovative attribute of the cabling system was the monitoring system, which surveys the cable patching, electronically processes orders, identifies IP devices, generates the alert and alarm event generation. The Dubai Exhibition Centre now offers a spatial design, flexible structure and a multi-functional networking platform at the core of the city.[45] From the above, it can be observed that at the start of any global construction project there will be a variety of information and views about the purpose of the project, life-cycle considerations, energy considerations, building materials and innovation to be used. In order to define a global construction project, it is essential to clearly and openly identify what the project is planned to achieve and what its scope will be. By using an integrated project management framework for project initiation, organisation, control and monitoring, this will ensure the timely and cost-effective delivery of deliverables. The way that a global construction project is managed and implemented is the key to its success. Despite the consequences, contractors working in a global environment will have to address the above factors.

6.14 Discussion questions

1. If you were appointed to manage the Dubai World Trade Centre project, what steps would you take to introduce whole life cycle project management methodology?
2. Why does WSP have a central role in this project?
3. How would you integrate the initiation, control, monitoring and closure of this project?
4. Based on the energy and environmental objectives of the project, discuss the apparent contradiction between the design of an energy-efficient project and an economy that has the highest carbon footprint per capita in the global economy.

6.15 References

1. European Construction Institute (2005). *Project Development and Definition.* Great Britain: European Construction Institute.
2. Morgan, B. V. (1987). Benefits of project management at the front end, *International Journal of Project Management*, **5** (2), pp. 102–119.
3. Pinto, J. K. (2010). *Project Management: Achieving Competitive Advantage.* 2nd edn. New Jersey: Pearson.

4. Diekmann, J. E. and Al-Tabtabai, H. (1992). Knowledge-based approach to construction project control, *International Journal of Project Management*, **10** (1), pp. 23–30.

5. Meredith, J. and Mantel, S. J. (2006). *Project Management: A Managerial Approach*. New York: Wiley.

6. Young, M., Owen, J. and Connor, J. (2011). Whole of enterprise portfolio management, *International Journal of Managing Projects in Business*, **4** (3), pp. 412–435.

7. APM (2006). *APM Body of Knowledge*. 5th edn. Hampshire: Hobbs the Printers Limited.

8. Archer, N. P. and Ghasemzadeh, F. (1999). An integrated framework for project portfolio selection, *International Journal of Project Management*, **17** (4), pp. 207–216.

9. PMI (2008). *A Guide to the Project Management Body of Knowledge*. 4th edn. Pennsylvania: Project Management Institute.

10. Morris, P. W. G. and Jamieson, H. A. (2004). *Translating Corporate Strategy into Project Strategy*. Newton Square, PA: PMI.

11. Jackob, K. (1999). Executing projects with portfolio management in large multi-business multi-functional organisations, *Proceedings of the 30th Annual Project Management Institute 1999 Seminars and Symposium*. Newton Square, PA: PMI, p. 1.

12. Markowitz, H. M. (1952). Portfolio selection, *Journal of Finance*, 7, pp. 77–91.

13. Dye, L. D. and Pennypacker, J. S. (2000). *Project portfolio management and managing multiple projects: Two sides of the same coin?* In Proceedings of the Project Management Institute Annual Seminars and Symposium, 7–16 September, Houston, Texas, USA.

14. Krebs, J. (2009). *Agile Portfolio Management*. Redmond: Microsoft Press, pp. 61–62.

15. Engwall, M. and Jerbrant, A. (2003). The resource allocation syndrome: The prime challenge of multi-project management? *International Journal of Project Management*, **21** (6), p. 403.

16. Nash, T. K. (2002). *Project portfolio management – PMO application*. In Proceedings of PMI Annual Seminars and Symposium, Texas.

17. Owen, J and Linger, H. (2011). Knowledge-based practices for managing the outsourced project, *Scandinavian Journal of Information Systems*, **23** (2), Article 4.

18. Aubry, M., Hobbs, B. and Thuillier, D. (2007). A new framework for understanding organisational project management through the PMO, *International Journal of Project Management*, **25**, pp. 328–336.

19. Meskendhal, S. (2010). The influence of business strategy on project portfolio management and its success – A conceptual framework, *International Journal of Project Management*, **28**, pp. 807–817.

20. Shenhar, A. J., Poli, M. and Lechler, T. (2001). A new framework for strategic project management. In Khalil, T. (ed.), *Management of Technology VIII*. Miami: University of Miami.

21. Jonas, D. (2010). Empowering project portfolio managers: How management involvement impacts project portfolio management performance, *International Journal of Project Management*, **28**, pp. 818–831.

22. Eve, A. (2007). Development of project management systems. *Industrial and Commercial Training*, **39** (2), pp. 85–90.

23. Srivannaboon, S. and Milosevic, D. Z. (2006). A two-way influence between business strategy and project management, *International Journal of Project Management*, **24**, pp. 493–505.

24. Longman, A. and Mullins, J. (2004). Project management: A key tool for implementing strategy. *Journal of Business Strategy*, **25** (5), pp. 55–60.

25. Fellows, R. F. and Liu, A. (2003). *Research Methods for Construction.* Oxford: Blackwell.
26. Bell, K. (1994). The strategic management of projects to enhance value for money for BAA Plc. Unpublished PhD Thesis, Herriot-Watt University.
27. Coxhead, H. and Davis, J. (1992). New development: A review of the literature. Henley Management College Working Paper.
28. Walker, A. (2002). *Project Management in Construction.* 4th edn. Oxford: Blackwell Science.
29. Struckenbruck, L. C. (1992). *The Implementation of Project Management: The Professional Handbook.* Massachusetts: Wesley Publishing Company Inc.
30. Kerzner, H. (1992). *Project Management: A Systems Approach to Planning, Scheduling and Controlling.* 4th edn. New York: Van Norstrand Reinhold Co.
31. Steyn, P. G. (2001). Managing organisations through projects and programmes: The modern general management approach, *Management Today,* **17** (3), April.
32. Murray-Webster, R. and Thiry, M. (2000). Managing programmes of projects. In J. Turner, R. and Simister, S. J. (eds), *Gower Handbook of Project Management.* 3rd edn. Aldershot: Gower, pp. 71–77.
33. Milosevic, D. Z. (2003). *Project management toolbox: Tools and techniques for the practicing project manager.*
34. Thiry, M. and Deguire, M. (2007). Recent developments in project-based organisations, *International Journal of Project Management,* **25** (7), pp. 649–658.
35. Kurtz, C. F and Snowden, D. J. (2003). The new dynamics of strategy: sense-making in a complex and complicated world, *IBM Systems Journal,* **42** (3), pp. 462–483.
36. Senge, P. (1994). *The Fifth Discipline Fieldbook: Strategies and Tools for Building a Learning Organization.* Doubleday.
37. Weick, K. E. (2001). *Making Sense of the Organization.* Oxford: Blackwell.
38. Dooley, L., Lupton, G. and O'Sullivan, D. (2005). Multiple project management: A modern competitive necessity, *Journal of Manufacturing Technology,* **16** (5), pp. 466–482.
39. Meetdubai (2011). Dubai rising. http://www.meetindubai.com/articles/show/dubai-rising [Accessed June 2011].
40. Mace (2011). Dubai World Trade Centre Project. http://www.macegroup.com/projects/dubai-world-trade-centre [Accessed June 2011].
41. Green Technologies FZCO (2011). Dubai World Trade Centre Development Phase 1. http://greentechno.ae/dubaiworld.htm [Accessed June 2011].
42. The National (2011). Dubai World Trade Centre work delayed. http://www.thenational.ae/business/property/dubai-world-trade-centre-work-delayed [Accessed June 2011].
43. Wikipedia (2011). Dubai World Trade Centre. http://en.wikipedia.org/wiki/Dubai_World_Trade_Centre [Accessed June 2011].
44. Foss Network Infrastructure Solutions (2011). Dubai World Trade Centre-New Exhibitions Halls (2011). http://www.foss.ae/case-study/dubai-world-trade-centre/ [Accessed June 2011].
45. Wikipedia (2011). Dubai World Trade Centre. http://en.wikipedia.org/wiki/Dubai_World_Trade_Centre [Accessed June 2011].

7

Managing Cultural Complexity in Global Projects

7.1 Introduction

Many multinational construction organisations operate internationally, moving resources to worldwide locations and having the capacity to work on a global scale. Construction is no longer a local industry and is becoming progressively more global in nature, which has brought unique challenges such as integration among project teams from different countries. Global construction organisations (GCOs) thus face a complex set of challenges characterised by cultural complexity, the management of which has become a core competency of senior managers. This chapter examines the nature of global construction project teams and their place in the global construction environment, with the main thrust being to explore cultural complexity within these teams. We will consider:

- multiculturalism;
- integration of multicultural project teams;
- current issues on internationalisation;
- managing change in project teams;
- communication across multicultural project teams;
- dealing with diversity and values;
- challenges of technological and organisational transfers between countries and cultures;
- national culture and project management;
- aligning organisations culture with strategy;
- organisational culture of sustainability;
- team integration principles.

7.2 Learning outcomes

The specific learning outcomes of this chapter are to enable you to gain an understanding of:

>> building effective multicultural teams at all levels and stages of the project organisation;
>> techniques for global team integration;
>> dealing with project complexity, cultural complexity and uncertainty.

7.3 Multiculturalism

As observed from the reviewed literature, European leaders have declared multiculturalism a failure.[1] According to Hammarberg,[1] policies to endorse multiculturalism have resulted in a separated society of parallel communities that do not relate to each other. UK Prime Minister David Cameron's recent discourse in favour of a more 'muscular' moderation and discarding of state multiculturalism has drawn both disapproval and support from all parts of the political continuum.[2] In October 2010, German Chancellor Angela Merkel affirmed that multiculturalism 'has failed totally'.[3] In Europe, multiculturalism has been examined at two different levels: the level of institutional decision and daily life. In North America, Asia Pacific, the Middle East, North Africa, Sub-Saharan Africa, South and Central America multiculturalism has been analysed in the form of assimilation of the society. One factor that cannot be ignored when linking multiculturalism with the global construction industry is cultural integration. The theory and practice of cultural integration has changed fundamentally over the last ten years, driven by the growing demands of multicultural teamworking and greater understanding of cultural issues and diversity in the internationalised global construction environment. The challenge to the global construction industry in both developed and developing countries is to address its performance on global people management by focusing on multicultural teamworking.

In the new global economy, multiculturalism has become an important focus in debates in construction management research.[4] An extensive literature reviewed highlighted that multiculturalism is often an indistinctly used term that has a diverse range of meanings, with very few empirical studies done on its role in construction project management. It could therefore be suggested that any construction project in which contractors bring different assumptions about working norms (either in design engineering or team behaviour) is a multicultural project. Even when all contractors are from one country, the construction project manager may still have to deal with cultural diversity. Ochieng and Price[5] established that some of the team differences are strictly cultural, while others stem from varied management styles and strategies, but all these differences will eventually show up during the project. Managing a global construction project team presents new challenges and opportunities to harness new skills, in particular language and cultural knowledge. In a global construction project environment, effective communication is an essential skill. It requires clients and construction project managers to acquire and promote knowledge and understanding of the cultures present, to understand their own attitudes and sometimes to adapt their working practices. Cultural differences and a lack of management talent can make it difficult for GCOs to attain their business objectives. According to Day,[6] organisations working with multicultural teams face a threefold multicultural challenge:

- enabling a mixed group to work towards a common goal;
- maximising contribution of each project team member;
- ensuring fair treatment for all, irrespective of background.

Whether the multicultural character of an organisation arises from its operation in various countries, or from the mixed backgrounds of a workforce in a single location, the client must address this diversity if it is to achieve these goals. Every multinational construction organisation has a strategic choice in how it will face this challenge, between a fundamentally defensive approach, and one that develops the individual and the group.

7.4 Integration of multicultural project teams

Multicultural project teams have become more common in recent years, and contemporary international management literature has identified that the management of multicultural teams is an important aspect of human resource management. Recent studies have focused on the positive effects of using multicultural teams. For example, Early and Mosakowski[7] stated that multicultural teams are used because they are perceived to outperform monoculture teams, especially when performance requires multiple skills and judgement. However, there has been little research into construction-specific multicultural teams, and many construction organisations, although expanding into global operations, do not fully appreciate the implications and are often unable to respond to cultural factors affecting their project teams.

In the last twenty years, project management has developed considerably with a much greater understanding of the key variables that lead to project success. Project team performance has been widely researched and the findings have clearly illustrated that best project performance is achieved when the whole project team is fully integrated and aligned with project objectives.[4,8,9,11] During this period, there has been a change in the way that many major construction engineering projects are delivered. This is especially noticeable in Western Europe, where local levels of investment have dropped and many project management contractors are now working on projects in other parts of the world.[12] The increased application and development of rapid worldwide electronic communications has led to a number of construction engineering projects being designed and developed in dispersed locations many thousands of miles away from the actual construction sites. In addition, there has been an inclination by clients to develop and undertake such projects in partnership with other companies as joint ventures, often collaborating with local companies based in the territory where the assets will be built. This has resulted in more teams with members from different cultures and backgrounds working together.

A number of authors, including Weatherly,[12] agree that project success is difficult enough to accomplish where the project team is located close to the construction project environment, and the situation is made considerably complex for teams that are widely separated geographically and have dissimilar organisational and regional cultures. The geographical division of multicultural project teams poses its own communication challenges. Emmitt and Gorse[13] have shown that, for factual data transfer, a number of communication problems have been addressed due to the development of rapid global information systems and telecommunications. However, when it comes to multicultural project teams many issues remain unresolved. For example, the loss of face-to-face communication can lead to misunderstanding, as can the loss of non-verbal signals – such as eye contact and body language. This can subsequently lead to difficulty in achieving mutual trust and confidence. It is difficult to manage or supervise teams without this face-to-face contact; likewise it is difficult to develop a working relationship.[12]

The success of managing a multicultural project team no longer lies in the simple delivery of the outcome. In a continually changing global context, senior project managers are required to deliver a project that will satisfy or exceed the client's needs and expectations at the time of delivery. There is a need for increased research efforts in understanding influential factors that affect multicultural project teams. There is mounting evidence and opinion indicating that integrated teamwork is a primary key in efforts towards improving product delivery within the global construction industry.[4]

7.5 Unifying multicultural thinking

The rapid globalisation of the world's economy has had significant impact on the way construction project managers work, bringing them frequently into contact with clients, suppliers and peers who they have not worked with before. In an era of globalisation, projects in the construction industry face unique challenges in coordinating clients, financiers, developers, designers and contractors from different countries. In addition, global construction project teams need to cope with the complexities of both local institutions and physical environments. Bartlett and Gosha[14] discussed the challenges facing organisations that are intending to work effectively across borders. They identified the major challenges as being able to develop practices that balance global competitiveness, multinational flexibility and the building of a worldwide learning capability. Achieving this balance will require multinational construction organisations to develop the cultural sensitivity and ability to manage and leverage learning in order to build future capabilities. While offering opportunities, globalisation also poses significant challenges, in that the management and development of people inevitably leads to considerations of diversity across different cultures, values, beliefs and practices.

For GCOs there is an increasing need to get groups of project managers from different nationalities to work together effectively, either as enduring management teams or to resource specific projects addressing key business issues. Many multinational construction organisations have found that bringing such groups of project managers together can be problematic, and performance is not always at the level required or expected. In addressing the issues relating to developing effective global construction project teams it appears that the following areas should be well thought out: communication techniques, smoothness of handover, teamwork, issue resolution, joint decision making, people selection and prioritisation.

7.6 Current issues on internationalisation

A number of multinational construction organisations have adopted a working definition of internationalisation as the process of integrating an international or intercultural construction team.[4] The success of the internationalisation progression is dependent on the involvement of the entire construction team, regardless of their roles or answerabilities. The acquirement of global team awareness and the diversity of cultures is an essential part of global construction project management, and getting teams to work effectively across international boundaries has become a major issue.[7,15,12] The trend is likely to continue and future business will increasingly depend on doing projects effectively in different cultural environments.[4,12,16,17] It has been widely recognised that global project teams have been common in recent years. Contemporary literature has identified the management of global teams as an important subject in human resource management. Most of the studies have focused on the positive effects of using global teams. For example, Early and Mosakowski[7] state that multicultural teams are used because of a belief that they outperform monoculture teams, especially when performance requires multiple skills and judgement.

Global team integration is a particular problem for clients and project managers. Once they are established, however, they are perceived to outperform monocultural teams in areas such as problem identification and resolution, by the sheer strength of their diversity.[18] The basic values, concepts and assumptions differ with each culture, and understanding these and enabling the 'settling-in' by recognising the cultural complexity is a required skill of a manager.[19,20] Choosing not to recognise cultural complexity limits the ability to manage the project. The fragmentation of a project delivery has been blamed on the cultural complexities that exist. Construction project managers of multinational organisations often make the common assumption that cultural differences are unimportant when individual members of different divisions in the same organisation are brought together as a team. The original research (Hofsteder[21]) suggested that 80 per cent of the differences in employee's attitudes and behaviours are influenced by national culture – and this still has resonance today.

Cultural differences reflect different expectations about the purpose of the team and its method of operation, which can be categorised into task and processes. The task area relates to the structure of the task, role responsibilities, decision making. The processes relate to team building, language, participation, conflict management and team evaluation. Culture is an issue with many different dimensions. Both Hofstede[22] and Trompenaars[23] discussed different levels of culture. The former mentions gender, generation, social class, regional and national and organisational levels. The latter presents national, corporate and professional levels of culture. The level that is important in this chapter is that of national and organisational culture.

It has been widely recognised that organisational culture is important in construction project management.[24,25] For example, contractors are usually drawn from a number of organisations, each with its own culture. To work as a team efficiently, it is essential to have some degree of cohesion of organisational culture. Most construction organisations have set ways of getting things done that can help or, in some cases, hinder a project.[26] It is essential, therefore, to introduce at the outset the organisational background and culture of all of the contractors involved in the project. Typically, leadership in global construction projects is complex, and critical to success in global team environments. For example, Weatherley affirmed that if management is *getting* the team to do what is required, global project leadership involves motivating the project team in such a way that they *want* to do what is required.[12] In this context, global project leadership is not so much about telling the project team what to do as leading by example and developing trust and confidence in the team to take the project forward. Indeed, this kind of team leadership is not just accidental but can be developed and is a required skill for creating successful multicultural project teams.[7]

In a number of countries, different ethical standards apply. This affects attitudes towards the law and, indeed, national laws can be very different in different territories. Weatherley claimed that in some nations, bribery and corruption have become institutionalised and are the only means by which some local officials can earn a living[12]. There may be an unwarranted predilection given to local suppliers or contractors, and planning laws and approvals can be very officious. In some nations, the government approval procedure can become the critical path on the project programme.

Project financing of global construction project teams can be difficult. Difference in currency rates can also play havoc with the cost management of a construction project. Due to this, a number of multinational construction organisations have difficulty in repatriating revenue from project work done outside their home country. Ochieng[4] established that there is also more hidden cost linked with multicultural projects such as customs, import duties, shipping, logistics and agent fees. A number of organisations now use low-cost design centres for the completion of detailed or standardised design and this can lead to a considerable reduction in costs due to the low currency rates in developing countries. Emmitt and Gorse noted that communication between the main project office and the low-cost design centre needs to be of high quality if complexity is to be reduced.[13] It is essential that these are used with care, since although the rates may be low, productivity is often also low, so the cost savings are much less attractive than might have first appeared. This undoubtedly impedes integration, since error rates tend to be higher for work farmed out in this way.[12] This presents a problematic context for achieving the integrated delivery of the project, since there can be a hostile response from project workers if there are different rates of pay for the same work.

Smith asserted that risk is present in all projects but becomes more definite in global projects where there are often new risks, predominantly if the project is being constructed in a country where security is an issue.[27] Edwin and Raymond affirmed that in some nations, contract law is not well instituted by other nationalities.[28] Emmitt and Gorse found that risks in communication and risks emerging from misunderstandings and misinterpretation are much greater.[13] There is also a danger of the expatriate project team 'going native' and becoming isolated from the project, and thus pursuing their own project goals rather than focusing on the overall project aim and objectives.[28] Another key issue that managers of multicultural construction project teams face is the assessment of skills and competencies of the project team.[4] For example, in a number of countries, the training and education standards and the relative value of qualifications can be very different. Weatherly also highlighted that job methods can be different because of specific local conditions such as working in heat, earthquake risk or local trade practices.[12]Due to the current political climate in the Middle East and North Africa, for example, it can be difficult to find individuals who can work effectively while away from their home country.

Langford and Rowland established that cultural differences are usually significant when managing project teams across different parts of the world.[29] Hofstede and Trompenaars classed the differences as national characteristics, ethnic differences, organisational culture or professional practices.[22,23] Indeed, if not addressed, these differences can lead to major divergence of working practice and can severely affect the project conclusion. For example, in some countries, status and hierarchy are very important. This can lead to a lack of empowerment of more junior project staff. Religious observance can also be very significant in getting the project completed successfully. For example, setting aside time for prayers during the working day might be required and, in a number of countries, there are religious festivals, fasts and feast days that are classed as non-working days.

Language is another factor that affects multicultural construction project teams.[15,29] Brett et al.[15] and Ochieng noted that destructive conflicts in a team can be caused through trouble with accents and fluency, direct versus indirect communication, differing attitudes towards hierarchy, and conflicting decision making norms.[4] The trouble with accents and fluency can occur when individuals who are not fluent in a team's dominant language may have difficulty sharing their knowledge. Direct versus indirect communication can transpire when some project workers use direct, explicit communication whilst others are indirect. Brett et al. further argue that team members from hierarchical cultures expect to be treated differently according to their status in the organisation.[15] With conflicting decision-making norms, project team members vary in how quickly they make decisions and in how much analysis they may require beforehand. Brett et al. asserted that an individual who prefers to make decisions quickly may grow frustrated with those who need more time. It is essential for the project manager to set out a common language so as to ensure a common understanding.[13] This arguably can lead to a substantial loss of efficiency for non-native speakers who are working in their second or third language, as well as increased risks of misunderstanding.

The realisation of an integrated global team as a single unit still remains the aspiration within the global construction sector. The various parties within the delivery team continue to face cultural issues. Egan stated that integrated teamwork is the key to construction projects that personify good whole life value and performance.[29] Integrated teams deliver greater process efficiency and, by working together over time, can help drive out the old-style adversarial culture and provide safer projects using a qualified and trained workforce. Teams that only construct one project at the client's expense would never be as efficient, safe, productive or profitable as those that work repeatedly on similar projects.[29] This, in particular in developing countries, has proven to be a long and complex process.[4] However, partnerships and cooperatives are being formed; and integration and collaboration are becoming generally accepted needs for individuals and companies to survive. Despite these difficulties, it is vital for the global construction industry to improve internationalisation. It is possible to get project teams from different countries and organisations to work together effectively. The task for global project leaders is to understand cultural issues and the secret of success so that more global construction project teams can be managed effectively to the benefit of the clients, in a way that properly rewards organisations involved in that delivery.

7.7 Managing change in project teams

In today's fast-paced, highly competitive world, change is inexorable. Multinational construction organisations must respond to change in order to remain competitive and client focused. The problem is that communications for implementing change often come from various sources and in many different formats, including change requests, defects, enhancement requests and problem reports. Projects are created to facilitate change, but by their very nature, are themselves incessantly changing. It is now projected that when considered in the context of global construction projects, change occurs in two places; internally or externally (in other words in the project organisation or project environment). It is organisational change that is being considered here and, in this context, a global construction project organisation is the same as any other organisation. Global project teams are conditioned to oppose change if it is seen as a threat rather than as an opportunity. It is vital for global construction project leaders to handle teams sensitively if they are to develop a cooperative culture – which delivers better value – in place of an adversarial culture. Constructive feedback must be sought at all stages from team members, and considered and acted upon in order to make the most of the benefits of integrated teamworking. There is a current drive within the industry to change to a value-based culture. Changes to projects may be proposed by the client or senior managers associated with it. In this case, it is essential that any proposed change to the project be formally controlled. Project leaders and clients need to review changes fully before they are approved and marked up for action. All approved changes should be fully documented and efficiently communicated.

The importance of communication in projects, in particular with regard to its influence on the acceptance of anything new, is well documented in the literature.[13] There are a number of reasons why communication is obligatory for the successful management of change to projects. These range from certifying an increased understanding of eradicating waste and motivating those involved in the change. In general, successful communication needs to be focused rather than broad-brushed, and timing is of crucial importance. There should be an effective change control system in operation and all senior managers should be familiar with its operation. It appears that if the above is met, the result will be a project within the allocated time and resources. Although much can be achieved by working with global construction project teams, the truly successful global construction firms are likely to be those that embed the change through integrated changes to cross-cultural team selection, joint decision making, communication, teamwork, effective people selection and project selection. In applying these factors, the value of global teamworking can be captured at many levels in the organisation, be they project based or permanent, and, furthermore, will allow global project teams to reach high-performance levels consistently.

7.8 Communication across multicultural project teams

Axley considered communication as a metaphorical pipeline along which information is transformed from one individual to another.[30] Thomason defined communication as 'the lifeblood of any system of human interaction as without it, no meaningful or coherent activity can take place'.[31] Nevertheless, defining communication is obscure, as it is such a multidimensional and imprecise concept. Despite the difficulties inherent in defining communication, it is essential that a working definition is developed to fortify the analysis of communication practice contained in this chapter. Here, communication is viewed as a professional practice where suitable tools and regulations can be applied in order to improve the utility of the data communicated, and is a social process of interaction between individuals. The problematic context of communicating in global construction project teams raises questions as to how project managers and clients can go about overcoming the structural and cultural conditions and constraints that define its operation, in order that it can develop an infrastructure that facilitates more effective communication in the future of the global construction sector. Moreover, it highlights that the construction research community and the industry need to find solutions of effecting change within the global sector in such a way as to overcome the present and future cultural constraints on the sector's development. In a theoretical sense, it could be argued that using effective communication tools should be fairly straightforward. However, the translation of theoretical perspectives actually into practice depends upon their interpretation by the individuals. Arguably, many of those with experience of working with global construction project teams have yet to develop skills to cope with such a challenging communication environment.

Given that global construction project teams involve people from a wide variety of cultures, there is no guarantee that the use of espoused good practices will result in successful project outcomes. Effective communication is about not only sending data or information; it includes ensuring that any message is received and understood by those team members to whom it is addressed. This is made easier by team members knowing each other, perhaps through relationships developed on previous projects. Effective communication on projects is aided by the early establishment of clear lines of responsibility and a clear robust issue resolution process within the integrated team. In order to achieve effective cross-cultural communication, adequate internal and external communication needs to be in place. It was established that effective communication is the key to managing expectations, misconceptions, and misgivings of global construction project teams. To develop as a global construction project team, it is essential that learning occurs. It is good practice not only to review the project objectives and deliverables at regular meetings, as a team, but also to conduct a process review. There is a need to communicate lessons learnt from previous company projects. Where suitable, such learning should be included or taken account of in the present project. Communicating learning can occur as the project proceeds, and it is essential that opportunity is given to review what is being done in order to pinpoint learning points and, if possible, to refine the way that the team is working. Clients have for some time interviewed people who have been selected to run projects. The main reasons behind this are to assess an individual's technical ability and to see if they can be part of a team.

When it comes to communicating project procedures, it is hard to get the message through to team members. For example, one former project manager, who is now a director, one time said that he used to spend 70 per cent of his time talking to people and the other 30 per cent at his desk. The surprising thing is that the director was one of the most successful project managers within the organisation. It is interesting to note that he rated communication as the most important tool when it came to managing projects.[4]

Thus it is clear that effective communication is not just about informing. A key aspect of communication is the ability of the client and project leader to listen, to give feedback and to respond to any project issues that might arise. On the other hand, good communication with a high level of trust, honesty and respect for others is significant in building and maintaining high team performance. Furthermore, communication must be maintained with members as individuals and as a team. Adequate internal and external communication systems must be in place. Senior managers must take an active role in keeping team members informed. Communication from the project leader to team members must be consistent regardless of their project location, and all team members must be aware that this communication. Expectations, misconceptions and misgivings from those outside the project team may increase with lack of information. Effective communication is the key to managing expectations and minimising misconceptions and misgivings. The issues raised have been summarised below as seven key dimensions of cultural differences on communication behaviours:

- establish clear lines of responsibility;
- institute team effectiveness (collectiveness);
- establish trust;
- implement honesty;
- encourage respect for others;
- introduce cultural empathy;
- implement value management techniques.

Project leaders need to establish clear project goals, team effectiveness and trust, and encourage respect between team members. In order to manage potential language barriers on global projects, project managers should have the ability to understand and clearly communicate team goals, roles and norms to other members of the global team. It is vital for a project manager to be cross-culturally and communicatively competent.

7.9 Working across different standards

As multinational construction organisations face constant challenges, it is increasingly essential that not only practising construction managers but also governments and educators in developing and developed economies understand how the global construction industry can build on its strengths. Niebles noted that, during the current challenging economic environment, the first steps that any construction firm must take are to understand its global exposure, quantify the impact on the business, and assess the various scenarios that might occur depending on the duration and severity of the economic crisis.[32] Once this is understood, there are a few key objectives that a business should look at in order to determine the best way to go forward and to maximise performance. The reality of today's global environment stipulates collaborative data sharing and problem solving, cooperative support and resource sharing, and joint action and implementation.[32] Working in a global construction project requires working with individuals of different nationalities and very different working styles. Cultural understanding is essential for senior managers, and the ability to work with and integrate a diverse project team has become a critical skill for all construction managers working in a global construction project environment. Multinationals recognise that their innovative potential depends on their capability to form a diverse global construction team.

7.10 Dealing with diversity and values

The management and development of project teams in the global context unavoidably leads to a consideration of diversity and related challenges. Within overseas construction projects, it is essential for multinational construction organisations to help their project managers to appreciate the international nature of the industry and to develop the ability to understand everyday job-site issues from different cultural perspectives. Multinationals must develop the cultural sensitivity and ability to manage and leverage learning to build future capabilities if they are to achieve this balance. Ely and Thomas and Jehn *et al.* demonstrated that diversity increases the number of different perspectives, styles, knowledge and insights that the team can bring to complex problems.[33,34] The world's most innovative firms, such as Microsoft, took advantage of this by intentionally introducing multicultural teamworking. Unfortunately, in contrast to sectors such as IT, manufacturing and aerospace, the construction industry has not taken into account the issue of cultural complexity and its influences upon different project teams. As established in the literature, all the evidence points to an assimilatory attitude, which largely ignores the needs of different project teams, expecting them to become accustomed to the dominant industry, national or organisational culture.[34,35] However, current thinking on global team integration requires multinational construction organisations to value explicitly global teamworking, to adapt to it and use it to generate improvements in project work performance and team effectiveness. Still, it should also be noted that linking different individual cultures to project outcomes is controversial. Although project teams from different cultures may well bring different perspectives and styles, the necessary conditions, likely consequences, and overall performance implications are yet to be universally accepted.

Existing literature on cultural diversity examines team members' demographical backgrounds and other factors relevant to their cultural characteristics, values and discernments.[36,37] The cultural diversity of a global construction project team has a number of benefits, including the variety of perspectives, skills and personal attributes that global team individuals can contribute.[38] As confirmed by Ng and Tung,[39] in brainstorming tasks diverse groups generate more ideas of high quality. Culturally diverse teams perform better than homogenous teams when it comes to identifying problems and generating answers.[39] According to Elron, organisations that utilise global teams make significant gains in productivity.[40] For example, Ng and Tung established that culturally diverse teams of a multi-branch financial services firm reported higher levels of financial profitability compared to their culturally homogenous counterparts.[39] More recently, Marquardt and Hovarth established that by assembling the energy and synergy of individuals from different backgrounds, organisations could generate creative approaches to problems and challenges that are faced by corporate teams in project-based operations.[19]

It has also been ascertained that communication in multicultural teams stimulates the formation of an emergent team culture. Unlike a homogenous or monoculture team, a multicultural team cannot refer to a pre-existing identity, because hardly any camaraderie exists among team members.[7] Thus, they develop and depend on a team culture of straightforward rules, performance expectations and individual perceptions. Earley and Mosakowski further confirmed that an effective multicultural team has a strong emergent culture, as shared individual prospects facilitate communication and team performance.[7] This suggests that the positive effect and trust generated by the perceived shared understanding can fuel performance improvement and boost team effectiveness. Most importantly, effective interaction among project team members can facilitate the formation of a strong emergent team culture.[41] Nonetheless, multicultural teams are particularly susceptible to communications problems that may affect team cohesion. Individuals in multicultural project teams have different perceptions of the environment, motives and behaviour intentions. Richardson argued that the effects of such differences could be visible in lower team performance due to impeded social cohesion.[42] Further research on team cohesion and team performance showed a positive correlation between these two variables.[43,44] Elron asserted that cohesive teams respond faster to changes and challenges and are more efficient.[40]

Managing cultural differences and cross-cultural conflicts is in general the most common challenge to multicultural teams.[18] However, there has been limited research on 'people issues' within multicultural teams in global construction management literature. The dominant focal point has been on 'research for management' rather than 'research of management'. Richardson noted that the recognition of techniques such as lean production and business process re-engineering are indicative of this point of view, as they mirror fashions in mainstream management, which are themselves based on a traditional culture of prescription and control.[42] People management in construction has become an important topic within the construction industry,[26] which needs to address its poor performance in people management by focusing on cultural issues.[26] Cultural issues among team individuals can cause conflict, misunderstanding and poor performance.[45] Five of the most distinctive challenges that managers face are: developing team cohesiveness; maintaining communication richness; dealing with coordination and control issues; handling geographic distances and dispersion of teams; managing cultural diversity, differences and conflicts.[18]

Construction project managers from different countries are likely to translate and respond differently to the same strategic issues or team tasks, because they have distinct perceptions of environmental opportunities and threats. An awareness of cross-cultural issues is therefore an essential competence of a manager's ability to address the common challenges faced by multicultural construction project teams.

While many researchers have investigated culture in construction, understanding of cross-cultural management factors on multicultural project teams is insufficiently developed.[24,26] Furthermore, the industry has not responded to cultural issues facing its workforce within the global construction industry. Multicultural project teams have merits from many points of view. International engineering construction projects involve multinational project teams from different political, legal, economic and cultural backgrounds. As the environment is becoming more complex and changes are faster, multinational construction organisations must improve their ability to address such external challenges. Multicultural construction project teams have a wider range of perspectives. Within this diversity of views, solutions for old problems are often found. However, this requires a certain level of integration. The sense of belonging to a group gives a feeling of safety and comfort to a team member.[46] This feeling gives the team better options for responding to project challenges. It also breaks the comfort zones and creates innovative solutions to project issues that might arise. Although concerns about cultural issues seem to be discussed widely by the construction research community, formal analysis on cultural issues affecting global projects teams emerge as a largely unexplored theme.[47,48] With the growth in globalisation, construction project managers will need to work with culturally diverse project teams. The good news is that these teams will bring fresh ideas and new approaches to problem solving.

7.11 Dealing with project complexity, cultural complexity and uncertainty

The escalating complexity of multi-purpose global construction project management emerged from the growing demands of clients and the increase of multidisciplinary and multi-supply chains that are gathered together to deliver the desired multi-million-value projects. According to Morris *et al.*, project complexity has amplified exponentially since the late 1980s.[49] Evidence shows that the augment in project complexity has many facets: mega projects, designs that approach the physical limits of construction materials and equipment, construction in remote sites, partnerships, data integration requirements, and various project delivery systems and contracting strategies.[50] Interestingly, globalisation and outsourcing also contribute to project and cultural complexity. As multinational construction organisations endeavour into new territories, factors such as geographical, social and political add to the complexity of the project. Managing global construction projects requires multinational construction organisations to ascertain how to work within an intercontinental environment with project teams that have different cultures, skill sets and language capabilities.

Global construction projects are invariably complex and have become more so. Baccarini stated that the construction process could be considered as the most complex undertaking in any industry.[48] However, Morris *et al.* emphasised that the construction industry has experienced great difficulty in handling the ever-increasing complexity of major construction projects.[49] It is essential to assert that the concept of project complexity has received little in-depth attention in global construction project management literature. A review of the literature showed that certain project characteristics present a basis for shaping the appropriate managerial actions needed to complete a project successfully.[51,52] Complexity is one such significant project dimension. As Bennett claimed, practitioners habitually portray their projects as simple or complex when they are discussing management issues.[52] This suggests a practical acceptance that complexity makes a difference to the management of global construction projects. As confirmed in the literature, the magnitude of complexity to the project management process is widely accredited, for example:

- complexity is a key decisive factor in the selection of an appropriate project organisational form;[53,54]
- complexity is often used as a criteria in determining a suitable project procurement arrangement;[55]
- as the Chartered Institute of Building (CIOB)[55] and Rowlinson[56] claimed, complexity affects project, cost, time, quality and objectives. Generally, the higher the project complexity, the greater the cost and the time required;
- project complexity hampers the clear identification of objectives and goals of construction projects.[57]

A common strategy of defining complexity is to quantify it in various dimensions such as number of stakeholders, number of units and amount of different resources. Baccarini's definition is centred on systems theory.[48] Baccarini's analysis of project complexity is based on two dimensions: organisational complexity and technological complexity. It is essential to underline that when project complexity is considered, project managers have to spell out to which of the project's dimensions they refer. Ochieng showed how differentiation and interdependency transpires through technological and organisational complexity.[4] Baccarini countered this argument by suggesting that organisational complexity based on differentiation can be either vertical or horizontal. Horizontal differentiation is determined by the number of organisational units and task structure (i.e. project job and specialisation), whilst vertical differentiation is the depth of the organisation's hierarchical structure. The other feature of organisational complexity in global construction projects is the degree of interaction between organisational elements and operational independencies. Technological complexity by differentiation is determined by the array of outputs, inputs, tasks and the number of specialities involved in a global project. As noted in the *Strategic Forum for Construction* report,[57] large projects are typically characterised by the engagement of diverse contractors and project teams. This leads to the formation of a temporary multicultural project structure to manage the global construction project.

From the above, a global construction project structure can be presented in two dimensions. The first feature is based on a relationship between complexity and uncertainty. The second feature involves the work of Baccarini.[49] Williams claimed that uncertainty is dwelling in the instability of circumstances and assumptions upon which the project is based.[58] As the project matures in real time, the uncertainty and hence project complexity is minimised.[59] Uncertainty can make project situations appear weighed down with danger. External factors can be looked at as the driving force for uncertainty in projects. Complexity will multiply through higher demands on the project's performance. The issue of complexity is the prime focus in today's global construction project management literature. Project complexity can be found in three dimensions: outside the project, inside the project and finally the environment outside the project. One of the key reasons why complexity varies is that clients have different goals, interests and expectations of the construction project that can originate from different levels. A second reason is that inside a global construction project, the project process is often the key focus. Moving from one phase to another probably means that the aim is on project result rather than process. Finally, a third reason is that in different project phases you find different driving factors. As a result, one could suggest that traditional forms of hierarchical team formation and leadership are being replaced by directed, self-managed team concepts.[60]

The continuous need for improved speed, cost, quality and safety, together with technological advances, environmental issues and fragmentation throughout the industry have contributed to the increased complexity of construction projects.[4] How to deal with complex global projects is of increasing concern throughout the industry. The few complexity-related studies that exist have tended to focus on either uncertainty and complexity in non-specific terms, or the experiences of individual organisations within the context of developed countries.[51] According to Cleden, uncertainty can be grouped into two main classes: variability uncertainty, which is behind a number of common problems in construction projects, and indeterminate uncertainty, which always leads to indistinctness in construction projects.[61] For example, uncertainty stemming from environmental issues, design and financial aspects of construction projects should at least be alleviated or ideally eradicated. Construction clients are intensely interested in realising some form of breakthrough that will lower team uncertainty on projects.[4]

For high uncertainty project teams clear project structures have to be in place, whereas low uncertainty global construction project teams demand less project structure. Uncertainty on projects is believed to be one of the principal casual factors underlying project team intercultural integration. An uncertain situation is dealt with by gathering more data to reduce the culture and information gap. Project leaders must try to be effective cross-cultural communicators in order to operate effectively and achieve high levels of team performance. Effective interpersonal skills, team effectiveness, ability to deal with cultural uncertainty, and cultural compassion towards others are learned behaviours that can be improved through multicultural training. Understanding how to develop the performance of a culturally diverse project team is a central goal of modern global construction management research.

Uncertainty is another side of cultural complexity that project leaders need to manage proactively. Uncertainty can have various impacts in global projects. They vary during different project phases and between projects. Impacts can be positive, negative and everything in-between. Uncertainty, together with other internal and external factors, causes challenges in global construction projects. Before uncertainty can be managed, the issue needs to be discussed.

The cultural weight that each contractor brings to a project is more often than not an intended action, reflecting aspects of culture that can be explained to others. However, individuals are not always aware that some of their actions and ways of thinking are dictated by more hidden or, in fact, unconscious values. For example, attitudes towards authority, approaches to carrying out a task, concern for efficiency, communication patterns and learning styles.[62,63] In most cases it is not a situation of one culture being right and another being wrong within a group, but rather there is a shared view of what is considered right or wrong, logical and illogical, fair and unfair. It is significant that cultural norms and values are passed on from generation to generation.[64,65] Cultural norms can affect the way global construction project teams communicate and behave within project environments. Human interaction does not occur in a vacuum or isolation; instead it takes place in a social environment governed by a complex set of formal and informal values, norms, uncertainties, changes, complexities, rules, codes of conduct, laws and regulations, policies and as well as in a variety of organisations.[22,23] Shaping, as well as being shaped by, these governing mechanisms is something that we often refer to as culture. Cultures materialise and evolve in response to social cravings for answers to a set of problems common to all groups.[22,23] In order to survive and to exist as a social identity, every construction project group, regardless of its size has to develop solutions to these problems.

All global construction industry players face similar problems and challenges. However, in developing countries these are exasperated by a general situation of socio-economic stress, chronic resources shortages, and a general inability to address the key issues within the industry. The problems have become more severe in recent years. The nature of complexity within construction engineering projects has been a subject of study with growing interest, especially since the EPSRC Networks-Engineering and Physical Sciences Research Council was set up in 2003.[51] However, it could be argued in research terms, that cultural complexity has been neglected both in terms of conceptualisation and empirical study. Given the supposed severity of cultural complexity and the obvious failings of the industry's approach towards people management, it is reasonable to assume that such an issue would provide a focus for research to improve practice.

7.12 Challenges of technological and organisational transfers between countries and cultures

Managing global construction project teams efficiently is crucial to project success in our globally connected and rapidly changing world. Pressures for working smarter, quicker and cheaper have increasingly led to searches for the best and most fitting resources across the globe.[68] With this concept shift, criteria for project success are also changing. Advances in technology and worldwide economic alliances have enabled multinational construction organisations to network their operations across wide geographic areas, taking advantage of the best fitting resources anywhere in the world. Partnership has become an important competitive tool in the globally connected construction environment. Multinational construction organisations likely to survive and prosper in the current amalgamated global marketplace will need to utilise new and innovative ways of integrating their resources and services more cost-effectively, timely and at higher value to their clients. Managing global construction project teams requires the partnership of multiple construction organisations and an interface among many stakeholders.[56] It requires cross-cultural experiences and continuing organisational development. As established by Ochieng and Price,[5] dealing with different cultures is a very important and perceptive issue that requires a careful and purposeful method.

Multinational construction organisations will need to ensure that there is a cultural balance of zest, knowledge, drive, organisation and control, which when brought together will ensure that construction projects are given the best possible opportunity to be successful. This can be achieved through multicultural team-building activities so as to create an environment that is both focused and efficient in producing the project deliverables. For today's global construction projects, success is no longer the result of skilled managers and multicultural teams. Rather global construction project team performance depends on multidisciplinary endeavour, involving diverse multicultural teams and support contractors working together in a highly complex environment. For multinational exertions, the project process requires pragmatic learning, cross-functional coordination and amalgamation of technical knowledge, information and components. Most construction project managers see their projects as a bleary process that cannot always be portrayed linearly or planned perfectly, nor can results be predicted with certainty.[5]

Global construction projects and project participants are all different and the big challenge facing clients and construction senior managers is the need for setting up a construction site team spirit almost immediately.[29] In a global project, the new team participants are not necessarily chosen as project team players but by the lowest price tag. They are not the project's employees but are leased from their home firm, which probably has other success criteria than the project in question. Since the project is new and the site is unbroken, nothing at all is as it was in the previous project. A second problem facing senior construction managers is that they need to act fast. In a project environment there is no such thing as a second try. The culture of teamworking must be introduced from the very beginning and kept all the time. In addition, service and support must be introduced in order to gain confidence, and the project's targets must be clearly communicated, particularly if the project is one in which recurrent changes may be expected.[13]

Flourishing global construction project management requires analysis of how cultural and project complexity affects the quality, cost, time, environment, and health and safety. We suggest that clients and project leaders in multinational construction organisations require this knowledge in order to manage the challenges of global projects. It is crucial that throughout the project life cycle clients and senior managers develop plans and standardise with the purpose of managing cultural complexity in the most efficient way. As stated by Emmit and Gorse, incessant communication and coordination during the project's life cycle facilitates effective management of cultural complexity. a perspective that is sustained by Baccarini and Williams.[13,48] However, it is essential to leave room for team adjustments within the standardised framework of construction project management. This allows flexibility for the project team to create project-specific solutions in order to maximise commitment on the individual level and thus increases project impetus and project success. In addressing the issues relating to developing effective global construction projects, the following areas should be considered:

- identifying a cross-cultural leadership style preferred by the project team so that the project leader's authority is respected;
- formalising team activities and workshops so as to enhance multicultural team maturity;
- understanding the nature and value of multicultural diversity;
- understanding factors relating to effective multicultural team maturity;
- recognising and leveraging cultural diversity and leadership;
- formulating processes for understanding, valuing and leveraging national cultural differences;
- classifying the nature and implications of national cultural differences within the multicultural construction project team.[67]

The strategies proposed in this section cannot be expected to resolve all the cultural issues and multicultural team-working issues in global construction projects. However, their use defines an approach that is superior to the traditional approaches typically adopted, and consequently merits far wider application. What does this mean for project leaders and international construction organisations? They must actively promote multicultural teamworking as the means of addressing poor performance on people management, and cultural issues on construction projects. In particular, if organisational change is to be effectively introduced globally, the organisations will have to ensure that their key decisions are being informed by the knowledge and experience of local or indigenous managers. This will require project leaders to have a better understanding of cultural change processes and procedures in countries. The proposed strategies present a better way of optimising the performance of project-based operations, thus enabling construction organisations to reform their poor performance on global projects and empower them to better manage emerging culture challenges in their future projects. In spite of the current difficulties faced by the industry, there is an increasing need to get multicultural construction project teams from different nationalities to work together effectively.

7.13 National culture and project management

The concept of culture can be defined at organisational, industry and national level, with all levels being relevant in the context of global project management research. Hofstede asserted that cultural differences can be construed as differences in shared values. National culture and organisational culture are different in nature.[22] Hofstede noted that national culture mostly stems from regularity in values, whilst organisational culture stems from evenness in practices. National cultures can differ in many ways. For instance, multicultural team members from different cultures vary in their communication behaviour, their motivation for seeking and disclosing information, and their need of self-categorisation. National cultures can be addressed in a global project environment if project leaders focus on five cultural orientations: uncertainty avoidance, communication richness, individualism, performance orientations and collectivism.[5]

Global team formation requires project leaders to be skilled communicators, to be able to give and receive constructive feedback, to openly discuss problems, and to communicate a desire for trusting relationships with global team members. It is essential that project leaders be able to recognise other project approaches towards work and decision making, and to adapt their project strategies based on their knowledge of other cultures. Many multinational construction organisations have found that global team integration can be problematic, and at times performance is not always at the level required or expected. With an ongoing increase of global construction project teams, project leaders in multinational construction organisations must be aware of national and organisational culture issues in order to function effectively and achieve high levels of team performance. National and organisational issues can be resolved if managed in a logical, uniform and ethical way. This will depend on the underpinning data of the existing national and organisational culture. An increasing number of global construction projects make knowledge and awareness of national culture important for team success.[4]

7.14 Aligning your organisation's culture with strategy

According to Weber and Dacin, cultural advances to organisations have experienced regeneration in recent years.[68] The rebirth of interest in cultural construction can be observed as a 'second wave' of cultural analysis in organisation science. The increasing global nature of construction projects has highlighted the importance of multiculturalism and the new challenges it brings to project execution. Global construction project management in Eastern Europe, Western Europe, Asia, the Middle East, North Africa, Sub-Saharan Africa, South and Central America is shaped by many social, political and economic factors interacting in complex and dynamic ways. These interactions are often specific to each country and often reflect its history, culture, legal systems, institutional frameworks and social capital.

Contemporary literature on best business practices introduces many different terms related to global project management, including management by projects, project-based organisations, project-oriented businesses and temporary project organisation structure. In recent years, the discipline of global project management has changed its application dramatically to accommodate emerging management processes and philosophies related to implementation of organisational development and strategic change. There are a number of reasons for such terminologies and they tend to reflect endeavours of modern multinational construction organisations to respond to the environmental changes by adapting specific patterns of coping behaviour. The implementation of strategic planning through projects makes the achievement of highest return possible

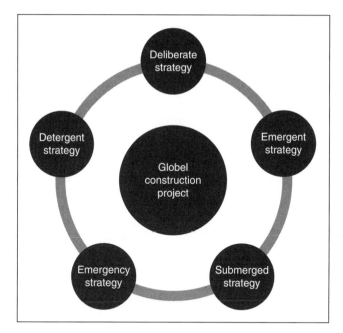

Figure 7.1 Strategy cycle
Source: Mintzberg et al. (1985).[70]

by encouraging optimal utilisation of available resources (including time, capital and people) more realistic.[69] Global construction projects should be defined with much more rigour than is usually the case. Equally, there needs to be a fully refined and continually steered project life cycle. In effect, these projects need to be guided much more sensitively towards their target than the more traditional 'fixed' notion of a project. Just as strategic planning has had to come to terms with greater fluidity and ambiguity, so must construction project management in emerging markets. Indeed the introduction of 'emergent' strategy in strategic management can be applied in an extended way. It is imperative to note that not only do these changes apply to project strategy but also project value – which can either be partly emergent or deliberate.

Figure 7.1 demonstrates an analytically useful approach to understanding global project strategy.

It may start off as deliberate but rapidly moves through other phases, which will include emergent strategy, submergent strategy, emergency strategy and detergent strategy. These phases can be characterised as follows:

- **Deliberate strategy:** a phase where the global construction project will have well-defined goals and a clear specific means of achieving these goals.
- **Emergent strategy:** where the global construction projects and goals are necessarily fluid and the means of achieving these goals can change in new and sometimes surprising ways.
- **Submerged strategy:** where the global construction project might be losing its way – its original goals now may seem distant and unreliable and project activities might begin to fragment.

- **Emergency strategy:** where the global construction project is truly fragmenting into near-random actions and the project as a whole appears to be overtaken by events.
- **Detergent strategy:** where the global construction project is recognised as off course and by now being steered back on its original track or onto a new track.[73]

The strategy cycle demonstrated in Figure 7.1 is not intended to be a deterministic series of phases to always go around clockwise. Indeed, global construction projects may interchange between emergent/deliberate modes as they are guided to target. Periodically, an emergent phase decays into both emergent/submergent during a project life cycle. Indeed, projects may continue to fly off course rather than being grasped firmly. The objective for integration of strategic and project management is for the introduction of an integral model that would essentially maximise the efficiency of strategy formulation and strategy implementation through projects.[71]

Multinational construction organisations working in global construction environments need to design an overall concept of management, i.e. strategic planning that will essentially maximise competitive power within organisations. To date, exploring the possibility of implementing strategies through projects mainly focuses on the junction between the two modes, i.e. between the completion of the strategy formulation process and the beginning of the project start-up.[72] The linking of strategic planning and global project management is a complex organisational problem for global construction project management practitioners. From the point of view of a successful project implementation, the solution lies in an overall project culture-oriented organisation per individual areas. The organisation of these areas should be subordinated to the principles of multi-project organisation changes. This should ensure that the global project management merges more and more in the existing management, in particular if the senior construction managers are committed to the task of culture development of the organisation.

Given that there are many different approaches to defining strategic process, care must be taken to ensure that any recommended processes are not prescriptive but permit a degree of flexibility that ensures the characteristics and needs of individual multinational construction organisations are taken into account. Operating in emerging markets, these organisations must also recognise the need to modify such processes to ensure that their individual construction organisation requirements are fully satisfied. There is a need for a cyclic process that is perceived on a regular basis to ensure that developments and performance can be closely monitored and emergent strategies are effectively managed.

7.15　Building an organisational culture of sustainability

Projects are a common way to carry out different types of tasks, which are in a number of ways unique. The continuous need for speed in global construction projects, alongside cost and quality control, safety in the working environment and avoidance of disputes, together with technological advances, environmental issues and the fragmentation of the global construction industry have resulted in a spiralling and hasty increase in the complexity of projects. It has today reached a level where senior construction managers must consider its influence on global construction engineering project success very seriously. It is crucial to highlight that construction engineering projects are made up of a multitude of interacting parts. Generally, it could be suggested that project management understands the project as an ordered and simple, and thus predictable, occurrence that can be divided into contracts, activities, work packages, assignments and so on, to be accomplished more or less independently. One could also see a project as a mainly sequential, assembly-like, linear process, which can be planned in any degree of detail through an adequate effort.[73] Consequently, one could indicate that global construction is generally complex in nature. For high project performance, multinational construction organisations need to set high specific goals for the project teams. Setting collective or individual goals within a project team is not enough for project performance and sustainability excellence, although it is essential to set collective sustainability priorities and project work plans. To certify the whole integrated project team is pulling in the same direction, and to make sure that the team is equally supporting each other, it is vital for senior managers to ensure that sustainability targets and team goals are aligned. This can be achieved through regular project team meetings. During the project meetings, senior managers should allow each team to test project goals against objectives. For each project goal, it is crucial to ascertain and, if possible, identify the extent to which it contributes to or achieves one or more project objectives. There are a number of reasons that multinational construction organisations recognise the need for a sustainability culture. As they contend in a global marketplace, contractors are looking for an edge in understanding the shifting client needs. Today's global construction marketplace is forcing a number of contractors to re-examine, restructure, and even work in partnership so as to keep up with the demand of clients and governments. While a number of global construction contractors have devoted years of research, schemes and solutions to sustainability, some are still trying to figure out why they should integrate sustainability principles.[4] Building a sustainable culture is a task that must be taken on from strategic, operational and project levels. The concerted endeavour of the construction industry, with support from governments and academia, is vital.

7.16 Proposing a competent organisational cross-cultural framework

Fellows and Liu,[74] suggest that a framework should capture the reality being modelled as closely as practical and include the essential features of that reality, whilst being reasonably cheap to construct and easy to use. It has been suggested by Bell, and Coxhead and Davis, that the use of a framework in a complex situation assists managers in minimising risk, imposing consistency, and provides a common generic and logical structure in decision making.[75,76] The framework proposed in this section delineates the key variables that influence the integration of global construction teams and highlights how cross-cultural issues can be managed. The emergence of a good global multicultural team is likely to depend on the establishment of a number of identifiable project level practices. These were found to fall within the following categories: leadership style,

Table 7.1 Cross multicultural project team performance variables

• **Leadership style**	• **Team selection and composition**
• Responsive leadership	**process**
• Inspirational leadership	• Picking people on value for money
• Authoritarian leadership	• Be based on ability individuals offer
• Charismatic leadership	• Capability to fit into the team
• Delegative leadership	• Ability to work in a team
• Participative leadership	• Respect between team members
• Organisational leadership	• Measure individual's beliefs
	• Be based on technical ability
	• Use of Meredith assessment and multicultural analysis
• **Cross-cultural management of**	• **Cross-cultural communication**
team development process	• Establish clear lines of responsibility
• Team building	• Cultural empathy
• Know individual's drivers	• Establish team effectiveness (collectiveness)
• Recognition and reward	• Value management
• Develop team loyalty (shared aims	• Establish trust
and objectives)	• Implement honesty
	• Encourage respect for others
• **Cross-cultural collectivism**	• **Cross-cultural trust**
• Good team organisation	• Good interpersonal relationships
• Participatory leadership (project	• Mutual respect between project leaders and
manager/ client)	team members
• Commitment from all team members	• Team-building activities
• Open decision making	
• Use of multicultural analysis	
• **Cross-cultural management**	• **Cross-cultural uncertainty**
• Keep project teams informed	• Articulation of project goals and objectives
• Interdisciplinary procedure should	• Clear project roles
be in place	• Managers need to be cross-cultural
• Verify project goals	communicators
• Encourage cooperative culture	• Effective interpersonal skills
• Promote constructive feedback	• Adopt project procedures that would apply
process	to everyone
• Open communication	• Gather more data to reduce culture and
• Build cohesion and stability in	information gap
teamworking	• Multicultural training

Source: Adapted from Ochieng and Price[67]

team selection and composition, team development process, cultural communication, cultural collectivism, cultural trust, cultural management and cultural uncertainty (Table 7.1).[4]

Multicultural teamwork requires greater fluidity and flexibility in responding to cultural issues on global construction projects. Integrating teams from different nationalities can be problematic and performance is not always at the level required or expected. In addressing this, there is a need to propose a framework that would:

1. highlight cross-cultural requirements for high-performing multicultural project teams;
2. establish how global project leaders can influence multicultural project teams to perform better;
3. illustrate how global project leaders can be cross-culturally competent.

Figure 7.2 summarises the main components of a multicultural framework. The framework for multicultural project performance needs to draw together:

- project purpose, objectives, values, roles, processes;
- cultural understanding through leadership;
- critical areas for cross-cultural action.

However, it is essential not to lose sight of what has been learnt in the broader sense about the factors associated with multicultural teamwork on projects and effective team performance. The culture of, and exhibited by, a project leader plays a major role in how members of a multicultural construction team perceive the project team performance framework proposed. The framework highlights the key factors of cultural complexity that have to be tackled within a multicultural construction project team in order to deliver a high-level performance. It presents a useful means to maximise the performance of these teams. However, much training would be needed before its full implementation. It also provides a ground work from which further research can be carried out on global teamwork. It is essential that researchers in construction management advance beyond the mere appeal of cultural diversity studies towards a more complete

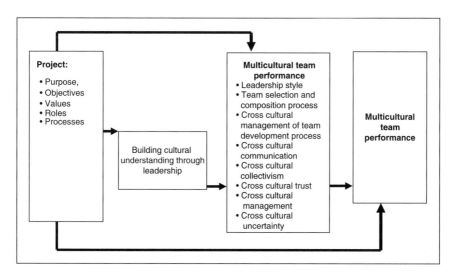

Figure 7.2 Summary of the main components of the above framework
Source: Adapted from Ochieng (2008).[4]

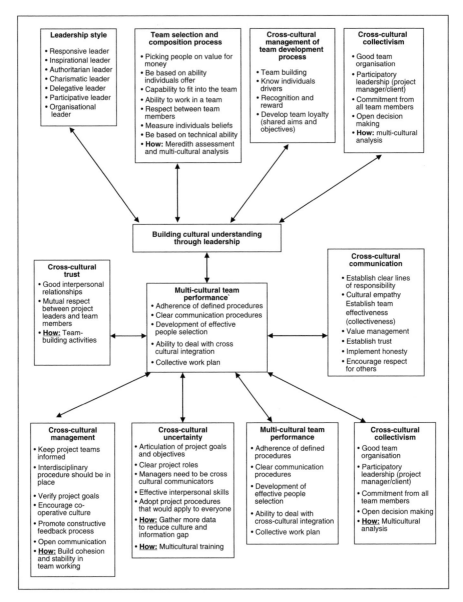

Figure 7.3 Cross-multicultural project performance framework
Source: Adapted from Ochieng 2008[4]

and detailed explication of multicultural team processes. A number of conditions and initiatives are conducive to good multicultural team performance for both the core team and the rest of the project team community. These can be summarised as:

- adherence of defined procedures;
- clear communication procedures;

- development of effective people selection;
- ability to deal with cross-cultural integration;
- collective work plan.

There is recognition that a climate of good team performance can be achieved if the five cultural effects are combined and managed effectively. The emergence of good multicultural teamwork is likely to depend on the establishment of a number of conditions and identifiable project level practices. The following diagram (Figure 7.3) is a representation of the cross-multicultural project performance framework proposed.

The utilisation of the proposed framework would not instantly transform multicultural construction teams into high-performing ones; however, it does identify eight key cross-cultural dimensions, which need to be considered. From the above, it is hopefully evident that in order to develop effective multicultural construction project teams it is necessary to create an environment that both acknowledges and values cross-cultural complexity. The primary function of culture in a project environment is to minimise uncertainty and ambiguity in everyday project team interaction and decision making by providing a framework for situational interpretation and limiting alternatives for appropriate behaviour and response. Cultures surface and develop in response to social cravings for answers to a set of problems common to all groups. In order for a project team to survive and to exist as a social identity, every project team, regardless of its size, has to come up with solutions to these problems.[22,23] These solutions then become characteristic of the group, which separate them from others.

7.17 Building effective multicultural teams at all levels and stages of the project organisation

The construction industry has a long-standing reputation for being adversarial, demonstrated by poor relationships between the client and project teams, which in turn lead to numerous problems, including poor project performance and a low number of long-term relationships between members of project teams.[77] These problems can be attributed to cultural issues between project teams. In a recent study that examined multicultural team integration on a global perspective, it was found that key determinants of team selection and composition process fall under the following variables:

- picking people on value for money;
- understanding client needs;
- initiatives to sustain involvement and wellbeing of all project workers;
- quality leadership.[4]

Project team members should be carefully selected for the ability they offer as well as their capability to fit into the project team. The client and project manager should allocate time so as working relationships can be developed within a project environment. This is particularly important where the project team do not know each other or have not previously worked together. Senior managers working with global construction teams must recognise and appreciate the value of getting the team to work together. Establishing the project team properly at the initiation phase helps in defining and setting team goals, which are united and can be measured to assure success for the team. Respect between team members has been highlighted by Cornick and Mather.[78] They conclude that project teams should have clearly defined project goals. This finding also emerged within the lay concepts of team effectiveness in the work of Akintoye and Guzzo.[79,80] Cornick and Mather and Egan showed that the construction project team has a purpose, composition and method of working that are unique to the industry it is formed to serve.[78,81] This uniqueness stems from the fact that its composition in terms of team members is not selected because of their ability to form an effective team in human terms but because they introduce the most technically and financially attractive competitive price for construction. The global project manager has to set a tone for the project, which will be conducive to the achievement of the project objectives. The creation of trust is more difficult to achieve,[82] and is one of the major challenges in cross-functional, distributed teams. In order to achieve trust, the client and project manager need to establish a common team culture from the outset and ensure that all team members are aware of their roles and responsibility in achieving the shared goals. Through client participation and project manager leadership, they need to show commitment and belief that it can be made to work.

To boost confidence and enhance cohesive teamworking, it is necessary that senior managers in multinational construction organisations build trust among themselves, especially in the early stages of the project. Trust is essential in cross-functional project teams: the higher the interdependence between disciplines, the more each team must depend on the functional expertise of other project teams. However, trust may be more difficult to create because the project teams are less familiar with the methods of team members from other disciplines and geographic location, thereby making it more exigent to form a shared understanding. As noted earlier on, it is vital for the client and project manager to set up project meetings on a regular basis to avoid complexities that could arise. This suggests that commitment and belief are key components of building trust. The challenges that global construction project teams face are compounded when team members are distributed and have few opportunities to interact face to face, rely heavily on technology to mediate their interactions, and have cultural or language barriers.

There is a growing recognition of the central position trust plays in integrated global construction project teams. For example, Thomas and Thomas stated that one of the basic elements of a cooperative relationship is mutual trust.[83] The authors further suggest that trust is a salient factor in determining the effectiveness of many relationships. Trust is crucial to affiliation, collectivism, leadership, participation and organisational culture. Many people instinctively feel that higher trust for one's project team members will result in better team affiliation. However, Ochieng established that the issue of trust affects other elements of project life.[4] It has appeared that, rather than having a direct impact on team affiliation, trust has a direct effect on other determinants of performance, such as leadership and participation.

7.18 Team integration principles

Integration means achieving coordination and collaboration among different work groups, responsibility centres and entire organisations.[84] The coordination needs of a one-time task may create unique interdependencies between various parts of the project, organisation or even several different contractors and agencies that have never before worked closely together. Project-related coordination requirements are often quite complex. Management must understand where these needs are, what kind of coordination is required, how best to achieve that coordination, and how these needs may change over time. An interesting feature of project integration needs is that they may vary significantly at different times in the project life cycle. The degree of integration can vary from merely sharing offices to using client engineers in one single project team. Integrating global project teams requires careful management and coordination. Some of the benefits of project integration include: better teamworking to meet project objectives, better cost estimation and control, improved risk analysis and management, quicker project start-up as contractors already understand the project.[85]

Multinational construction organisations have to ensure that there is a cultural balance of enthusiasm, experience, drive, organisation and control, which when brought together ensures that projects are given the best possible opportunity to succeed.[85] The underlying message that underpins the project team should be to try to enjoy the day-to-day activity on a project. This can be achieved through team-building activities so as to create an environment that is both focused and efficient in producing the project deliverables. The Procurement Guide report demonstrates that a supply chain is made up of all parties responsible for delivering a specific project.[86] In the context of a global construction project, there may be a number of specialised supply chains delivering design services, construction services, manufacturing and assembly of products that are fabricated off site and so on. It is imperative for each chain team member to be accustomed to working together as part of a fully linked chain. Supply chains move from project to project; they are brought together as an integrated supply team to meet a particular business need.

An integrated global construction project team is made up of the clients, project team and the supply team of consultants, contractors and specialist suppliers, as demonstrated in Figure 7.4. It brings together the design and construction activities, with maintenance considered as well, whether or not the integrated project team will be responsible for the ongoing maintenance of the project.

Large construction projects operating in emerging markets can have major financial consequences not only for those who manage or use them but also for the organisations that build and finance them. What is patently lacking is a strategic approach that gives construction managers in emerging markets an understanding of how to tackle complexity and diversity of global team integration.[86]

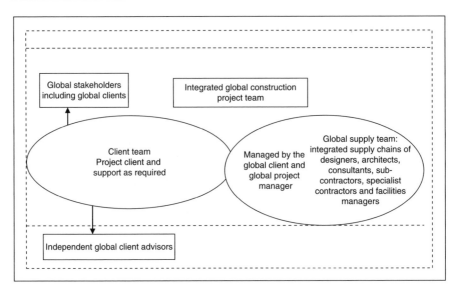

Figure 7.4 An integrated global project team
Source: OGC 2003[86]

7.19 Techniques for global team integration

The global construction industry is multifaceted and multidimensional, and major international construction projects in emerging markets are often carried out with multicultural project teams from developed countries. International alliance can be of particular benefit to less-developed and developing countries. International construction projects are normally fast-paced and require a longer span of control compared with national projects, and more parties are involved. Association between the concerned project teams requires clear project definition, and each set of project objectives under definition may be subject to changes as the project evolves. Project teams on international construction projects are also concerned with the clarity of local laws and the interpretation of those contracts governed by local laws. Trans-global alliance calls for cultural understanding and sensitivity in terms of people management by the concerned parties.

For today's construction projects, success is no longer the result of skilled project teams and leaders.[5] Rather, project team performance depends on multidisciplinary endeavour, involving diverse teams and support contractors working together in a highly multifaceted global construction environment. The process requires pragmatic learning, cross-functional coordination and amalgamation of technical knowledge, information and components. In spite of these challenges, a number of construction project team's work highly effectively across international borders, producing great results within agreed budget and schedule constraints. In terms of success for team integration, it is important to identify the critical factors. Global construction project management could be characterised by a twofold approach. Projects are embedded in their social context and they need to be managed accordingly. Success factors will depend on: defining the team structure, assembling a high-performance image, utilising tools and techniques to local culture, coalescing management processes and encouraging a culture of continuous support.

Multinational construction organisations have discovered that integration must be considered seriously, not only because they are dealing with an increasing number of international clients but also because more alliances are being formed with organisations based in other countries. Collaboration between internationally dispersed contractors is becoming commonplace in the contemporary construction environment. Managing projects across international borders is difficult and challenging. Common understanding of team values must be attained if a diverse project team is to work effectively, as lack of a coalescing management process to cultural differences can result in the raising of barriers that can inhibit teamwork and ultimately lead to a belligerent environment. Although the outcomes of cross-cultural teamwork can be creative and beneficial, poor team integration can lead to benefits being left unrealised or the outcomes being destructively conflicted. The challenge is for multinational construction organisations to learn to be conscious of the roots of cultural differences, to assess their impact, and to build structures, procedures and working environments that promote cultural synergy. This process is potentially costly and time consuming, but conversely, not dealing consciously with diversity and differences causes delays, and sometimes thwarts a creative and effective team.

This section has provided only a glance at the enormous problems facing international construction firms in team integration. Further studies are required in order to identify how knowledge sharing can be promoted in global projects.

7.20 Chapter summary

One of the more significant findings to emerge from this chapter is that global team integration is a particular problem for clients and project managers. Multinational construction organisations must respond to change in order to remain competitive and client focused. It is vital for global construction project leaders to handle teams sensitively if it is to develop a cooperative culture, which delivers better value than an adversarial culture. Cultural understanding is essential for senior managers who work on global construction projects. To develop as a global multicultural construction project team, it is essential that learning occurs.

7.21 Case study: China's infrastructure footprint in Africa

Underpinning China's entry into the African continent has been a calculated effort to forge close economic alliances with local state owned entities. An emerging pattern of China's expanding investment in Africa is that it has made an effort to work closely with African based construction organisations. This has been done to:

- Access new technologies
- Access new markets
- Fast track political networking.[87]

It is worth noting that, for Western organisations operating in Africa, the Chinese factor is set to weigh heavily on future investments. In simple terms, they do not have adequate response strategies to deal with China's all-inclusive strategy in accessing construction projects in competitively tight markets and therefore face a continual threat of being blind-sided by an increasingly finely honed Chinese construction project acquisition policy. According to *Executive Research Associates*,[87] China's integrated approach towards foreign investments in Africa allow Chinese organisations to enjoy lower political and economic risk entry levels than those faced by Western organisations. The Chinese risk framework underpinning the way it conducts business and delivers projects is fundamentally different from that held by Western companies.

China is currently involved in delivering infrastructure projects in 35 African countries. A concentration of infrastructure projects is to be found in Angola, Nigeria and Sudan. However, China is planning a new range of infrastructure projects in other parts of Africa, especially in the Democratic Republic of Congo. It is worth highlighting that the country's activities have been divided into two main segments: power generation (especially hydropower), and transport (especially railways), followed by ICT sector (mainly equipment supply). One way to measure the international competitiveness of the Chinese construction industry is to look at the performance of Chinese firms under open tenders.[87] For instance, multilateral agencies, such as the African Development Bank and World Bank, require unrestricted International Competitive Bidding to take place on important contracts that they finance. Procurement information from these two agencies can be used to gauge the share of contract value going to Chinese companies bidding for projects in different sectors of the market. This in turn provides an objective indication of the competitiveness of Chinese construction organisations.

About 70 per cent of the contract value won by Chinese companies under multilateral projects has been accounted for by four nations: Ethiopia, Mozambique, Tanzania and the Democratic Republic of Congo. One could suggest that this is quite different from the geographical spread under Chinese funded projects, where more than 55 per cent of contract value is accounted for by Sudan, Nigeria and Angola. This determines that Chinese construction firms have significant presence and experience in a number of African countries that have not yet featured prominently in Chinese financing deals. Examining China's footprint in Africa's infrastructure development trail, the following picture arises:[87]

- *Dams*: A number of dam projects undertaken by Chinese firms have a hydropower dimension to them.
- *Power*: The sector attracting the largest investment from Chinese firms has been the power sector with more than US$5.3 billion in cumulative at

present. Natural resources are being used to secure some of the financing. For instance, the Congo River Dam in the Republic of Congo and Bui Dam in Ghana are being financed by the China Ex-Im Bank loans backed by guarantees of crude oil in case of the Congo River Dam, and cocoa, in case of Bui Dam.

- *Rail*: Chinese venture into Africa began in the 1970s, with the construction of the Tanzania-Zambia railway, which came to signify China's contribution to African economic development. In recent years, China has made a major comeback in the African rail sector, with financing commitments on the order of US$4 billion for this sector. The largest investments have been in Nigeria, Gabon and Mauritania.
- *Roads*: The road ventures that Chinese companies have undertaken have been comparatively small compared to average project sizes in other sectors, and a number of them are financed by grants from the Ministry of Commerce.
- *Water and Sanitation*: This sector accounts for a moderately small share of China's total financial commitments to African infrastructure development.

Chinese project managers working in Africa are now expected to lead multi-regional projects. This adds the element of different cultures, language barriers, variations in currency exchange rates, ethnic differences, professional differences and organisational differences. From the above, the Chinese government and Chinese firms have shown that it is possible to get project teams from different companies and diverse cultures to work together effectively.

7.22 Discussion questions

1. What are the cross-cultural factors that would determine the success of Chinese firms working on infrastructure projects in Africa?
2. What are some of the challenges that Chinese project managers are likely to face at any project level (*strategic, operational, and technical*) in leading different sets of cultures in Africa?
3. Imagine that you were appointed the Project Director of Chinese firms in Africa. How would you accustom to a management model and a business model that suits the local culture in Nigeria, Gabon, Mauritania, Sudan, Angola and Democratic Republic of Congo.
4. Among the most challenging individuals to work on a global infrastructure project are project managers who are uninformed about differences in leading across cultures. As a project director how would you go about selecting project managers who are able to adapt their behaviour to different cultures and contexts?

7.23 References

1. Hammarberg, T. (2011). The real meaning of 'multiculturalism' is positive-and part of European values. http://www.neurope.eu/blog/real-meaning-multiculturalism-positive-and-part-european-values [Accessed March 2011].

2. Kelly, P. (2011). Multiculturalism is not a coherent policy that can be abandoned; David Cameron's speech reveals more continuity with Labour's 'British national identity' project than a radical departure from his 'liberal conservatism. http://www2.lse.ac.uk/researchAndExpertise/Experts/p.j.kelly@lse.ac.uk [Accessed July 2012].

3. Friedman, G. (2010). Germany's Chancellor Merkel declares multiculturalism a failure. http://www.thecuttingedgenews.com/index.php?article=21701&pageid=&pagename [Accessed June 2011].

4. Ochieng, E. G. (2008). Framework for managing multicultural project teams. Unpublished PhD Thesis, Loughborough University.

5. Ochieng, E. G. and Price, A. D. F. (2010). Managing cross-cultural communication in multicultural construction project teams: The case of Kenya and UK, *International Journal of Project Management*, **28** (5), pp. 449–460.

6. Day, R. (2008). Facing the challenge of managing a diverse multi-cultural workforce. http://www.cnplus.co.uk/facing-the-challenge-of-managing-a-diverse-multi-culturalworkforce/908807.article [Accessed June 2010].

7. Earley, P. C. and Mosakowski, E. (2000). Creating hybrid team cultures: An empirical test of transnational team functioning, *Academy of Management Journal*, **43** (1), pp. 26–49.

8. Baiden, B. K. (2006). Framework for the integration of the Project Delivery Team. Unpublished PhD Thesis, Loughborough University.

9. Cheng, J., Proverbs, D. G. and Oduoza, C.F. (2006). The satisfaction levels of UK construction clients based on the performance of consultants. *Engineering, Construction Architectural Management*, **13** (6), pp. 567–583.

10. Chervier, S. (2003). Cross-cultural management in multinational project groups, *Journal of World Business*, **38**, pp. 141–149.

11. Kumaraswamy, M. M., Ng, S. T., Ugwu, O. O., Palaneeswaran, E. and Rahman, M. M. (2004). Empowering collaborative decisions in complex construction project scenarios, *Engineering, Construction and Architectural Management*, **11** (2), pp. 133–142.

12. Weatherley, S. (2006). ECI in partnership with Engineering Construction Industry Training Board (ECITB), *ECI UK 2006 Master class Multi-cultural Project Team Working*, London: http://www.gdsinternational.com/infocentre/artsum. asp?lang=en&mag=182&iss=149&art=25863- [Accessed December 2006]

13. Emmitt, S. and Gorse, C. A. (2007). *Communication Construction Teams*. Oxon: Taylor and Francis.

14. Bartlett, C. A. and Gosha, S. (1989). *Managing across Borders*. Boston, MA: Harvard Business School Press.

15. Brett, J., Behfar, K. and Kern, M. C. (2007). *Managing Multi-Cultural Teams*. Boston, MA: Harvard Business School Publishing Corporation, available online at: www.hbrreprints.org [Accessed January 2008].

16. Earley, P. C. (1993). East meets west meets Mideast: Further explorations of collectivistic and individualistic work groups, *Academy of Management Journal*, **36** (2), pp. 319–348.

17. Earley, P. C. (1994). Self or Group? Cultural effects of training on self-efficacy and performance, *Administrative Science Quarterly*, **39** (1), pp. 89–117.

18. Peterson, M. F., Smith, P. B., Akande, A. and Ayestaran, S. (1995). Role conflict, ambiguity, and overload: A 21-nation Study', *Academy of Management Journal*, **38** (2): pp. 429–452.

19. Marquardt, M. J. and Hovarth, L. (2001). *Global Teams: How Top Multinationals Span Boundaries and Cultures with High-Speed Teamwork*. Palo Alto, CA: Davies-Black.

20. Kang, B. G., Price, A. D. F., Thorpe, A. and Edum-Fotwe, F. T. (2006). Ethics training on multi-cultural construction projects, *CIQ Paper 201, CIOB, Englemere*, **8** (2), pp. 85–91.
21. Hofstede, G. (1980). *Culture's Consequences: International Differences in Work-Related Values*. London: Sage Publications.
22. Hofstede, G. (1991). *Cultures and Organisations: Software of the Mind, Intercultural Co-operation and Its Importance for Survival.* New York: McGraw-Hill.
23. Trompenaars, F. and Hampden-Turner, C. (1997). *Riding the Waves Culture*. London: Nicholas Brealey Publishing.
24. Kandola, R. and Fullerton, J. (1998). *Diversity in Action. Managing the Mosaic*, London: Institute of Personnel and Development.
25. Meek, V. L. (1998). Organisational culture: origins and weaknesses, *Organisation Studies*, **9** (4), pp. 453–473.
26. Bathorpe, S., Duncan, R. and Miller, C. (2000). The pluralistic facets of culture and its impact on construction, *Property Management Journal*, **18** (5), pp. 335–351.
27. Smith, N. (1999). *Managing Risk in Construction Projects*. Oxford: Blackwell Science.
28. Edwin, H. W. and Raymond, Y. C. T. (2003). Cultural considerations in international construction contracts, *Journal of Construction Engineering and Management*, **129** (4) pp. 375–381.
29. Langford, D. A. and Rowland, V. R. (1995). *Managing Overseas Construction Contracting*. London: Thomas Telford.
30. Axley, S. (1984). Managerial and organisational communication in terms of the conduit metaphor. *Academy of Management Review*, **9**, pp. 428–437.
31. Thomason, G. F. (1988). *A Textbook of Human Resource Management*. 4th edn. London: Institute of Personnel Management.
32. Nibles, E. (2009). Management: Tough times make it time to reappraise systems. Available online at: http://www.nce.co.uk/market-data/consultants-file/management-tough-times-make-it-time-to-reappraise-systems/1995523.article [Accessed June 2011].
33. Ely R. J. and Thomas, D. A. (2001). Cultural diversity at work: The effects of diversity perspectives on work group processes and outcomes, *Administrative Science Quarterly*, **46** (2), pp. 229–273.
34. Jen, K. A., Northcraft, G. B. and Neale, M. A. (1999). Why differences make a difference. A field study of diversity, conflict and performance in work groups, *Administrative Science Quarterly*, **44**, pp. 741–763.
35. Loosemore, M. and Al Muslmani, H.S. (1999). Construction project management in the Persian Gulf – inter-cultural communication, *International Journal of Project Management*, **17** (2), pp. 95–101.
36. Loosemore, M. and Chau, D.W. (2002). Racial and discrimination towards Asian workers in Australian construction industry, *Construction Management and Economics*, **20** (1), pp. 91–102.
37. Watson, W. E., Kumar, K. and Michaelson, L. K. (1993). Cultural diversity's impact of Interaction process and performance: Comparing homogenous and diverse task Groups, *Academy of Management Journal*, **36**, pp. 590–602.
38. Townsend, A. M., DeMarie, S. and Hendrickson, A. R. (1998). Virtual teams: Technology and the workplace of the future, *The Academy of Management Executive*, **12** (3), pp. 17–29.
39. Ng, E. S. W. and Tung, R. L. (1998). Ethno-cultural diversity and organisational effectiveness: A field study, *The International Journal of Human Resource Management*, **9** (6), pp. 980–995.

40. Elron, E. (1997). Top management teams within multinational corporations: Effects of cultural heterogeneity, *The Leadership Quarterly*, **8** (4), pp. 355–393.

41. Evans, C. E. and Dion, K. L. (1991). Group Cohesion and Performance, *Small Group Research*, **22**, pp. 175–186.

42. Richardson, B., (1996). Modern management's role in the demise of sustainable society, *Journal of Contingencies and Crisis Management*, **4** (1), pp. 20–31.

43. Shenkar, O. and Zeira, Y. (1992). Role conflict and role ambiguity of CEOs in international joint ventures, *Journal of International Business Studies*, **23** (1), pp. 55–75.

44. Schein, E. (1985). *Organisational Culture and Leadership: A Dynamic View*. San Francisco, CA: Jossey-Bass.

45. Kwan, A. Y. and Ofori, G. (2001). Chinese culture and successful implementation of partnering in Singapore's construction industry, *Construction Management and Economics*, **19**(6), pp. 619–632.

46. Thomas, R., Marosszeky, M., Karim, K., Davis, S. and McGeorge, D. (2002). The importance of project culture in achieving quality outcomes in construction. In Proceedings of 10th Annual Conference on Lean Construction, 6–8 August, Gramado.

47. Cleland, I. C. and Bidanda, B. (2009). Project Management CIRCA 2025. Atlanta, USA: PMI Publications.

48. Baccarini, D. (1996). The concept of project complexity a review, *International Journal of Project Management*, **14** (4), pp. 201–204.

49. Morris, P. W. G., Patel, M. B. and Wearne, S. H. (2000). Researching into revising the APM Project Body of Knowledge, *International Journal of Project Management*, **18** (3), pp. 155–164.

50. Turner, J.R. (1998). *The Handbook of Project-Based Management*. 2nd edn. London: McGraw-Hill.

51. Winter, M., Smith, C., Morris, P. and Cicmil, S. (2006). The main findings of UK government-funded research network, *International Journal of Project Management*, **24**, pp. 638–649.

52. Bennett, J. (1991). *International Construction Project Management: General Theory and Practice*. Oxford: Butterworth-Heinemann.

53. Morris, P. W .G. and Hough, G. H. (1987). *The Anatomy of Major Projects: A Study of the Reality of Project Management*. Chichester: John Wiley & Sons.

54. Wozniak, T. M. (1993). Significance vs Capability: 'Fit for Use' Project Controls', In Proceedings of the American Association of Cost Engineers International (Trans) Conference, Dearborn, Michigan, A.2.1–8.

55. Chartered Institute of Building (CIOB) (1991). Procurement and project performance. Occasional Paper No. 45. Chartered Institute of Building, Ascot, England.

56. Rowlinson, S. M. (1988). An analysis of factors affecting project performance in industrial building. Unpublished PhD Thesis, Brunel University.

57. Strategic Forum for Construction (2002). Rethinking construction: Accelerating change. Consultation paper, Strategic Forum for Construction, London.

58. Williams, T. M. (1999). The need for new paradigms for complex projects, *International Journal of Project Management*, **17** (5), pp. 269–273.

59. Meredith, J. R. and Mantel, S. J. (1995). *Project Management–A Managerial Approach*. New York: John Wiley and Sons Inc.

60. Thamhain, H. and Wilemon, D. L. (1996). Building high performing engineering project teams. *Technology Management*, **5** (2), pp. 203–212.

61. Cleden, D. (2009). *Managing Project Uncertainty*. Surrey: Gower Publishing Company.

62. Johnson, S. K., Bettenhausen, K. and Ellie Gibbons, E. (2009). Realities of working in virtual teams: Affective and attitudinal outcomes of using computer-mediated communication, *Sage Journal: Small Group Research*, **40** (623), DOI: 10.1177/1046496409346448.

63. Tirmizi, S. A. (2008). Effective multicultural teams: Theory and practice: *Advances in group decision and negotiation*, **3**, pp. 1–20.

64. Elisa Mattarelli, E. and Maria Rita Tagliaventi, M. R. (2010). Work-related identities, virtual work acceptance and the development of glocalized work practices in globally distributed eams, *Journal of Industry and Innovation*, **17** (4), pp. 415–443.

65. Dekker, D. M., Rutte, G. C. and P. T. Van Den Berg (2008). Cultural differences in the perception of critical interaction behaviors in global virtual teams, *International Journal of Intercultural Relations*, **32** (5), pp. 441–452.

66. Nuick, A. and Thamhain, H. (2009). Managing global project teams: http://www.projectsatwork.com/content/articles/249601.cfm [Accessed June 2010].

67. Ochieng, E. G. and Price, A. D. F (2009). Framework for managing multicultural project teams, *Engineering, Construction and Architectural Management*, **16** (6), pp. 527–543.

68. Weber, K. and Dacin, T. (2011). The cultural construction of organizational life. *Organization Science*, **22** (2), pp. 286–298.

69. Thompson, J. L. (1998). *Strategic Management: Awareness and Changes*. 3rd edn. London: International Thomson Business Press.

70. Mitnzberg, H. and Waters, J. A (1985). Of strategies, deliberate and emergent, Strategic Management Journal, **6**, pp. 257–272.

71. Cleand, D. I. and Ireland, L. (2002). *Project Management: Strategic Design and Implementation*. New York: McGraw-Hill.

72. Kalos, A. and Diertrich, P. (2004). *Strategic Business Management through Multiple Projects. The Wiley Guide to Managing Projects*. John Wiley and Sons Inc.

73. Kokela, L. and Howell, G. (2002). The underlying theory of project management is obsolete. In proceedings of the 2002 PMI Conference, Seattle.

74. Felows, R. and Liu, A. (1997). *Research Methods in Construction*. Oxford: Blackwell Publishing Limited.

75. Bel, K. (1994). The strategic management of projects to enhance value for money for BAA Plc. Unpublished PhD Thesis, Herriot-Watt University.

76. Coxhead, H. and Davis, J. (1992). New development: A review of the literature. Henley Management College Working Paper.

77. Murray, M. and Langford, D. (2003). *Construction Reports 1944–98*. Oxford: Blackwell Science.

78. Conick, T. and Mather, J. (1999). *Construction Project Teams: Making Them Work Profitably*. London: Thomas Telford.

79. Akintoye, A. (2000). A survey of supply chain collaboration and management in the UK construction industry, *European Journal of Purchasing and Supply Management*, **6**, pp. 159–168.

80. Guzo, R. A. (1995). Introduction: At the intersection of team effectiveness and decision making. In Guzzo, R. A. and Salas, E. (eds), *Team Effectiveness and Decision Making in Organisations*. San Fransico, Jossey-Bass, pp. 1–8.

81. Elgon, J. (1998). *Rethinking Construction*. London: Department of the Environment, Transport and the Regions.

82. Bishop, S. K. (1999). Cross-functional project teams in functionally aligned organisations, *Project Management Journal*, **30** (3), pp. 6–13.

83. Thomas, G. and Thomas, M. (2005). *Construction Partnering and Integrated Team working*. Oxford: Blackwell Publishing.
84. Project Management Institute (2000). *A Guide to the Project Management Body of Knowledge*. Newtown Square, PA: PMI.
85. Verma, V. K. (1997). The human aspects of project management managing: Managing the the project team. Pennsylvania: Project Management Institute.
86. OGC (2003). Achieving excellence in construction, 'The integrated project team'. Procurement guide: Office of Government Commerce.
87. Executive Research Associates (Pty) Limited (2009). China in Africa: A strategic overview. http://www.ide.go.jp/English/Data/Africa_file/Manualreport/pdf/china_all.pdf [Accessed February 2013]

8

Partnering and Alliancing in Global Projects

8.1 Introduction

Global construction projects are often unique, on a large scale and highly complex. They involve a high degree of technical innovation and are delivered to tight cost and time schedules by large multidisciplinary teams drawn from very different organisations. In order to deal with these issues, partnering principles, frameworks and alliancing contracts have become established and very important features of the global construction sector. These terms cover a vast range of arrangements, including standard forms partnering, frameworks and alliancing contracts such as PPC 2000, Joint Contracts Tribunal (JCTs), constructing excellence, Engineering and Construction Contracts (EEC), supply chain for contractors. Partnering agreements are drawn up at an early stage of a construction project, setting out agreed dispute avoidance procedures and related issues. Partnering and alliancing strategies are fundamental to the successful management of global construction projects. This chapter explores all aspects of this, including drivers and prerequisites for partnering and alliancing in global construction projects. The chapter includes the following sections:

- inter-organisational excellence;
- changing perceptions of quality within construction;
- improving project performance;
- relationship contracting;
- drivers and prerequisites of partnering and alliancing;
- project alliance.

8.2 Learning outcomes

The specific learning outcomes are to enable you to gain an understanding of:

>> quality and performance concepts;
>> relationship management;
>> partnering and alliancing.

8.3 Inter-organisational excellence

8.3.1 The need to improve construction project performance

The construction industry has considerable bearing on the wealth and quality of life of any nation. It is central to national economies and delivers new homes, regional regeneration, modern transport facilities and a wide range of other public infrastructure. The provision and maintenance of buildings accounts for more than 10 per cent of the total UK gross domestic product (GDP); construction and associated manufacturing industries employ about 14 per cent of the UK workforce; there is currently a trade deficit of £1.8 billion on imported construction materials and components. There have been many calls for the UK construction industry to improve its productivity and performance in order to exceed clients' ever-growing expectations and needs.[1] One obstacle to achieving the required improvement has been the construction industry's reluctance to learn from other industries, hence causing considerable time lags between the development and implementation of new technologies and management techniques. The traditional method of construction procurement has been to produce a specification and award the contract to the lowest tender. However, this process has proven problematic and encouraged a claims-driven industry resulting in adversarial relationships throughout the supply chain. In order to improve the situation, many projects are now being awarded on a best value basis, thus resulting in more long-term partnering and less adversarial relationships throughout the supply chain. These characteristics substantially differentiate the construction industry from others; unfortunately, this uniqueness is all too frequently used as an excuse for not adopting new technologies and management techniques that have been successfully implemented elsewhere. As a result, the construction industry has traditionally tended to lag behind other industries, such as manufacturing and automotive industries, in the adoption of new technologies and management techniques for performance improvement.[2]

8.3.2 The Latham Report

The construction industry has been of considerable concern to several UK governments. The last two decades have seen successive governments support several initiatives and commissioned reports aimed at improving construction performance. For example, the final report of the Technology Foresight Panel on Construction (1995) recognised the lack of modern business and management processes in construction and described developments in improving managerial processes as slow and evolutionary.[2] In 1994, the UK government and the construction industry jointly commissioned Sir Michael Latham to independently review what was generally accepted as an underperforming construction industry. The main recommendations within the resulting report, *Constructing the Team: The Final Report of the Government Review of Procurement and Contractual Arrangements in the UK Construction Industry* (1994), were that the client should be at the focus of the construction process and there should be greater use of teamwork and cooperation. Project partnering was one of the specific methods recommended.[3] The report highlighted many problems associated with construction and the proposed solutions relate closely to Total Quality Management (TQM) philosophies. TQM is a comprehensive management paradigm that can help improve organisational performance and competitiveness. There is more on this later in the chapter.

Several recommendations were made in the final Latham report, including a construction cost reduction construction target of 30 per cent. Latham also highlighted that, despite widespread agreement on three previous construction-related reports by Simon, Sir Harold Emmerson and the Banwell, there had been little in the way of follow-up action. As a result of the Latham Report (1994) and the Efficiency Scrutiny of Government Construction Procurement Report (1995), HM Treasury issued a series of government guidance documents.[4] These provided best practice advice at a strategic level relating to the clients' role in the construction procurement process. Emphasis was placed upon: roles and responsibilities; training and skill development; achieving value for money; project management. The issues of partnering and incentives were also covered.

8.3.3 Egan and the construction task force

Although the Latham Report created an impetus for change, concerns still existed regarding the lack of progress and, in 1997, the Construction Task Force chaired by Sir John Egan was commissioned by the deputy prime minister: to report on the scope for improving the efficiency and quality of delivery of UK construction; reinforce the impetus for change; and make the industry more responsive to customer needs.[5] The resulting report, *Rethinking Construction*, was published in 1998. Although the task force stated that UK construction industry at its best was excellent and its capability to deliver the most difficult and innovative projects matched that of any other construction industry in the world, concern was expressed regarding: the under-achievement of the industry as a whole; unacceptable level of defects; lack of predictability; lack of contractor profit; need for customer feedback; lack of investment in capital, research, development and training; dissatisfaction amongst the industry's clients. The report highlighted issues such as teamwork, partnering and long-term relationships, all of which depend upon appropriate and effective empowerment of individuals and teams. The Egan Report also recommended that the industry should create and adopt integrated project processes and teams. The main targets from the Egan Report are summarised below:

- **Committed leadership:** requiring top management commitment to leading an improvement agenda involving whole organisation, cultural and process changes.
- **A focus on the customer:** recognise the importance of identifying the customer's needs and focus on the value the product delivers to the customer. Eliminate activities that do not add value to the customer.
- **Integrate the process and the team around the product:** the most successful enterprises do not fragment their operations – the process and the production team are integrated to deliver value to the customer efficiently and eliminate waste in all its forms. Construction has traditionally approached

the project process as a series of sequential and largely separate operations undertaken by individual designers, constructors and suppliers who had no stake in the long-term success of the product and no commitment to it. Changing this culture was thus seen as fundamental to increasing efficiency and quality in construction.

- **A quality-driven agenda:** move away from traditional perceptions of quality (i.e. zero defects, delivery on time and to budget) and pay greater attention to after-sales care and reduced whole life costs.
- **Commitment to people:** commit to the training and development of and respect for all participants in the process, within a no-blame culture based on mutual interdependence and trust.

Additionally:

- Targets for improvement included: annual reductions of 10 per cent in construction cost and construction time; defects in projects to be reduced by 20 per cent per year.
- The industry needed to: provide decent and safe working conditions and improve management and supervisory skills at all levels; design projects for ease of construction.

To achieve these objectives, the task force recommended the following:

- The industry needed to make radical changes to the processes through which it delivered its projects. These processes needed to be explicit and transparent to the industry and its clients.
- The industry should create an integrated project process around the four key elements of product development, project implementation, partnering the supply chain and production of components.
- Sustained improvement should be delivered through use of techniques for eliminating waste and increasing value for the customer.
- The industry needed to replace competitive tendering with long-term relationships based on clear measurement of performance and sustained improvements in quality and efficiency.

Over the last 20 years, new management tools and methods have been developed to help improve performance in delivering products and services to customers. TQM techniques have helped to provide new performance tools and have found many successful applications within many industrial sectors. This has led to awareness that TQM has the important cultural philosophy that customer satisfaction is inseparable from business goals. Although the construction industry is often reluctant to adopt external models, there have also been significant changes in the management of construction processes and the industry has become more conscious of performance in terms of safety, cost and quality. Four factors frequently cited as being central to TQM and continuous improvements are leadership, teamwork, client focus and effective communications. The Latham and Egan reports both identified fundamental factors that were well aligned with the key TQM factors and are applicable to both construction and other business concerns.

8.3.4 Long-term arrangements

The recent focus on the effective integration and management of construction supply chains has emphasised the need to ensure that all key stakeholders contribute fully to the achievement of project objectives.[5] In recent years, many project-specific partnering arrangements have emerged both in the public and private sectors. Strategic alliances and long-term partnering arrangements are now starting to develop throughout construction supply chains.[6,7,8] To be effective, these new long-term arrangements must involve increased empowerment throughout the supply chain, thus creating new roles and responsibilities based on mutual trust and respect. The process is effectively empowerment of teams across organisational boundaries and offers considerable potential cost savings through leaner supply chains and reduced checking and monitoring costs.

Effective alliances can provide a seamless supply network, offering better business relationships, integrated solutions, reduced costs, higher quality and shorter delivery times; however, despite the shift in relationships and responsibilities that such changes are likely to engender, the process of empowering (i.e. delegation of power and responsibility) organisations and individuals within this framework has been largely ignored. Egan alluded to this in his 'Rethinking Construction' report, when he identified the need for a commitment to people as a key requirement for change and performance improvement within the industry.[5] There is more on empowerment in 8.5.2 below. In a keynote address to the Movement for Innovation's (M4I) conference Profiting from Innovation, Construction Minister Nick Raynsford called for the industry to now focus on addressing the respect for people theme (2000).[9] He also recommended the industry to do more to improve the work environment so that the most able employees will be attracted to the industry and remain within it.

8.4 Changing perceptions of quality within construction

8.4.1 Introduction to quality and performance concepts

Evidence from the British Properties Federation Survey (1997) suggested that one-third of the major construction clients were dissatisfied with the performance of the industry as a whole.[10] Egan stated that, if the construction industry is to improve client satisfaction, there needed to be considerable improvement in terms of quality and efficiency.[5] Although the construction industry has become more competitive and considerable steps have been taken to improve quality and achieve excellence, clients' expectations of construction performance have increased over recent years and customer satisfaction keeps moving out of reach. In the drive to improve both the quality of finished products and the efficiency of the processes through which they are produced, many construction organisations have introduced TQM-based strategies, resulting in the development of integrated supply chains, long-term partnering and lessons-learnt procedures. However, the construction industry has tended to lag behind other industries when it comes to achieving the performance benefits that can be achieved through the implementation of TQM.[11] In addition to traditionally low profit margins, the construction industry has also experienced relatively low levels of efficiency and is often responsible for considerable wastage of resources. It is therefore vital that the industry continues to review and improve its processes and deliverables in the light of ever-changing perceptions of quality and performance. Despite some improvements and success stories, such as the Toyota Production System (i.e. an integrated socio-technical system comprising management philosophies and practices), which have led the way in the comprehensive development of lean production, construction still has a lot to learn from other industries.[12]

8.4.2 TQM and inter-organisational collaboration

If a TQM philosophy is to be fully embraced within an organisation, it should not be considered as a single company in isolation but must take due account of the whole supply chain. Zhao explored how advanced TQM approaches can lead to best performance and inter-organisational excellence by developing a theoretical framework on the application of TQM concepts.[13] Within a multi-organisational project-based environment, teamwork must transcend organisational boundaries with the objective of achieving excellence through inter-organisational collaborations. There are many reasons for adopting inter-organisational collaborations, for example:

- to increase core competencies and capability;
- to gain a competitive advantage;
- to reduce the cost of quality;
- to reduce or improve the management of risks;
- to build a multidisciplinary team capable of delivering complex products;
- to encourage a long-term perspective and more innovative approaches;
- to access local knowledge or technical expertise.

Such collaborations need strong leadership with support and commitment demonstrated by senior management cascaded through the organisations. Organisational objectives must be aligned around a common goal with high levels of trust and openness. Power traditionally embedded within a single organisation has to be transferred to the project entity (such as Eurotunnel) around which the inter-organisational collaboration is based.

8.4.3 TQM and the project alliance

Walker and Keniger described the TQM system and innovative approach to project alliancing that was adopted on the National Museum of Australia project.[14] The project demonstrates how the integration of selection criteria and performance measures can be used to develop an effective risk and reward structure. The resulting cross-team relationship helped to achieve the desired quality culture resulting in high levels of quality and a team highly motivated to deliver best value. Mellat-Parast and Digman proposed a framework for adopting quality management practices within a strategic alliance based on integrating core elements of quality management and strategic alliances with the contribution of trust and learning being emphasised.[15]

8.5 Improving project performance

8.5.1 Integrated teams

The importance of integrated teams was highlighted and proposed in the late 1970s as a way of improving construction project performance. However, the mechanisms needed to develop high-performing integrated teams was not fully understood and the industry has been through a learning curve with some major global construction projects highlighting the benefits and barriers to successful integration as briefly summarised below:

- 1970s: the economic crisis during the premiership of UK Prime Minister Harold Wilson resulted in considerable conflict within the construction industry and a claims conscious environment.
- 1985: the Broadgate Project in the City of London adopted an integrated team approach and was an early example of construction management. The client adopted many ideas from the US, including an integrated design and construction process.
- 1987: the Channel Tunnel between England and France was perceived as a perfect integrated team. The joint venture had a revolving chairperson but experienced a very aggressive style of management resulting in considerable conflict and 100 per cent cost growth with a reduced specification.
- 1993: the Canary Wharf project in London developed the integrated design and construction team approach further by developing direct links with suppliers and introduced processes to support integrated team decision making, which resulted in a 15 per cent cost reduction, bespoke contracts and a high degree of trust with no checking.
- 1999: the Petronas Towers project in Kuala Lumpur, Malaysia, achieved full integration supported by improved knowledge sharing.
- 2008: the Channel Tunnel rail link, taking passengers from central London to the English end of the Channel Tunnel, took the concept of integrated teamwork further through the adoption of risk and reward sharing.

The Latham and Egan reports have been referred to in this chapter to highlight the emergence of a new mindset and the recognition in the 1990s of the need for innovative procurement approaches that improve performance and reduce confrontation. Since these reports were published, there has been reports on the construction industry that, although dealing with other topics, have also highlight the need for new procurement models to drive change and improve supply chain efficiency through increased integration and collaboration, as demonstrated below:

- The Office of Government Commerce (2003) Achieving Excellence suite of procurement guides replaced the Construction Procurement Guidance series and reflects developments in construction procurement over recent years. *The Achieving Excellence in Construction Procurement Guide 05: The integrated project team: teamworking and partnering*, explains how to work together as an integrated project team to enhance whole-life value.

- The Low Carbon Report produced by the Innovation and Growth Team in 2010, although providing a series of recommendations aimed at reducing carbon and energy, also highlighted that integration across the supply chain still needs to be improved along with greater sharing of construction risks and rewards.
- The need for integration of the whole supply chain was a common theme among those interviewed during the development of the Infrastructure UK (2010), *Infrastructure Cost Review: Main Report.*
- In 2011, the Cabinet Office announced the government's new construction strategy and the government's intention to require collaborative 3D building information modelling (BIM) on its projects by 2016, thus pushing forward an approach that supports the integration and collaboration agenda.

8.5.2 Empowerment

Recent trends towards integrated teams and supply chain management within construction have provided the opportunity for key stakeholders to better contribute to the achievement of project objectives. The increased application of partnering and strategic alliances has resulted in a shift in relationships and responsibilities and the changes have empowered organisations and individuals. The industry has thus become more reliant on virtual teams, fragmented work groups, subcontracted labour and multi-organisational project delivery structures, along with a more general awareness of the significance of teams for organisational performance. The level of fragmentation and autonomy that the emerging structures promote arguably provide an ideal climate for the systematic implementation of performance-enhancing empowerment, leadership and teamwork strategies.[16] The importance of empowerment was explicitly supported in 2005 by the Egan implementation task force examining *Respect for People*.[17] The task force stated that significant improvement was only likely to come about when those working in the industry were empowered to drive the performance agenda. Empowerment has thus been an important element in achieving the Rethinking Construction agenda:

> Respect for people means that all workers need to be consulted, involved, engaged and ultimately empowered in a spirit of partnership – not just management. The workforce is a rich source of ideas to improve the way work is carried out. And involving the workforce will not only demonstrate that they are respected and valued, but will improve productivity and quality.
> – *Rethinking Construction report on Respect for People (2000).*[18]

Empowerment can be considered positively in terms of improving business performance and enhancing innovation. Developing and applying appropriate empowerment strategies throughout the project delivery system is essential if the potential benefits of self-managed teams is to be exploited. Understanding and commitment to teamwork and empowerment is essential if managers are to: improve motivation; design effective self-managed teams; develop individual skills; better manage team interaction, communication and decision making. Teamwork is fundamental to effective empowerment. All parties need to be clear regarding the aims and objectives if teamwork is to be effective. Sharing personal and organisational objectives helps to ensure that relationships are honest and open. Trust and good communications are key attributes for success and help to foster an equitable approach across teams. Honold and Johnson[2] highlighted the importance of having an effective leader as crucial to creating a common goal.[19,20] Communication is a key component of empowerment and by communicating effectively and creating supportive environments, leaders can successfully empower their teams. Existing organisational structures and cultures are potential barriers to empowerment and can hinder the implementation of empowerment and change, which is a slow process. Effective long-term partnering must ensure that all parties fully appreciate the new roles and responsibilities that emerge. This could result in the client relinquishing some of its traditional roles and responsibilities, and the contractor being expected to take them up. The full implications of this process of organisational empowerment are not fully understood by all and can often result in the failure to achieve the full potential that effective partnering arrangement can offer.

The increased use of partnering has led a greater need for understanding how to develop and maintain long-term relationships. Effective integration and management of construction supply chains requires all key stakeholders to fully contribute to the achievement of project objectives.[5] Many project-specific partnering arrangements have been adopted both in the public and private sectors, leading to strategic alliances and long-term partnering arrangements being adopted throughout construction.[6,7,8] However, to be effective, such arrangements rely on appropriate empowerment throughout the supply chain, thus creating new roles and responsibilities based on greater mutual trust and respect. The process is one of empowering teams across organisational boundaries and offers considerable potential cost savings through leaner supply chains and reductions in costs associated with checking and monitoring. The awareness of supply chain members within projects, regarding the importance of cultivating long-term relationships and empowerment of the supply chain through teamwork and leadership concepts needs to improve.

Although partnering can be applied alongside most project delivery systems, it has greater potential to deliver benefits on fast-track, complex or uncertain global projects, which tend to have many high-level inter-organisational interactions, for example between client and contractor. Partnering benefits such projects as it requires:

- inter-organisational interactions that are conducted in a non-confrontational way and issues resolved in the best interests of mutually agreed objectives;
- open and horizontal organisational infrastructure, which facilitates direct and uninhibited communications thus reducing the cost and increasing the effectiveness of each interaction.

8.5.3 Supply chain perspectives

The Rethinking Construction Task Force emphasised the importance of the construction supply chain and its importance to driving innovation and improving performance. Partnering should not be entered into by believing it is an easy option, as it can often be more demanding than more traditional forms of contracting, especially in the early stages. It generates interdependence between the involved parties, which requires open relationships, trust, effective performance measures, commitment to continuous improvements, and opportunities for all parties to share the rewards for improved performance and the pain resulting from poor performance.

8.6 Relationship management and contracting

8.6.1 Relationship management

'Relationship management' can be considered as an umbrella concept that emphasises communications, teamwork and cooperation. According to Galbreath: 'for the most part [customer relationship management] CRM, human resources management (HRM), enterprise resource planning (ERP), supply chain management (SCM), partner relationship management (PRM) and similar programs have paid very little attention to the relationships that underpin those processes, or to the intangible – relationship – assets embedded in them.'[21] Although new forms of relationship management have emerged over recent years, it is not an entirely new concept and can be applied in many different ways to address specific situations and organisational objectives. The first that comes to mind is customer relationship management (CRM) in which organisational competitiveness is improved by focusing on customers' needs. Partner relationship management (PRM) can take the form of focusing on strategies and processes for improving communications and relationships between a manufacturer or producer and its channel partners who work with it to market and sell the manufacturer's products, services or technologies. Enterprise resource planning (ERP) refers to the management of internal operations such as manufacturing, finance, sales, resources and distribution.

8.6.2 Relationship contracting

Relationship-based approaches to project management have developed over the past two decades as part of a general trend and in response to problems experienced throughout the construction industry. The key elements of relationship management – teamwork and cooperation – have been used across the full range of construction project delivery systems, including: Build Operate Transfer; Private Finance Initiative (PFI); and Design and Build. Most types of relationship management in the construction sector take the form of partnering or alliancing. Two useful definitions of defined relationship contracting are provided below:

> A term broadly adopted to describe new styles of contract that attempt to achieve outcomes that are acceptable to all parties involved in the delivery of construction projects
>
> Gunn (2002),[22]

> A flexible approach to establish and manage relationships between clients and contractors and to implement proven practices and techniques to optimise project outcomes
>
> The Australian Constructors Association (1999).[24]

For relationship contracting to be successful it needs:

- project objectives to be well aligned and agreed to by all parties;
- risk allocation to be fair and equitable, taking into account who is best placed to manage them;
- the project scope to be clearly defined with complete and unambiguous objectives;
- a form of contract that is appropriate and encourages relationship development;
- an integrated project management team provided with sufficient authority, responsibility and accountability to achieve the project objectives;
- a long-term approach to shared *gainshare and painshare* (i.e. a sharing of a project's profit and losses between alliance partners) based on the project's objectives and encouraging innovation and attainment of best value;
- an open and honest approach to effective communications.

8.7 Partnering and alliancing

8.7.1 Definition of partnering

Partnering is a voluntary non-contractual commitment between a project's client and contractor(s) to actively cooperate and work together in meeting well-aligned project objectives. As such, there is no partnering contract, only a partnering charter, which is an agreement signed by the relevant parties expressing their commitment to teamwork and collaboration on a project. Partnering has been defined as:

- 'a structured management approach which encourages teamwork across contractual boundaries';[23]
- 'two or more organisations working together to improve performance through agreeing mutual objectives, devising a way for resolving any disputes and committing themselves to continuous improvement, measuring progress and sharing the gains';[5]
- 'a commitment by those involved in a project or outsourcing to work closely or cooperatively, rather than competitive and adversarial';[22]
- 'a co-operative relationship focused on aligned objectives between two or more parties, which may include an operator, to deliver enhanced business results for all parties and a mechanism for sharing risk and reward'.[24]

Organisations can use partnering to maximise performance through collaborative business affiliations based on best value. Partnering requires substantial effort to set up, but a number of construction clients who have experienced partnering have found it rewarding. Benefits of partnering can be summarised as:

- increased client satisfaction;
- better value for client;
- creation of an environment that promotes innovation and technical development;
- better understanding between partners and driving down of real costs;
- stability that provides more confidence for better planning and investment in project staff and resources;

- improved 'buildability' through early participation of contractors;
- acknowledgement and protection of profit margin for suppliers and contractors;
- shorter overall deliver period;
- better predictability of time and cost.[25]

8.7.2 Different types of partnering

Partnering started as a one-off project-by-project approach. In time, parties became more confident of this new kind of working relationship, and trust grew, resulting in more long-term partnering arrangements, which afforded more benefits and avoided some of the early set-up costs. Long-term partnering emerged to provide clients, contractors, suppliers and the industry at large opportunities to draw upon the benefits associated with long-term relationships. However, risks materialise throughout the process and their management can be a major challenge to parties involved. It has been defined as: 'the development of sustainable relationships between two or more organisations, to work in cooperation for their mutual benefit in the requisition and delivery of works, goods and/or services over a specified period to achieve continuous performance improvement'.[26]

Vertical partnering is based on a relationship between purchaser and supplier, for example:

- supply chain partnering to deliver better value to the client over the long term by improving quality through better systems, skills, communication and relationships;
- one-off project partnering based on competition/negotiation, strategic/full or long-term partnering;
- post-award partnering;
- one-off competition partnering usually on public sector projects;
- preselection partnering on a one-off long-term basis.

Horizontal forms of partnering include organisations performing similar roles within the supply chain but usually with different but complementary competencies and skills, such as a major contractor and a specialist subcontractor. They generally tend to develop into a more formal contractual relationship, thus shifting away from the informal approach that characterises partnering.

Networks can be considered as a form of partnering and tend to be open-ended and provide a vehicle for debate, driven by the members' own agenda. Such networks can take many forms, as demonstrated by the following examples:

- **Networking forums**: established to enable members to learn from each other by benchmarking their performance and practices.
- **Business associations**: with the remit of promoting members' interests.
- **Sector groups**: established to improve performance of an industry by developing sector strategies and advising government agencies.
- **Cluster groups**: to improve performance in related industries.

8.7.3 Early partnering developments

The initial successes of post-award partnering were in the US building industry, where project alliancing in the oil and gas industries had become more common and soon filtered through into the UK civil engineering and building industry in the late 1980s during a period of economic buoyancy in which construction clients were investing heavily. There was considerable early success in project-specific partnering on major projects, where there usually tends to be a well-defined project scope. However, partnering was less successful within long-term frameworks where client and contractors often have conflicting long-term objectives. Partnering arrangements were encouraged during a period of public sector deregulation and were encouraged to adopt a more commercial approach to procurement, involving a move away from lowest cost to best value where contractor is selection-based not on price alone but on the contractor's ability to provide best *value-for-money*. However, this requires the process for determining value to be predetermined, usually based on a range of qualitative and quantitative criteria (not lowest cost), often considering: technical expertise and experience; safety and quality records; approach to risk management, relationships and innovation; financial standing. Consequently, features of relational contracting, such as shared objectives, financial incentives and benchmarking, started to emerge with long-term frameworks.

During the early 1990s, the recession resulted in a lack of work within agreed frameworks, which to a large extent undermined many long-term relationships. However, this tended to lead to more collaborative arrangements being developed. During the early stages of partnering in the public sector, there was considerable concern that the selection process may not include sufficient elements of competition, however, this did not prove to be the case. The European Construction Institute 'Partnering in Europe' task force comprised 37 organisations from clients, contractors, consultants and academic institutions. The task force produced guidelines for the application of project-specific partnering arrangements in Europe, which identified the reasons to adopt and circumstances under which they might apply. It suggested that the implementation of long-term partnering and strategic alliances should be developed. There has been high take-up in the UK, Australia and Hong Kong, through various government initiatives. However, this is not the case throughout mainland Europe. Continuous improvement is one of the main drivers for partnering, leading to greater efficiency and competitive advantage. In order to demonstrate improvement, relevant indicators must be determined and performance in terms of value measured. These can then be compared with internal or external benchmarks.

8.7.4 Drivers for partnering and alliancing

Partnering was first applied in the construction industry by the US Army Corps of Engineers in the late 1980s. Like many other countries, the US suffered from problems where traditional forms of contract were one-sided with little control, often resulting in considerable cost and schedule growth, increased litigation and a highly confrontational approach to relationships. The new approach was initially a post-award one to partnering, where the client and successful contractor would discuss their mutual expectations, share issues of concern and develop well-aligned mutually beneficial objectives. The project participants would then sign a partnering charter.[27] One of the first UK applications of the project alliance concept was in the North Sea oil industry, where it was used to achieve significant time and cost savings on major projects. The main driving force was to significantly reduce construction costs to make marginal oil fields financially viable. The Andrews Oilfield is the best-known early example. It was completed six months early with a final cost of £39 million below the original agreed cost of £373 million, which itself was £77 million below the client's original estimate.[28] In addition to many TQM tools such as workshops, benchmarking, process mapping, team charters, partnering also needs effective dispute resolution mechanisms. Partnering has proven to be effective on many projects by increasing productivity, reducing cost and time, improving quality and certainty, and increased client satisfaction.

During the 1970s and 1980s construction experienced considerable increase in litigation on major projects. There was a tendency for clients to minimise their own risk exposure by transferring more contractual risk to the contractor. The updating and general reworking of industry standard forms of contract issued in 1999 only partly mitigated this trend.[27] At the same time, project partnering became well established as a non-adversarial and performance-enhancing approach to contracting in several countries, including Australia, the UK and the US. However, the simple and often imprecise language used to describe partnering sometimes makes it difficult to differentiate between partnering and other forms of relationship contracting, thus driving the need for considerable changes in practice and cultures through rigorous implementation processes, followed by effective training and performance monitoring.

8.7.5 Strategic alliances

In non-equity alliances, the partners agree to work as a stakeholder in the objective of the project, whereas equity alliances and joint ventures (JVs) tend to be more formal arrangements in which the partners become both stakeholders and shareholders by contributing equity capital to fund the alliance.[25,29] The four main types of strategic alliance are:

- A *business co-operative*, which is usually set up to reduce costs or create marketing opportunities, often by joint purchasing or joint marketing. It can involve many members and is usually a delegated management function.
- *JVs*: according to Khemani and Shapiro (1993),[30] a JV is 'an association of firms or individuals formed to undertake a specific business project. It is similar to a partnership, but limited to a specific project (such as producing a specific product or doing research in a specific area)'. Unlike strategic alliances, where firms agree to cooperate with each other towards a mutually agreed objective, a JV involves participating companies forming a new and separate legal entity in which they will invest capital and share ownership. The JV will thus own its assets independently from the

parent companies. Although the partner companies will actively participate in developing the JV's strategic direction, the JV will be staffed by a separate management team. The JV is a popular approach for entering into international markets, especially those that do not allow wholly owned subsidiaries. A JV is selected where traditional subcontracting is not appropriate. There are many difficulties that come with JVs as they require considerable effort to establish, manage and operate. Before entering into a JV approach, due consideration should be given to a traditional subcontract approach, which can provide many advantages as the liabilities, responsibilities and payment methods are well established and require less administration. However, JVs can provide many benefits, such as reducing the risks associated with very large projects; local participation requirements being satisfied; all parties being granted equivalent status, which is not possible through traditional subcontracts; combined partners' resources; partners' bonding requirements shared; increased prequalification credibility; technology transfer.

- A *strategic alliance* is set up to improve project performance and gain an advantage on competitors. It is similar to a JV but on a longer term and more open basis (e.g. access to financial information). There are many different types of strategic alliance throughout the supply chain: pooled purchasing to increase a firm's purchasing power; supplier partnering to improve quality, reduce cost and increase speed by establishing long-term relationships.

- *Project alliancing* is a highly evolved form of partnering embedded within a contract and focused on client and contractor relationships, however, it is not a joint-investment type of relationship. Where project partnering is usually considered to be a voluntary agreement to work collaboratively together, this is not the situation with a project alliance, which involves a formal contract where the parties agree to work cooperatively in the best interests of the project and share the project risks and rewards based on specific predefined performance measures. A project alliance is thus a mechanism that supports relationship management and project delivery. The client and contractor create an integrated project team in which individual organisational commercial interests are aligned with the project objectives. NASA[26] defined a strategic (project) alliance as: 'a business relationship in which firms combine complementary strengths and share risk by pooling financial, physical, and personnel resources regarding a specific business venture; each firm remains a distinct entity, separate from its strategic alliance partner'.

8.7.6 Prerequisites to successful partnering in global construction projects

There are prerequisites to successful partnering (leading to project alliancing) that need to be in place either before the process starts or implemented at the very early stages. As this involves several parties agreeing to align their objectives and working practices, workshops are critical to success not only at the start in developing the desired teamwork approach but also for maintaining relationships over considerable periods of time. Some of the prerequisites to successful partnering are summarised in Figure 8.1:

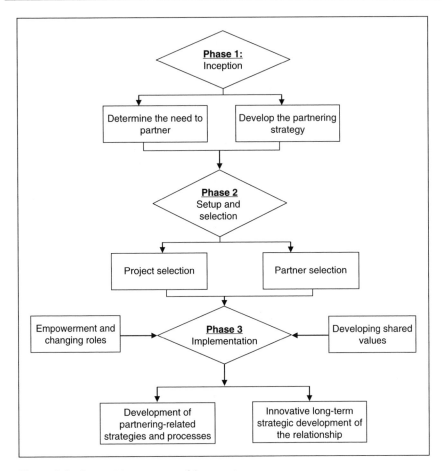

Figure 8.1   Prerequisites to successful partnering

Phase 1: Inception (deciding if and how to partner)

- **Determine the need to partner**
  - There has to be recognition of the need and a desire to change with top management providing a committed individual to take the initiative, and champions to lead the process by cascading the partnering philosophy throughout the organisation and help to change culture and process in line with the intended partnering relationship.
  - Top management should *walk the talk* to encourage change and ensure the availability of resources.
  - The client must decide what forms of relationship to enter into and if there is sufficient volume of work where partnering is to be adopted to warrant a change in the business model.
  - Decide with whom it would be best to have overarching partnering agreements and at an executive level agree to engage in specific partnering processes on future contracts that the parties enter into.
  - Determine budget constraints and provide sufficient resources to partner. This should include the direct and time-commitment costs associated with

workshops, training personnel and communications. These are critical to the success of the process and tend to be underestimated, so it's best to err on the high side.

- **Develop the partnering strategy**
  - The client must establish a partnering strategy that is clear about what type of relational contracting is being considered and what proportion of future work is to be partnered.
  - SMART objectives need to be developed and a time for implementation of the partnering agreed.
  - The champion needs to be provided with the required support and resources, including an implementation team. It may be best to appoint two champions, one at executive level and one at operations level.
  - There needs to be an internal workshop to develop or validate the above.

Phase 2: Setup and selection phase
- **Project selection**
  Assess if the project or work is suited to partnering:
  - Determine the degree of complexity as this determines the level of partnering communication needed.
  - Consider the duration of the contract (relationship) as long-term arrangements provide the greater benefits.
  - Determine how to deal with intellectual property issues.
  - Assess how important the project is in terms of top level management time and involvement.

- **Partner selection**
  - Check compliance with current legislations.
  - Develop and apply a relationship value-based partner selection process.
  - Introduce the client's interest and commitment to partnering in the solicitation documents and provide information on the partnering process and philosophy for potential bidders not familiar with the process.
  - Ensure that the commitment to partnering is emphasised at the prebid meeting.
  - Ensure executive level cooperative relationship and commitment.
  - Select an appropriate partner, bearing in mind potential for development of long-term relationships.
  - Agree and formalise the relationship. All processes must be mutually agreed to by all parties – initial contacts should be at the executive level.

Phase 3: Implementation
- **Empowerment and changing roles**
  - The client has to be comfortable with abandoning the micro-management of specific matters and passing an increasing amount of control to the contractor. The contractor needs to be able to take on this additional responsibility.

- **Developing shared objectives**
  - The contractor's and client's business and project goals need to be well aligned.

- The contractor must be committed to helping the client achieve its business objectives, and vice versa.

- **Developing shared values**
  - All the parties need to be mature enough to adopt partnering and must share the same values regarding honesty, respect and trust.
  - There has to be a desire by all the parties to partner and adopt a positive attitude towards cooperation. This desire should not just reside with those developing the strategy and process but also strong leadership should ensure this is cascaded down to those responsible for implementing these on a day-to-day basis.

- **Development of partnering-related strategies and processes**
  - Develop and implement the processes, capability and commitment needed to collect and analyse performance-related data.
  - Develop and implement a communication strategy with mechanisms for two-way communications.
  - Develop and implement mechanisms for early problems (dispute) resolution.
  - Develop, agree and sign a partnering charter.
  - Hold joint workshops to develop and maintain the above.

- **Innovative long-term strategic development of the relationship**
  - Consider adopting a strategic equitable pain/gain share alliance approach that aligns the long-term commercial objectives of the partners.
  - Ensure all relationships have a no-blame culture that encourages innovation.
  - Develop whole-life value approaches and value-adding strategies, such as lean, value-engineering and sustainability.
  - Continuously benchmark and monitor key performance indicators to maintain a competitive advantage.
  - Run joint non-optional partnering training programmes for partners on large projects.

8.7.7 Cost and benefits of partnering and alliancing

Alliancing tends to require more resources than partnering, which is more resource intensive than traditional contracting. The additional costs tend to be associated with the early stages of a project, such as the selection process and the need for information throughout the project. Partnering and alliancing usually require change and the introduction of new ways of working, which have to be learnt over a period of time. Such change needs early senior management involvement and commitment, which can incur considerable direct costs especially when establishing the partnering or alliance strategy. As the project progresses, direct costs will also be incurred for workshops, monitoring, evaluation and staff training.[26] However, there are efficiencies to be made throughout the project as a result of effective teamwork, increased trust, less checking and increased innovation resulting in long-term improvements. Incentives need to represent an equitable sharing of contract risk and provide real incentives for all partners to improve project performance and not just create an additional cost burden for the contractor, for example:

- the sharing of cost savings from value-engineering reviews; or
- bonuses aligned with the achievement of project milestones.

Project incentives should be designed to result in:

- adoption of good practice and aligned objectives;
- improved project performance terms of cost, time and quality;
- reduced litigation, administrative and legal costs;
- increased innovation;
- improved whole-life value.

8.7.8   Public Private Partnership (PPP) and Private Finance Initiative (PFI)

Public Private Partnerships (PPP) and the Private Finance Initiative (PFI) provide excellent opportunities to apply project partnering during the procurement phase of a project. A PPP is where a government service is funded and operated through a partnership between government and the private sector. PPPs include all types of collaboration between the public and private sectors. In the UK, the Private Finance Initiative (PFI) is the most common form of PPP used in the UK to deliver policies, services and infrastructure. Other forms of PPP include: Strategic Service Delivery Partnerships (SSDPs); Concessions; Strategic Infrastructure Partnerships; and JV Companies. One of the main advantages of PFI is the ability to access private-sector funds for new or improved capital assets.

8.7.9   Typical problem and potential solution

Example of a typical problem
- The client has estimated that the project will be delivered six months behind schedule and is under considerable pressure to have the project delivered on time but has a contingency of £10 million still unspent.
- The contractor believes that the project is five months behind schedule. Estimates that they will lose £10–15 million on the project with a large proportion of the costs arising as materials have been purchased at much higher rates than originally planned. As a result, the contractor reduced spending due to cash-flow problems, which could add further delays.
- The subcontractor estimates that the project is four months late and that his losses on the project will be £1–2 million. The subcontractor is experiencing delays due to poor sequencing of work but is also suffering from poor productivity.

Solution: adoption of partnering and alliancing
- Partnering would have encouraged better information flows, which in turn would have helped to improve poor sequencing of work, thus improving productivity – 'working out of sequence'. The client would also have been in a better position to know the extent of the problem and identify potential solutions.

- Alliancing with a *painshare and gainshare* based on the schedule would have motivated all three parties to work together more effectively with the shared objective of minimising delays.
- Alliancing with a painshare and gainshare based on cost would have allowed the contractor to share the impact of cost increases with the client and avoid the delays associated with cash-flow problems.

8.8 Chapter summary

The prerequisites to successful partnering and alliancing have been set out in Section 8.7.6 along with the steps that need to be undertaken. Given the size and cultural complexity of global construction projects, the potential risk and consequences of failure tend to be magnified. This section summarises some of the key strategies that need to be developed in order to mitigate these risks. A good starting point is to explore various TQM concepts and philosophies in relation to the proposed project and existing approaches. These will provide some direction to the type of procedures and organisation cultures that need to be adopted. The size and complexity will invariably result in global construction projects being built, owned and operated by a special-purpose vehicle formed by several organisations that provide their own individual expertise to create a strong project team. There are different approaches that can be taken to the relationship contracting, and a strategy needs to be developed in line with the needs of the project. This will lead further to the development of strategy for supply chain management, which will have to take into account the multinational characteristics of the supply chain. A long-term procurement strategy needs to be developed that will ensure the most appropriate partners to be selected. Given the number of organisations and individuals to be involved in the project, an effective communication and engagement strategy needs to be developed that takes due account of cultural complexities. The project governance strategy needs to emerge (see Chapter 2), once the various partners have been selected. This should facilitate effective matrix management, align the goals of the key partners and ensure project loyalty. Although the philosophies and approaches will have already taken shape, there is a need to further consider how the risk and rewards associated with the project will be shared between the various parties through the development of a painshare and gainshare strategy. This will also influence the degree of innovation that is adopted on the project. These strategies have been summarised below:

1. A **quality and performance strategy** based on well-established TQM concepts to inform the selection of project processes and culture.
2. A **relationship strategy** that informs the type of relationship contracting to be developed.
3. A **supply chain management strategy** in line with the adopted approach to TQM.

4. An **engagement and communications strategy** to ensure organisations and individuals are part of the process.
5. A **long-term procurement strategy** that includes partner selection.
6. A **project governance strategy** that helps to align partner objectives.
7. A **painshare and gainshare strategy** that encourages innovation and continuous improvement.

8.9 Discussion questions

1. Discuss how TQM principles have been incorporated into partnering and alliancing.
2. Discuss the evolution of integrated teams and their role within global construction projects.
3. What is relationship contracting and how does it apply to global construction projects?
4. Explain how partnering can improve the performance of global construction projects.
5. Discuss the role of supply chain management within a project alliance.
6. Discuss the main prerequisites to strategic alliance.
7. Explain the difference between partnering and alliancing.
8. What are the costs and benefits of partnering and alliancing?
9. Compare the different types of strategic alliance.
10. Discuss partnering from the perspective of the cost of quality.

8.10 References

1. DTI (1997). Competitiveness through partnerships with people. London: Department of Trade and Industry.
2. Technology Foresight Panel (1995). Final report of the Technology Foresight Panel. London: HMSO.
3. Latham, M. (1994). *Constructing the Team*. London: HMSO.
4. Bennett, J. and Jayes, S. L. (1995). *Trusting the Team: The Best Practice Guide to Partnering in Construction*. London: Thomas Telford.
5. Egan, J. (1998). *Rethinking construction: The report of the Construction Task Force*. Norwich: HMSO.
6. Allen, J. D. (1999). Minds over natter, *Construction News*, 19 Aug (No. 6629), pp. 19.
7. Leitch, J. (1999). Steering in a new direction, *Contract Journal*, **400** (6239), 18–19.
8. Barlow, J., Cohen, M., Jashapara, A. and Simpson, Y. (1997). *Towards Positive Partnering*. Bristol: The Policy Press.
9. Raynsford, N. (2000). *Keynote address: Profiting from innovation*. In Proceedings of the M4I Annual Conference, 22 May, NEC, Birmingham. http://www.m4i.org.uk/ conference/index.htm [Accessed May 2000].
10. British Properties Federation Survey (1997). Survey of major UK clients. In Egan, J. (1998). *Rethinking Construction*. London: HMSO, p. 11.

11. Loraine, B. and Williams, I. (2000). *Partnering in the Social Housing Sector: A Handbook.* London: Thomas Telford.

12. Howell, G. A. (1999). *What is lean construction?* In Proceedings Seventh Annual Conference of the International Group for Lean Construction (IGLC-7), 26–28 July, Berkeley, CA .

13. Bresnen, M. and Marshall, N. (2000). Partnering in construction: A critical review of issues, problems and dilemmas, *Construction Management Economics,* **18** (2), pp. 229–237.

14. Walker, D. T. and Keniger, M. (2002). Quality management in construction: An innovative advance using project alliancing in Australia, *The TQM Magazine,* **14** (5), pp. 307–317.

15. Mellat-Parast, M. and Digman, L. A. (2007) A framework for quality management practices in strategic alliances, *Management Decision,* **45** (4), pp. 802–818.

16. Price, A. D. F, Bryman, A. and Dainty, A. R. J. (2004). Empowerment as a strategy for improving construction performance, *Leadership and Management Engineering,* **4** (1) pp. 27–36.

17. M4I (2000). A commitment to people 'our biggest asset', Report of the Movement for Innovation's Working Group on Respect for People. http://www.m4i.org.uk [Accessed July 2001].

18. Construction Excellence (2000). *Rethinking construction report Respect for people.* London: Construction Excellence.

19. Honold, L. (1997). A review of the literature on employee empowerment, *Empowerment in Organisations,* **5** (4), pp. 202–212.

20. Johnson, P. R. (1994). Brains, Heart and Courage: keys to empowerment and self-directed leadership, *Journal of Managerial Psychology,* **9** (2), pp. 17–21.

21. Galbreath, J. (2002). Success in the relationship age: Building quality relationship assets for market value creation, *The TQM Magazine,* **14** (1), pp. 8–24.

22. Gunn, J. (2002). *The Effective Use of Partnering and Alliancing.* Australia: Minter Ellison.

23. Construction Industry Board (1997). *Partnering in the Team.* London: Thomas Telford.

24. Broome, J. (2002). *Procurement Routes for Partnering: A Practical Guide.* London: Thomas Telford.

25. European Construction Institute (2003). *Long-Term Partnering: Achieving Continuous Improvement and Value.* London: Thomas Telford.

26. NASA (2012). Strategic alliance partnering-some guidelines. http://sbir.nasa.gov/SBIR/alliances.htm [Accessed May 2012].

27. Jones, D. (2002). The Development of PPPs in Australia. [2001] ICLR333-347.

28. Bakshi, A. (1995) Alliances change economics of the Andrew field development. Offshore Engr, Jan. 1995, pp. 30–34.

29. Black, C., Akintoye, A. and Fitzgerald, E. (2000). An analysis of success factors and benefits of partnering in construction, *International Journal of Project Management,* **18** (6), pp. 423–434.

30. Khemani, R. S. and Shapiro D. M. (1993). Glossary of Industrial Organisation Economics and Competition Law, compiled and commissioned by the Directorate for Financial, Fiscal and Enterprise Affairs, OECD.

9

Financing Global Construction Projects

9.1 Introduction

Most construction companies are very different in their asset base to those in the manufacturing and service sectors, mainly due to their production activities usually being located at many different and temporary sites rather than a fixed permanent location. Contractors' main physical assets tend to be equipment that is highly mobile and rapidly depreciates due to the harsh treatment it is usually subjected to. These assets usually provide a very poor collateral base when considered alongside the risks associated with construction projects and the potentially one-sided contracts that frequently result in delayed payments. Lenders in any industrial sector will consider the associated risks before providing any loan, which tends to involve: an assessment of how the money is to be used; the creditworthiness of the borrower, including how well the company is managed and its profitability; the value of assets to be used as collateral to which the lender has recourse. Considering these factors, the argument for financing construction projects through conventional means is substantially weakened. This chapter discusses the various sources of construction-related finance, including the concept of project finance with reference to global construction projects. The focus is on sources of global project finance, financial institutions, minimisation of commercial risks, project conformity and contracts. The chapter also examines how construction organisations work with global financial institutions. Finally, the procedures for investment appraisal and cost modelling are presented.

9.2 Learning outcomes

The specific learning outcomes of this chapter are to enable the reader to gain an understanding of:

>> corporate financing;
>> bilateral loans;
>> syndicated loans and commercial bonds;
>> global project finance;

>> indirect guarantees;
>> global countertrade;
>> development bank project finance procedures;
>> financial risk assessment and management.

9.3 Global corporate financing

9.3.1 Corporate financing

Corporate financing, often referred to as on-balance sheet financing, involves using a company's internal capital to directly finance a project, or using company assets as collateral to obtain loans, which can be categorised as: short-term finance (up to two years); medium-term finance (two to five years); long-term finance (over five years). Project investors, sponsors and other equity providers try to minimise the equity they invest in a project in order to increase the rate of return and reduce exposure. In contrast, lenders must ensure that equity providers have a high financial stake in the project in order to increase commitment and reduce the risk of the equity providers walking away from the project if it proves to be unsuccessful, thereby reducing the lender's risk. Before deciding whether or not to finance a project, and the level of interest to be charged, lenders will evaluate the financial viability of the project, the risks involved and how the project is to be managed. The various parties will work together and develop a mutually acceptable package. There is usually a trade-off between the cost of borrowing and the cost of the adopted risk management strategies. According to Swiss Re (1999), in industrialised countries most construction projects comprise: 10–30 per cent equity; 60–90 per cent senior debt; 0–15 per cent junior debt. However, in developing countries there tends to be more equity and less debt. The main advantages of corporate financing are discussed below:

- **Access:** companies' internal capital approval procedures tend to be quicker than those of external lenders. The approval of loans based on a company's creditworthiness and assets is also quicker than undertaking the due diligence associated with loans based on the project's projected cash flows and assets. In some cases, corporate financing may be the only option available where projects are too small to be considered for project finance.
- **Confidentiality:** the internal financing of a project helps to ensure commercially sensitive project information does not reach competitors or other third parties.

The main disadvantages of corporate financing are:

- **Funding limits:** the amount that can be borrowed may be limited due to: a company's internal budgetary constraints; a company's ability to repay; the need to keep the cost of loans down by maintaining a high credit rating; the lender's gearing requirements (e.g. 50–80 per cent of a project to be funded by company assets).
- **Risk and liability:** there are fewer opportunities to transfer risk, and the company's internal capital and assets are at risk if the project fails.

9.3.2 Short-term finance

Working capital is the amount of money a company has at its immediate disposal and is needed to fund a business's daily transactions. It can be defined as the sum by which current assets exceed current liabilities. A company's current assets appear on its balance sheet and comprise stock held, cash in the bank, debtors and short-term investments. Current liabilities are the payments that have to be paid within one year, this includes creditors, Inland Revenue payments, shareholder dividends, short-term loans and bank overdrafts. Short-term finance tends to be used to overcome immediate cash-flow problems such as the purchase of materials, plant hire, payment of employees and subcontractors. As described below, this may include the use of short-term loans, overdrafts, trade credit, revenue payments and dividends to the shareholders.

1. **Bridging loans** are used to cover periods when other forms of finance are not available. They are a source of short-term finance that can be used to access cash at short notice but at a relatively high interest rate. The principal balance is the amount owed, which in most other forms of loan is paid off during the loan period and reduced to zero at the end of the loan. This is often not the case with bridging loans used to finance the construction phase of a project, which is where most of the risks occur. Construction bridging loans, therefore, tend to be a short-term loan covering the construction phase and attract high interest rates. Once the construction phase is complete, the loan is transferred to a prearranged long-term loan with a lower interest rate. On more risky projects, the short-term bridging construction loan may also include an initial part of the operation period, in order to ensure that the project has stabilised and the risks are reduced.

2. **Overdraft facilities** are considered as short-term finance and have the advantage of being flexible as money can be withdrawn as the need arises up to a pre-agreed limit. However, the limits tend to be low and interest rates on overdraft are high when compared to longer types of finance and are calculated on a daily basis. In order to satisfy immediate liabilities a company can: request longer to pay, borrow money or sell some of its fixed assets. However, all three approaches have their drawbacks: the first damages the company's reputation and creates the risk of being blacklisted by suppliers; the second increases costs; the third reduces operating capacity. Delaying payments of trade credit debts (see point 3 below) is one way of increasing a company's working capital over the short term. Many contractors use this approach but run the risk of being blacklisted by their suppliers.

3. **Trade credit**, according to Christy and Roden (1973),[1] 'arises when a business receives goods or services from a supplier without the requirement of an immediate cash payment'. It is still considered a source of finance although it provides goods and services on credit rather than actual cash. Trade credit can take the form of a discount applied to the invoice price; or a delay in payment due, starting from when the goods were received to the time payment is due. Companies receiving trade credit need to ensure prompt payment, as any delay could pull down their credit rating.

9.3.3 Medium-term finance

There are three main types of medium-term finance: finance institutions; bank loans; sale and leaseback. Finance institutions provide funds in the form of hire purchase, and leasing and loan agreements to the industrial, commercial and consumer sectors. Under most construction equipment hire purchase arrangements, the contractor is required to provide an initial cash deposit with the finance institution providing the remaining balance, which is paid back through regular monthly instalments over a pre-agreed time period (usually two to three years). The equipment is owned by the contractor but the financial institution will have a claim on it should the contractor default on the instalments. The advantage of hire purchase is that companies can still obtain it even if they have no access to a bank overdraft, or if their overdraft limit has been reached. However, hire purchase tends to be an expensive source of finance. In contrast, in leasing arrangements, the finance institution purchases the equipment but leases it to the contractor who pays for it through monthly instalments over a specified time period of usually two to three years. No cash deposit is required from the contractor, which improves short-term cash flow. There are also some tax benefits associated with leasing as it can be considered as an off-balance sheet debt. The contractor does not own the equipment at the end of the lease period but could purchase it from the finance institution. Sale and leaseback can be used when a company wishes to sell some assets because the company is in financial difficulty, requires additional capital to fund future growth, or takes a strategic decision to concentrate on its core business activities. Cash can be raised by remortgaging, which may raise only a proportion of the assets value, or by selling the property on the open market, which would tend to imply a new occupier of the asset. However, sale and leaseback arrangements maximise the cash generated whilst ensuring that the existing owners continue in occupation. Many sale and leaseback arrangements are made with insurance companies, financial trusts, property investment companies and pension funds looking for long-term property investments not occupation.

9.3.4 Long-term finance

Most limited construction companies use a combination of finance sources (usually share capital, long-term loans and medium- to short-term facilities) to finance the purchase of long-term assets and fund day-to-day business operations. Long-term finance can be used to help the company expand, for example, through the purchase of equipment and plant. It is considered to be risky because of the timescale involved but this can be reduced if there is recourse to assets (e.g. via property deeds). Short- and medium-term finance is less risky but is sometimes unsecured and also relatively more costly to administer and thus tends to be more expensive than long-term finance. As noted below:

- **Retained profits** are an attractive form of long-term finance because they do not require any interest payments on the money raised. At the end of the financial year, any annual profit is used to pay corporation tax and shareholders' dividends with the remainder being reinvested in the business. Most companies will have a corporate policy or strategy, which influences how much profit is paid out as dividends and how much is retained.

- **Share issues** are used to generate substantial equity and to provide access to external sources of finance. There are two main types of shares: ordinary and preference, as defined by Rutherford:[2] 'A portion of the financial capital of a limited company which gives the holder an entitlement to a fixed return, in the case of preference shares, or, in the case of ordinary shares, a variable dividend decided by the board of directors. Ordinary shares are also known as equities'. Ordinary shareholders own the company and have full voting and dividend rights, but only receive dividends after payments have been made on debentures, loan stock and preference shares.

9.4 Bilateral loans

9.4.1 Grants and loans

A grant is a sum of money provided to support a project in order to meet the stated objectives of a third party and does not have to be repaid so long as the objectives have been satisfied. Loans, however, relate to a sum of money that has to be repaid, plus interest at an agreed rate, at a later date. Bilateral loans are where there is one borrower and one lender with interest rates being set by adding an amount to the specified index rate. This is the prime rate that is usually set by individual banks and is charged to its more creditworthy borrowers. A bank 'base rate' is the lowest rate it will lend at and this rate will usually be referred to in the loan documentation. The London Interbank Offered Rate (LIBOR) is the index usually referred to on large commercial loans. It is lower than the prime rate as it is the interest rate at which banks lend money to one another on the London interbank market. Loans are usually made to a time and credit limit, which may be renegotiated if difficulties are encountered. However, most overdrafts are 'repayable on demand'. The following steps may be taken by a bank should a borrower fail to repay a debt when it is due:

- The best option is to negotiate a new solution, such as reducing monthly repayments and increasing the duration of the loan, which would ensure the completion of the project and the bank being able to recoup all debt including any interest due.
- When the loan has been secured, for example, with land deeds deposited with the bank, the assets can be sold to recover the debt. However, this can be very damaging to the borrower, and the sale of the assets on the open market may not ensure that the back recoups all of the debt due.
- If the loan is unsecured, the bank can sue for bankruptcy. In construction, the borrower is unlikely to have enough cash and the bank would lose its only real security, which is a successful completed project.

9.4.2 Senior or junior loans (debt)

Most loans provided by large international banks will be considered as 'senior' debt, which must be serviced before any other debt or equity. Most senior loans tend to be secured against physical assets and carry the lowest risk of the commercial financial mechanisms. They are thus the cheapest source of capital, with interest rates determined by the interest rates associated with the loan currency plus a margin based on the perceived risk associated with the project. Junior loans are considered to be subordinate and are therefore serviced after senior debt but before equity. They can also be secured against physical assets, but with the senior debt having the first claim. Junior loans cost more than senior loans due to the increased risk.

9.4.3 Loan conditions

Traditional bank loans are drawn down once and have a fixed maturity with a variable (floating) interest rate, adjusted at pre-set intervals of say 6 or 12 months and quoted as a percentage above the London Interbank Offered Rate (LIBOR) or the US prime rate. Payments are made to a pre-agreed schedule, although 'repayment holiday' periods may be included. Fixed-rate loans are available but usually at a higher interest rate, with other conditions such as arrangement and withdrawal fees, and limited time periods. Additional facilities can be introduced to improve the loan's value. Revolving credit acts very much like a credit card, and permits the borrower to draw the loan (or part of it) and make repayments at its discretion, or to an agreed programme. This will have a higher interest rate but may have a lower overall cost as the sums borrowed can be kept to a minimum but can be changed to meet short-term needs. Standby facilities permit the borrower to delay taking the loan until it is needed, but a contingency fee is charged until the loan is drawn down and interest is paid. Multi-option-facilities tend to be much more complex than traditional loans. They can be tailored to suit individual borrower's requirements and thus represent good value. Interest rates are usually determined by adding an amount to the specified index rate. This is usually the prime rate set by individual banks and is charged to its more creditworthy borrowers. Many construction projects have long periods during which there is high expenditure with little or no income generated. Any interest accrued during the construction period is usually added to the amount borrowed and paid once construction has finished and income is being generated during the operation period. The ability to draw down the loan when required throughout the construction period helps to keep interest payments low.

9.5 Syndicated loans and commercial bonds

9.5.1 Syndicated loans

On large loans, say over US$75 million, it is unlikely that a bilateral loan, where only one bank lends the money, would be used, due to the degree of risk exposure to the bank. It is in the interests of both the borrower and the lender to ensure that no single lender has too much risk exposure with one borrower. Syndicated loans are perhaps the most important risk-sharing device for large loans, including those for global construction projects, as they can be used to spread debt among a number of lenders. They involve a formal subcontracting of the loan and its associated risk to several lenders, thus avoiding overexposure by a single lender. They help lenders to better manage their portfolios by reducing risk, increasing fee business and improving returns. As well as being an excellent risk-sharing device for lenders, syndicated loans also have other key advantages for the borrower, mainly due to the number of lenders participating, such as: the overall risk is reduced thus leading to a reduction in the overall cost of borrowing; the speed of execution once drawdown starts; the large funds available; reduced administration; increased flexibility of timing.

A syndicated loan is typically structured and priced by the lead manager or agent, who then sells portions of the deal to other lenders under terms negotiated by the agent. The borrower negotiates the loan conditions with a lead manager (agent) who will structure and price the syndicated loan. The 'lead manager' will sell parts of the loan to other lenders who are prepared to offer on common terms (e.g. a medium-term, syndicated or rollover loan) to a single borrower. Although there are many types of syndicated loans, they are typically medium-term loans with durations of three to fifteen years and interest rates quoted as a percentage above the London Interbank Offered Rate (LIBOR) or the US prime rate, and adjusted at pre-agreed intervals of say six or twelve months. The syndication process has become more complex over recent decades. This has partly been due to an increased tailoring of the loan to the specific needs of the borrower, but there has been a significant reduction in the number of banks to sell debt to.

However, there has been a growth in non-bank investors, such as finance companies and insurance companies, which has been essential to supplement the reduced number of commercial banks. Corporate bonds are longer-term debt instruments with a maturity of at least a year after being issued by a corporation to raise money and expand its business. Instruments with a shorter maturity are usually referred to as 'commercial paper'. Corporate bonds and syndicated loans are used extensively by Europe's largest firms to support their investments, sometimes simultaneously. Altunbas et al.,[2] in his report to the European Central Bank, explored the determinate of borrowing via the syndicated loan market and compared syndicated loans with the main alternative: the corporate bond market. This choice of debt instrument is often linked to a range of factors, including: economies of scale; transaction costs; the potential need for future debt renegotiation. Altunbas et al.[2] highlighted the recent changes in the syndicated loan market – mainly increased capacity and transparency – that have moved syndicated loans closer to corporate bonds and further away from bilateral bank lending, thus making it the main substitute to corporate bonds.

Over the past three decades, there has been considerable speculation over the impending demise of the syndicated loans market. During the 1980s, there was a reduction in the number of syndicated loans being arranged. This was due to fewer high leverage buy-outs and multi-billion-dollar loans required for US takeovers; reduced real-estate-related lending in the US and UK resulting from falling property prices; support for syndicated loans being substantially dampened; recession. However, when the conditions are favourable, remarkable oversubscriptions and innovative approaches have been taken by project owners in order to secure the funds they require. Chui et al.[3] examined developments in the syndicated loan markets and explored the collapse of international bank finance during the recent financial crisis. Their investigation suggests that balance sheet constraints of international banks significantly contributed to the collapse of syndicated lending, which coincided with a sharp decline in global economic activity. The supply of syndicated loans tends to be more sensitive to bank balance sheet constraints than bilateral lending, which tends to involve more long-standing relationships.

9.5.2 Parties to a syndicated loan

The roles that the various parties to a syndicated loan perform can be categorised as:

- The *borrower* must agree to certain conditions prior to drawdown, ensure that there is no default on existing loans and guarantee that the nature of business will not change. The borrower will provide the lead manager with full financial statements throughout the loan period.
- The *lead manager* is the bank responsible for negotiating the terms of the loan and, if appointed, will act as the *agent* bank who will: collect funds from the syndicate members and distribute when needed; collect and distribute interest and repayment from the borrower.
- The *managers* help the lead manager to organise the participants and fully or partially underwrite the loan.
- The *participants* usually include smaller banks who have been asked to share the loan and non-underwriting risk.

9.5.3 Steps in completing a syndicated loan

Although there are different approaches that can be taken when arranging a syndicated loan, the process has to be based on well-founded procedures. As the loan needs to be individually tailored according to the borrower's needs, each loan is different and the lead bank needs clear instructions from the borrower as to the actions needed. A good syndication strategy has to be agreed between the lead bank and the borrower if the borrower's needs are to be satisfied. There are usually four main steps required to a syndicated loan:

1. Discussion between the client and the lead bank, who will need to agree on key issues such as: the number of banks within the underwriting group; ratio of relationship to and non-relationship banks; conditions attached to the loan; launch date.
2. The lead bank to produce a prospectus and prepare the loan syndication.
3. Negotiate the details of the loan agreement.
4. Finalise the loan.

9.5.4 Loan documentation

The loan conditions, roles and duties of the parties involved are detailed in the 'loan agreement', which is the contractual basis of the loan. Lenders and borrowers have become more experienced and the complexity of the legal documents has increased. Over the past forty years, loan agreements have increased in size from only six or seven pages to between forty and ninety pages today. The loan agreement must state the choice of law and the legal jurisdiction in order to avoid later uncertainty if a dispute arises at a later stage. The lead manager, in determining an acceptable interest rate for the loan, has the responsibility to the borrower to ensure that the loan is competitive, whilst also ensuring that the loan is attractive to potential lenders. The cost of a loan includes the fees and the spread/margin. The lead manager receives a single management fee for arranging the loan; the agent receives an annual fee for administering the loan.

Participating banks will charge a fee to cover parts of the loan not drawn down. The spread/margin tends to be based on the interbank rate for the currency of the loan and is quoted as a percentage over LIBOR, US prime rate, or SIBOR (Singapore Interbank Offered Rate).

9.6 Global project finance

9.6.1 Advantages of project finance

Although there are many types of financing options available, most lenders require some form of security should the borrower default on the loan. When the borrower is unable or unwilling to provide recourse through traditional corporate assets, project financing provides a mechanism whereby the lender has recourse to the project assets and the loan's security is based wholly or primarily on the project's cash flow, which is used to service the loan repayment. Early examples of project financing include global construction projects such as the Panama Canal, and the railroads and oilfields in the US and UK, all being very large capital-intensive projects with potential good payback periods. Project finance has become the preferred approach for financing most large infrastructure, energy and public service projects. A special-purpose vehicle (SPV) is usually set up to undertake the project by entering into contracts with key stakeholders who can provide the required construction, operation and other services. The terms 'non-recourse finance' and 'off balance sheet finance' are often used interchangeably with the term 'project finance', although this can be misleading as these terms represent different levels of recourse back to the project sponsors. Off balance sheet finance means that the finance is not provided through loan secured against the parent company's balance sheet assets, whereas in non-recourse finance there are no loan repayment guarantees provided by the parent company.

Project finance was defined by the US Financial Standard FAS as follows: 'The financing of major capital projects in which the lender looks principally to the cash flows and earnings of the project as the source of funds for repayment and to the assets of the project as collateral for the loan. The general credit of the project entity is usually not a significant factor, either the entity is a corporation without other assets or because the financing is without direct recourse to the owner(s) of the entity.' Owing to the complex nature of most projects, it is rarely possible to achieve pure project finance, and lenders must check that the project will generate sufficient cash flow to service the debt and pay for other liabilities (for example, running costs, taxes). The lender will thus need to thoroughly evaluate the project's operating viability in considerable detail, rather than the activities and assets of the parent company. The principal advantages of project finance are that they enable the sponsor to:

- undertake a project that would have otherwise been impossible and to benefit from a large capital asset without having to finance it;
- minimise their equity contribution by using the project finance secured loan to attract additional investors;
- keep the project loan off the parent company's balance sheets, thus providing a higher level of gearing and maintaining a high credit rating;

- shift some of the risk away from the parent company on to the project cash flows by limited recourse to the parent company's assets;
- help the sponsor to raise loans on better terms than would have been achieved through direct borrowing;
- secure large amounts of debt for capital-intensive projects.

9.6.2   Disadvantages of project finance

The principal disadvantages with project finance are: the high set-up costs usually associated with establishing the project finance structure can generally be only justified for large projects, usually valued at over US$20 million; the additional project-specific risks that both lenders and investors must pay close attention to and agree how best to manage these. This is more complex than conventional loans, where the lender is primarily reliant on the overall creditworthiness of the borrower. In addition to the risks associated with the construction phase of the project, the project-specific risks have to take account of the whole life of the project in relation to operating revenue and costs. However, this should involve some of the risk moving from the tradition client to a consortia or special-purpose company who would be better placed to manage this risk.

9.6.3   Build–operate–transfer

There has been a substantial drop in construction activity over the last five years. To compete in the current climate, contractors cannot afford to be too choosy, and are often willing to pay high fees to banks capable of producing a competitive financial package. Contractors on many global construction projects have had to take a greater proportion of the risk involved in the form of build–operate–transfer (BOT) projects, which as a means of project financing is a relatively new concept and falls under the general heading of concession financing. BOT involves the creation of a company to handle not only the construction of a project, but also to operate the new facility for a predefined time period. The arrival of BOT projects stems from governments wishing to avoid the risks associated with the financing of major construction projects. The BOT model is often selected because it prevents losses and reduces the danger of accumulating long-term debts. BOT has also been promoted as a way for lesser-developed countries to build infrastructure projects without having to finance the scheme from public funds. BOT is often the temporary privatisation of public construction projects. The scheme is built and operated by a private developer, and finally transferred back to the government after being operated over a substantial time period. The developer is given a certain number of years of positive revenue to pay back any investment before the government takes over. Most time periods are over 20 years, however, Eurotunnel was granted a 55-year (later increased to 60-year) concession on the Channel Tunnel project before it reverts back to the French and UK governments. The important steps that are required to safeguard investors' money are:

1. identify and justify the need for the project;
2. consider various options and select the most appropriate;
3. assess if the project is economically viable based on estimates of project costs and income;

4. negotiate and agree the concession period;
5. identify key risks and select the most approximate risk-offsetting mechanisms;
6. negotiate and agree safeguards (i.e. the government could make up any toll revenue shortfalls);
7. form a project company or special-purpose vehicle and agree equity contributions from members;
8. negotiate a debt package with a bank and bank syndicate;
9. appoint an underwriting group for local and international equity investors;
10. sell equity in the project as required during the construction period.

Most BOT projects will involve an initial element of equity used to establish a consortium comprising the contractor, the operator, sponsors, banks and the buyer project's products or services (i.e. a special-purpose vehicle. The equity offers an incentive to the contractor to complete the project on time and within budget, demonstrates the sponsoring consortium's belief in the economic viability of the project, encourages lenders to provide additional finance, and provides some security in case the project is not as economically viable as initially expected.

9.7 Indirect guarantees

9.7.1 Single-purpose entities and guarantees

The owners of many construction projects are special-purpose vehicles (SPVs) or single-asset entities that have no significant assets other than the project. SPVs protect the sponsors' parent companies from project-related liability but they also protect the project should the owner run into financial difficulties outside of the project. The SPV can take the form of general partnerships, limited partnerships, limited liability companies and trusts, depending on liability concerns and tax considerations. Due to the limited liability of many SPVs, lenders often require some form of personal *guarantees* from the owners or principals of the entity, not only to provide an additional source of loan repayment should the project not meet its expectations, but also to provide access to the required technical, financial and management skills.

9.7.2 Take-or-pay and throughput contracts

Under project finance, where the loan is secured against a project's future cash flows, it is usual to seek indirect guarantees that provide some reassurance that the future cash flows will be generated. However, 'take-or-pay' or 'throughput contracts' are increasingly being used as indirect guarantees to ensure steady cash flow of funds once construction has been completed. These types of arrangements help to assure lenders of the project's viability and should be disclosed in the company's consolidated balance sheets. They can take the form of: a take-or-pay contract in which suppliers agree a fixed price for raw material, or customers agree to take an agreed quantity of goods at a fixed price over a predefined time period: or a throughput contract in which, for example, sponsors agree to use or pay for the services offered by a processing facility, which may include an oil refinery, gas pipeline or a transport facility. They will often

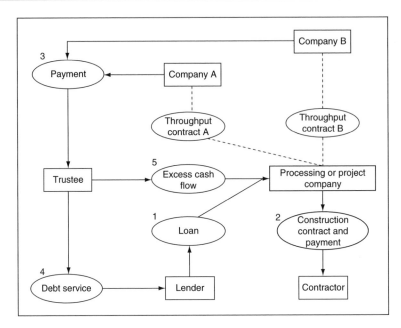

Figure 9.1 Throughput contract cash flow

receive a discount for the goods or service provided, but should they not wish to take the goods or service a charge will still need to be paid. An example of an indirect guarantee, in the form of a throughput contract used to secure a loan to fund the construction phase for a processing plant, is illustrated in Figure 9.1.

There will also be a construction contract between the processing company and the contractor who builds the facility. The processing company will usually be required to sign a loan agreement and assign the throughput contract to a trustee who will receive payments due from Company A and Company B. The trustee will first use these payments to service the debt and the remaining funds will be paid to the processing company.

9.7.3 Owner/sponsor guarantees

The owner, in the form of a parent company of the completed project, is perhaps the most obvious guarantor for the project funding. However, the owner may not have sufficient resources or may not wish the liability to be on the company's balance sheets. If a subsidiary is established to run a project it will have no established track record and probably be undercapitalised, thus resulting in a poor credit rating. Project lenders thus require additional guarantees from more creditworthy sources, often involving the parent company. This type of debt guarantee will usually appear on the consolidated balance sheet of the owner or sponsor as a liability. However, there are other forms of direct and indirect undertakings that, if structured correctly, the sponsor could treat as off-balance sheet liabilities. Organisations who need the facility to be built but do not wish to have an equity share in it can act as indirect guarantors.

9.8 Global countertrade

9.8.1 Countertrade

Nations have traded silk, spices, cloth and animals of all kinds since early times. Nowadays, many nations trade defence equipment, raw materials and food. Although the products may have changed, the basic concept is still the same. Countertrade is usually referred to as a reciprocal and compensatory trade agreement between nations involving the exchange of goods and/or services for different types of goods and/or services. Although no money is exchanged, monetary evaluations are usually needed for purposes of taxation, accounts, sales tax and import/export duties. Countertrade has emerged as a sophisticated marketing and financing device to: fund global construction projects; access new markets; maintain a balance of payments; transfer technology; facilitate foreign investment. Most countertrade arrangements are: based on a single contract between the direct parties; completed in very short periods of time in order to reduce the impact of future price fluctuations; made on a government-to-government basis; supported by bilateral clearing agreements. Most countertrade deals are used to maintain a country's balance of payments and avoid transferring foreign exchange reserves with one another. Where there is a trade imbalance, the accounts can be settled annually. Most transactions nowadays are done either in a real or electronic form of money, hence there appears to be little place for true barter. However, very innovative approaches have emerged and at least 80 countries still regularly use countertrade. There is considerable disagreement as to how much of world trade involves countertrade, with estimates varying between 5 and 30 per cent. However, based on expert opinion, Okaroafo[5] estimated countertrade accounted for 20 to 25 per cent of world trade.

9.8.2 Forms of reciprocal trade

Countertrade encompasses all forms of reciprocal trade, the main variants of which are discussed below:

- **Barter** is probably the simplest form of countertrade and involves a one-off direct and simultaneous exchange of services and/or goods without the exchange of money. This is a very narrow definition as money is usually involved at some stage. To reduce any impact of interest-rate fluctuations and changes in market demand and supply, the exchange process needs to be completed within a short period of time. In some cases, the supply of the principal goods is delayed until sufficient revenue has been generated from the sale of the bartered goods. In practice, there may be a time lag between the exchange of goods or services. There are two different approaches that can be taken if long delays occur: adjust the ratio of goods exchanged to take into account inflation, or purchase goods as they become available at pre-agreed prices.
- **Counterpurchase** tends to last for one to five years and is favoured by less developed countries. It involves an importing country or company purchasing goods and/or services from a foreign country or company that promises to make a future purchase of specific products from the initial purchaser. The

initial exporter will either buy the goods/services or guarantee to find a buyer for the initial importer of its own goods.

- **Compensation** (or **buyback**) occurs when a company builds a facility or supplies technology, equipment, training or other services to the country and agrees to take a certain percentage of the exported products' output as full or partial payment for the goods initially provided. When the repayment is in the form of related goods, the process is referred to as 'buyback'. Compensation deals are more frequently being used to finance large turnkey factories. The contract value tends to be higher than other forms of countertrade (excluding offsets) and cover periods can vary between five and twenty years.

- **Offset** involves one country agreeing to purchase a hard currency product from a foreign company that agrees to purchase raw materials or components from the buyer of the finished product, or the assembly of such product in the buyer country. Offset arrangements usually have similar characteristics, often associated with counterpurchase and compensation. They also tend to involve the sale of high-tech products, such as defence equipment, from multinational companies based in one of the more industrialised countries and a foreign government. Some deals will need the home government's approval to export certain high-tech information. There are several Offset variants such as: technology transfer; co-production; joint ventures or subsidiary arrangements; licensed production; subcontract production.

9.8.3 Switch trading

Switch trading or countertrade is used where there is a mismatch between the needs for goods and/or services between the parties of a countertrade agreement. It is a process that involves one party selling purchase options at a discounted price (sometimes as high as 40 per cent depending on demand) to a third party who then sells the goods and/or service to other parties. Some companies specialise in switch trading in order to generate a market where several countertrade contracts can be indirectly linked, thus providing a better match between buyers' needs and the goods available. Switch trading effectively opens up countertrade to companies that can only act as either a supplier or receiver of goods.

According to an official US statement: 'The U.S. Government generally views countertrade, including barter, as contrary to an open, free trading system and, in the long run, not in the interest of the U.S. business community. However, as a matter of policy the U.S. Government will not oppose U.S. companies' participation in countertrade arrangements unless such action could have a negative impact on national security.' (Office of Management and Budget; 'Impact of Offsets in Defence-related Exports', December 1985[4]).

Although countertrade has its advantages, there are many difficulties encountered when using this route to finance construction projects as: only very large organisations can fully exploit countertrade; they are extremely difficult to arrange and can take a long time to finalise; definite starting dates are needed for construction projects; countertrade places so much responsibility on third-party organisations that establishing a deal can be too time consuming. It is usually best to avoid countertrade unless there are clear market advantages, or conventional financing methods are not available. However, countertrade is here to stay and many construction organisations have no choice but to get involved.

9.9 Development bank project finance procedures

9.9.1 Development projects

A development project was defined by Baum and Tolbert as:[5] 'a discrete package of investments, policy measures, and institutional and other actions designed to achieve a specific objective (or set of objectives) within a designated period'. There are many international development agencies that provide extensive funding. However, external donors set limits and conditions on the funds made available. Although intended to be of benefit to the development of some of the world's poorest countries, project costs can often be higher than expected and some of the anticipated benefits may not be fully realised, thus resulting in the donor's limit being exceeded and placing considerable debt repayment burden on the local economy.

9.9.2 Development banks

Development banks can be defined in a variety of ways and in practice have little in common apart from the basic concept. There are many types of ownership, objectives, sources of finance, policies and degree of government involvement that can be adopted. A development bank is considered by the United Nations to be 'an institution concerned primarily with long-term loan capital'. However, this vague definition can be applied to most banks. In practice, the main objective of a development bank must be to speed up the development process. In most cases, these institutions have been established by various governments mainly to encourage growth within the private sector of their own economy. International development banks can be characterised on a geographical basis: national; regional (e.g. African Development Bank (AfDB), Asian Development Bank (AsDB); or universalist (World Bank)). The ownership and membership of regional development banks is not restricted to the regional states, hence entities other than states can also be included. The objectives of most regional development banks are generally restricted to economic development and integration. Some, such as the AfDB, are also concerned with the encouragement of social progress. In order to achieve these objectives, sections are often established within the bank that have one or more of the following functions:

- to facilitate economic construction or reconstruction;
- to help development planning and programming;
- to cooperate with other institutions with similar goals;
- to provide funds for capital investment;
- to encourage and promote both public and private investment from abroad;
- to generate international capital flow.

9.9.3 The World Bank Group

The World Bank Treasury has more than 60 years' experience in financing major projects and mainly comprises: the International Bank for Reconstruction and Development (IBRD), which lends to governments of middle-income and creditworthy low-income countries; the International Development Association (IDA), which provides interest-free loans and grants to governments of the poorest countries. There are three other affiliates to the IBRD, namely: the International Finance Corporation (IFC), which provides loans, equity and technical assistance to the private sector in developing countries; the Multilateral Investment Guarantee Agency (MIGA), which guarantees against losses caused by non-commercial risks to investors in developing countries; the International Centre for Settlement of Investment Disputes (ICSID), which provides conciliation and arbitration facilities for investment disputes.

9.9.4 The World Bank project cycle

As with most other development banks, the World Bank lends money to low- and middle-income countries in order to support development and change. The final financial package used for a project is often unique to that project, as each project and executing agency has their individual needs that have to be satisfied, consequently several routes will have to be explored involving considerable backtracking and iteration before the final package is arrived at. The procedures and policies of the funding agencies change with time, but most of the development banks are mandated or have a charter that governs their actions. In order to ensure that funds invested in development projects reach their intended target, borrowing countries have to follow certain rules and procedures. To assist in the assessment and monitoring of investment and project performance, projects are usually broken down into various stages, often referred to as the 'project cycle'. The term project cycle highlights the close relationship between the various stages of a project, and that the final stage should provide a feedback basis for planning future projects. There are many different approaches to defining a project life cycle. However, the approach taken by the World Bank has been fairly constant over the past 30 years and comprises the following:

1. **Country strategy and identification:** the first phase of the cycle involves the World Bank, the borrowing country's government and other stakeholders to identify ideas that appear to represent a high-priority use of the country's resources to achieve an important development objective. After considerable fact finding and analytical work, strategies and priorities for reducing poverty and improving living standards will emerge. Any emerging project needs to meet an initial test of feasibility and provide assurance that technical and institutional solutions can be achieved at costs commensurate with the expected benefits. Important documents to be produced include: a project concept note, which presents objectives, imminent risks, alternative scenarios and timetable; the project information document; the integrated safeguards data sheet, which identifies key issues.

2. **Project preparation:** once a project idea has been identified, further assessment and evaluation is needed before a firm decision can be made whether or not to proceed with it. This requires a considerable period of consultation with key stakeholders, and progressive refinement of the project definition,

taking due consideration of the technical, economic, financial, social and institutional issues. It can take several years to complete the required feasibility studies and prepare the required engineering and technical designs, which are only a few of the documents required. An Environmental Assessment Report and an Indigenous Peoples Plan need to be developed during this phase for integration within the design of the project and used as part of an early screen process to determine if the proposed project has environmental or social impacts as highlighted in the World Bank's Safeguard Policies.

3. **Project appraisal and approval:** this phase provides potential stakeholders with opportunities to review the project design and resolve any outstanding questions. The World Bank works with the government to review the identification and preparation phase and confirm the expected project outcomes, intended beneficiaries and evaluation tools for monitoring progress. Before approving a loan, external agencies normally require a formal process of explicit appraisal that assesses the overall soundness of a project and its readiness for implementation. The project appraisal document (for investment lending) and the programme document (for development policy lending) are prepared and presented to the bank's board of executive directors for consideration and approval. The project can be sanctioned once funding has been approved and conditions for effectiveness satisfied.

4. **Implementation:** this phase includes: the development of the project design details, specifications, procurement of goods, works and services; the actual development or construction of the project, up to the point at which it becomes fully operational; extensive monitoring of the project's progress, outcomes and impact on beneficiaries; evaluating the ultimate effectiveness of the operation and the project in terms of results.

5. **Project completion:** a project is completed and closed at the end of the loan period, which can be between one and ten years. A World Bank operations team compiles an implementation completion and results report, which evaluates final project outcomes based on a documentation of the results, problems encountered, lessons learnt and knowledge gained.

6. **Post-evaluation:** the post-evaluation of about one in four completed projects is performed by the World Bank's Independent Evaluation Group and seeks to determine whether the original objectives were achieved and to draw lessons from the project that can be applied to similar projects in the future. The post-evaluation is the last part of the project cycle, and if the project has run for some period, the evaluation can be based upon actual project costs, operating costs and early benefits. The evaluation should also determine whether or not the project has been worthwhile and also indicate if:

- the original objectives were properly defined and achievable;
- the appropriate technical choices were made;
- the appropriate procurement procedures were adopted;
- the project's target group was correct;
- the local conditions were correctly identified;
- there were excessive cost overruns;
- project institutions were strengthened.

9.10 Financial risk assessment and management

9.10.1 Certainty, risk and uncertainty

All investment decisions involve an element of predicting the future and it is possible to classify them into certainty, risk and uncertainty. However, there will always be some form of overlap between the types of decision. Lenders and investors will assess project risk and work together with sponsors to develop a mitigation strategy. Certainty regarding an investment refers to where nothing short of a major international disaster would impact on the predicted outcome, but it is not possible to be 100 per cent certain about an investment. Investment into government bonds that provide a fixed rate of return is a good example of an investment decision made under the conditions of certainty. However, recent events have demonstrated that there will always be doubt regarding the security of an investment. The conditions of certainty usually only provide modest returns. There is a general consensus that there is a direct relationship between risk and reward: that more risk should yield a higher return. There is an element of risk associated with all commercial transactions. The amount of risk that is acceptable will relate to the anticipated profits. An increase in the uncertainty around the investment is generally reflected by an increase in the level of expected reward. Investment decisions made under the conditions of risk (i.e. where historical data exists) can be made on a rational or intuitive basis involving assessing the probability of different outcomes occurring. A good example of an investment decision made under the conditions of risk predication is the estimation of a project's duration where previous similar projects are available to provide a good indication of project duration, taking into account major delays that may occur. A risk assessment of an investment decision can only be performed if there has been sufficient similar investments previously made to enable a trend to be established.

Most global construction projects tend to be unique in that little or no historical data exists. Consequently, investment decisions are being made under the conditions of uncertainty. Indeed most innovative products or processes have no historical performance data to draw upon, and resulting investments are made under the conditions of uncertainty. Although the probability of failure is high, the potential rewards associated with successful outcomes can also be very high. Such decisions should also involve some assessment of the probability of future events: due to a lack of historical data this will involve the development of a range of scenarios to predict future trends and markets. Companies that develop and introduce new products are often referred to as market leaders and will operate under the conditions of uncertainty – and are often unable to draw on historical data relating to their products. Changes in government can create a new environment and context where previous historical information is no longer relevant.

9.10.2 Risk management

This section discusses the potential effects of financial and business strategy risks on global construction projects. The identification of potential risk is the first and most important step towards managing such a project. Effective risk identification is essential and must be ongoing throughout the project if a risk is to be discovered and managed before it develops into a loss. Risk management is a five-step decision process comprising:

- risk identification;
- risk assessment;
- evaluation of risk alternatives;
- selection of an alternative or risk control measure;
- ongoing monitoring of the risk management programme.

9.10.3 Risk identification: project-related risks

Construction contracting is a high-risk business due to the numerous factors that influence it but remain beyond contractors' control, such as the weather, economic climate and changes in regulations or specifications. The main types of project risk are grouped and summarised below. This provides a starting point for interested parties – such as banks, clients and contractors – to develop a check to support risk identification during key investment decision-making processes leading to the development of effective risk analysis methods suited to that particular line of business. Global construction projects need to be closely monitored during the construction phase and supported by performance bonds or guarantees. Most of the financial risks manifest themselves during the construction phase and this is the greatest period of risk exposure for all parties involved with the project because once construction has started, the financial expenditure and commitments rapidly increase.

Planning stage

- **Feasibility risk:** where the project evaluation demonstrates that the project is not viable due to financial, technical, environmental, social or political reasons. Although not necessarily costly, termination at this stage could be politically embarrassing and lead to an important development need not being met.
- **Permit/licence risk:** where planning permission or access permits are refused. This does not always result in the project not going ahead but substantial changes may be required, thus resulting in considerable delays and additional costs.

Construction phase

- **Estimate risk:** where the contractor fails to allow enough margin for overheads and profits.
- **Reservoir risk:** on very large projects it is important to ensure that there are sufficient local raw materials and labour that will need to be consumed during construction. Access and transportation are also important considerations to be taken into account, as material or labour that has to be brought in or transported from outside the local area tend to rapidly escalate project costs. High local demand can also push up costs during the construction period.
- **Completion risk:** although the risk of schedule and cost growth, and delayed completion, need to be minimised on all construction projects, it is especially important where project finance is used as the loans have been secured against future income generation, which usually does not start until the construction phase has been completed. Schedule growth is associated with cost growth, hence there are also increased borrowing requirements, with

potential penalties in the form of increased interest rates when combined with a delay in income generation and loan repayment.

Operating phase

- **Technological risk:** where new and innovative technologies are introduced there will always be an increased risk on technological failure – thus close monitoring and evaluation is needed.
- **Financial and currency risk:** flaws in the financial model may come to light during the operational period, especially where the currency of payment differs from the loan currency.
- **Political, legal and regulatory risk:** whereby there is risk of civil disorder and outright expropriations without compensation.
- **Market price risk:** predicting the correct price to charge for goods or services is one of the major risks associated with project finance: if prices are too high, the generated volumes may be insufficient to make a profit; if the prices are too low, a high volume may be generated but the overall income may not be sufficient to service the debt and provide a decent return on equity.
- **Operating risk:** predicting the running and maintenance costs is not easy, especially where project finance is used. Global construction projects tend to be on–off, so there is little historical data available upon which to base future predictions over the long time frame usually associated with these types of projects.
- **Decommissioning risk:** although it may be possible to sell off the asset at the end of the project life, there could well be a need to refurbish and hand back the project in an agreed state, or to decommission the project and reinstate the land back to its original form. Prediction of the associated costs (or income) at the end of a project with a long duration is also highly problematic.
- **Counterparty risk:** where competitors provide a better alternative that is more attractive to potential customers.

Lenders tend to find construction contractors to be high risk-takers who are difficult to evaluate in terms of creditworthiness. There are two approaches that lenders can take to add additional security to recovering a loan: the first is to tie the lay repayments through project income that flows directly to the lender, with surplus money going to the contractor; alternatively through recourse to valuable (saleable) assets should the contractor default on agreed loan repayments. An asset's value can be determined either in terms of its market price or alternatively its potential earning capability; these two values may differ considerably and may influence the course of action that the lender takes to recoup its loan. Prior to lending money, banks need to evaluate the risks involved, which requires a good understanding of the determinants of contractor success or failure. A good starting point is outlined in the following two sections, in which ten factors that contribute to the risk of contractor failure (Schleifer[5]) are categorised as relating to business strategy or accounting considerations.

9.10.4 Risk identification: business strategy

Schleifer[6] identified five determinants of failure relating to business strategies or practical considerations. However, most of these factors cannot be avoided as they form an important part of company growth and strategic develop-ment. Contractors should not fear growth, expansion into unfamiliar locations or the adoption of new construction methods, but must recognise that poor decisions relating to these factors have resulted in the failure of many con-tractors. Although not always straightforward, this may take some considerable time and be expensive to rectify; the main way to reducing the risks associ-ated with these factors is to ensure that the quality and suitability of project staff is improved, either through a rigorous training programme or new key appointments, which may even be insisted upon by the lending bank. Global construction organisations should consider the following factors when making decisions:

- **Increase in project size**: as with many companies in other sectors, rapid expansion and a scaling up of a contractor's operations is perhaps the main cause of failure. Very often the existing business model is not scalable and moving to a higher level of operation requires new skills and competencies that are not apparent until it is too late. Lenders and borrowers need to determine the size and capacity of the contractor, especially during a period of expansion. A common risk is that contractors often reduce their margins to increase the volume of work and maximise revenue, thereby running the risk of making no profit or perhaps operating at a loss. A significant and rapid increase in project size is the most common cause of failure. During profitable years, contractors often take on larger projects, but problems start to develop before the first of the larger projects is completed. Many contractors need to take on larger projects as part of their planned growth. However, such growth needs to be gradual and in relation to the company's capabilities and mean size of its current projects. Also, some contractors may be profitable when executing projects of a certain size, but this does not mean that it will also be able to make a profit on significantly larger projects.
- **Unfamiliarity with new geographic areas:** moving to new locations where a contractor does not normally work is also a common action that precedes project or even company failure. Although there may be sound business rea-sons for a contractor to expand into new geographic areas – such as growth, lack of work in its primary area or more opportunities in the new areas – there are considerable risks involved due to the significant change in context that may not be initially apparent, such as potential changes in legislation, condi-tions of contract and labour relations, in addition to the increased complexity of the logical requirements.
- **Moving into new types of construction:** contractors may change the type or introduce new types of projects or construction methods. This may be the result of changes in: business strategy including mergers and acquisitions; legislation and regulations; safety strategy leading to adoption of increased off site; the types of equipment available through a process of innovation. Changing from one type of civil-engineering work to another, or from build-ing housing estates to high-rise buildings, involves many technical risks and can result in extensive addition costs associated with training and the pur-chase of new equipment. There will also be a considerable learning period, during which the organisation adjusts to the new type of work. This is costly and many new types of project will be undertaken at a loss before profitable

projects start to emerge. This often proves to be considerably more expensive than expected, thus leading to some companies going bankrupt in the process. Most contractors tend to be more specialised than they like to admit: they often bid for different types of project but their track record results in most of the awarded projects being of a similar type.

- **Replacing key personnel:** to succeed, contractors have three core functions that have to be well managed with a high-calibre senior manager responsible for each, these are: planning and estimating; contact administration and accounting; construction operations. The loss of senior managers in these areas will put any profit-making construction company at risk, especially during times of rapid expansion in project size or geographical location.
- **Lack of management maturity:** lack of management maturity needs to be carefully analysed as it is the most comment factor relating to contractor failure and can contribute to some of the other determinants of contractor failure at both project and company level. Successful project completion, on time and within budget, is usually a result of having skilled and mature project managers. Company organisational changes involving new process, technologies or cultures are a necessary part of growth periods. However, knowing when and how to make such changes needs skilled and mature managers.

9.10.5 Risk identification: management or accounting

The second group of risk identification includes five determinants of contractor success that relate to the management or accounting processes adopted. These tend to be unacceptable situations that can be easily rectified if the situation is identified early enough. The adopted management and accounting processes can easily be benchmarked against other organisations thus leading to the early identification and rectification of any deficiencies. There are many inherent dangers associated with these factors, hence thorough planning and a good understanding of the risks involved are needed in order to mitigate their potential impacts, as elaborated below:

- **Poor use of accounting systems:** given the very large sums of other people's (e.g. subcontractors and material suppliers) money that passes through contractors' accounts over very short periods of time, very effective accounting systems and procedures are needed to capture information and process it very quickly and efficiently. Also contractors can often be at risk to cash-flow problems, for example, during the early stages of construction where the initial on-site costs have been very high but payment for which has been spread over the whole construction period. Contractors' with weak invoicing and debt collection procedures are at increased exposure especially in construction, where unlike other industries, full payment for completed construction work is not made until after a pre-agreed retention period has expired.
- **Failure to evaluate project profitability:** many contractors work on very low profit margins but with a relatively high level of turnover. However, there are many factors that make it hard to determine a project's profitability until some time after completion. Very few projects are tendered for, awarded and completed within a single financial year, which would expedient verification project profitability. Uncompleted and ongoing construction

projects are not easy to accurately measure as large proportions of many activities are partially complete. Most parties involved in the process of determining percentage completion benefit from an overestimation and tend to err on the higher side, thus leading to an overestimation of a project's early profitability.

- **Lack of equipment control:** although the costs associated with hired equipment are additional costs and easily accounted for, the costs associated with company-owned equipment is not so easily accounted for as equipment tends to be used on several projects and may even already have been paid for. However, all construction projects will incur costs for equipment, irrespective of it being company-owned or hired, and company-owned equipment is best accounted for by applying a hire rate for all equipment on site. To do otherwise could result in a situation where all projects report a profit, but the company reports a loss. How idle equipment on a project is accounted for also affects project profitability, as costs are often being incurred whether the equipment is being used or not.
- **Poor billing procedure:** the regular monthly progress payments that are commonplace throughout the construction sector make it very easy for businesses to start up with a very small amount of capital. However, such low-capitalised contractors with inefficient valuation and collection procedures are at considerable risk, given the terms of payments in construction contracting, which compel contractors to work to tight budgets, and given that the late approval and collection of a large progress payment could jeopardise the contractor's planned cash flows.
- **Transition to or problems with computerised accounting:** it is essential to keep accurate and timely records of the numerous transactions and variation orders that occur on each contract. The need to convert to new computerised accounting systems increases as projects become more complex and the company grows. The process of moving to a new IT system often presents many problems, and lenders need to consider the probability associated with this transition taking place in the near future.

9.10.6 Evaluation of risk

Risk evaluation has to take into account many variables and a systematic approach has to be applied, wherever possible based on reliable current and historical information. Following a period of risk identification, the identified risks need to be reasonably evaluated with due diligence. A sensitivity analysis should be used to determine the significance of each risk in relation to the extent to which resulting losses affect the project. Risk evaluation usually involves careful consideration of three characteristics:

- how often the losses will happen;
- how severe they will be if they happen;
- the ability to foresee their happening.

In addition to project-related risks, business strategy risk, and management and accounting risks, lending banks also need to take into account more specific factors, such as:

- the bank's extent of unsecured exposure and recourse to assets;
- the company's audited balance sheet, and profit and loss account;
- the company's debt–equity ratio;
- the project and company management structure and the board of directors;
- project and company insurance including comprehensive insurance of assets at all times;
- project planning, monitoring and control processes;
- the project's primary and secondary sources of income;
- the project's condition of contract;
- the introduction of covenants.

9.10.7 The use of bonds to offset risk

There are many different types of bonds that can be used to provide some assurance or compensation in relation to a contractor's performance or project completion as specified in the construction contract. A performance bond tends to be an agreement between three parties: the client, the contractor and the surety who issues the bond. The surety effectively 'bonds' itself to the contractor, and guarantees the contractor's performance to the client. If for some reason the contractor does not perform (e.g. fails to complete the contract on time), the surety steps in and pays the client a sum of money equal to the loss incurred, usually in the region of 10 per cent of the contract price. However, the client needs to decide if the size of the bond is sufficient. Performance bonds are not the same as insurance, they are completely different, for a variety of reasons. Insurance tends to be an agreement between two parties (the insured and the insurer); however, bonds tend to involve three parties (the contractor, the client and the surety) and offer no security to the contractor. With performance bonds, the surety does not totally release the contractor from all responsibilities, the surety is there to provide assurance to the client that the work will be completed. If the surety has to pay compensation to the client, the surety can reclaim the money from the contractor through the courts or by some other means. Performance bonds include the following:

- **Bid or tender bonds** assure that the contractor's tender is a serious bid. If the bid is accepted but the contractor does not go through with it, any costs incurred whilst the client finds a new contractor will be paid by the surety.
- **Advanced payment bonds** enable contractors to purchase goods and equipment before construction has started, with the bond protecting the client against losses resulting from the contractor's early default.
- **Retention bonds** may be issued in lieu of money retained, in order to ensure that the contractor rectifies any mistakes that manifest themselves during the retention period.
- **Maintenance bonds** can be used to fund the correction of construction defaults discovered once construction has been completed. Performance and retention bonds are often converted into maintenance bonds once the construction phases has finished.

9.10.8 The use of guarantees to offset risk

Guarantees have an important role to play within project-related finance, as they enable specific risks to be shifted to interested parties best placed to manage the risks. The guarantees may be direct in the form of guaranteeing loan repayments, or indirect where the guarantee is to supply or purchase goods or services at an agreed rate, thereby providing a degree of certainty to future cash flows. There are four main types of project-related guarantees:

1. **Supplier's guarantees** are used where a supplier is sufficiently motivated to guarantee that if a manufacturing facility is constructed, material will be supplied at an agreed rate and price in return for a share of the new market.
2. **User's guarantees** are used where a user of a project's products is sufficiently motivated to guarantee that if a manufacturing (or process) facility is constructed, the user will purchase goods (or services) at an agreed rate and price in return for guaranteed volumes, usually at a discounted price.
3. **Contractor's guarantees** mainly relate to the previous discussed bonds, such as bid bonds enclosed with tenders, performance guarantees, advance payments and retention bonds.
4. **Government guarantees** may be provided by government agencies that wish to have some control on the project but not own or finance it.

9.10.9 Insurance to offset risk

With the best risk management programme, insurance cover provides the cornerstone to financial protection. However, insurance does not eliminate all the risks involved, but transfers most of them to a professional risk bearer. Insurance provides contractual security against an anticipated liability, whereby the insurer agrees to indemnify the insured for a stipulated premium. There are many different policies, some are mandatory and others are optional. The main types of contractor insurance policies include: Comprehensive General Liability Insurance; Employers' Liability; Builder's Risk Insurance; Contractor's Plant Coverage and Property Insurance.

9.10.10 Modelling and monitoring project cost

To demonstrate due diligence, investors need to assess a project's financial viability, the ability of the management team to undertake the project and the technology involved. They also need to monitor the project's post-financing performance. The financial assessment process usually involves:

- developing the project's financial model;
- determining and analysing the key financial indicators;
- performing a sensitivity analysis;
- performing a risk assessment and mitigation strategy.

A project's financial model is thus an important tool to be used within the financial assessment process and will typically include (see Figure 9.2) the following:

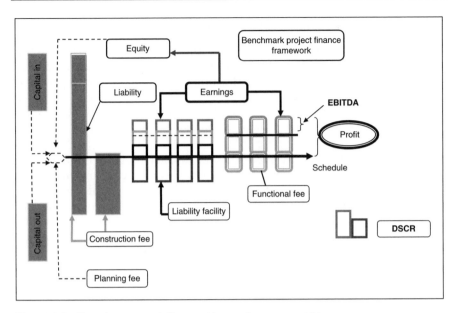

Figure 9.2 Typical project cash flows and key performance variables

- The main assumptions made regarding the model's input variables and the basis for the assumption.
- The analysis based on a number of scenarios (e.g. best, worst and most likely case).
- The outputs, which will include a cash-flow statement, profit and loss account, balance sheet and key financial indicators, such as:
  - **Earnings before interest, tax, depreciation and amortisation (EBITDA)**: project revenue minus operating costs;
  - **Interest cover ratio:** EBITDA divided by interest payments;
  - **Debt service cover ratio (DSCR):** the ratio of EBITDA to all debt servicing requirements (CDM[7]).

9.11 Case study: Channel Tunnel

One of the best examples of a BOT with concession arrangement is the Channel Tunnel, which was funded on a limited recourse basis with contractors being paid and the debt serviced by the cash flow generated by the completed project. The Channel Tunnel is a 50 kilometre undersea rail tunnel between Folkestone, Kent, in the United Kingdom and Coquelles in northern France. The tunnel is used by high-speed Eurostar passenger trains, Eurotunnel Shuttle roll-on/roll-off vehicle transport and international rail freight trains. The project is an entirely private sector affair, with the French and British governments both stating that they will not help if the project gets into trouble. However, the British and French governments retained control over the main engineering and safety decisions. The Channel Tunnel Group (TML) designed and built the tunnel with financing through a separate legal entity – Eurotunnel, which absorbed TML. A throughput contract in the form of a Railway Usage Agreement was signed between Eurotunnel, British Rail and the Société Nationale des Chemins de fer Français, thus ensuring future revenue in exchange for half of the tunnel's capacity. Eurotunnel is a private company created for the project and granted a 55-year (increased to 65-year) concession on the tunnel traffic before the British and French governments take over. However, this was only awarded after the winning consortium took substantial equity in the project, thus demonstrating their faith in the economic viability of the project. This equity also created a big incentive to complete the project on time and within budget.

The Eurotunnel venture is very risky because of the nature of the project. Normally, if something goes wrong the project or its assets can be sold off to recoup some of the investment. However, if the tunnel fails, it will be difficult to sell a hole between two countries. The cost of financing the project was initially estimated to be over £6 billion, making it the largest venture ever in the UK. Five billion of this was in the form of a syndicated loan provided by 40 banks, led by National Westminster Bank in the UK and Credit Lyonnais in France. The loan was issued in three tranches, subject to the raising of $1 billion equity. The first tranche of £41 million was provided by the 14 original project promoters and the main lending banks. The second tranche of £200 million was raised in October 1986, with £70 million coming from French and British markets. The remaining £760 million was obtained towards the end of 1987. Eurotunnel was financed mostly in francs and sterling, as the eventual revenues will come from French and British currencies. This approach applies to most global construction projects. There was an 80 per cent cost overrun, partly due to enhanced safety, security and environmental demands, with costs being 140 per cent higher than forecast.

9.12 Chapter summary

Whilst many of the key actions listed below apply to most construction projects, when dealing with the size and complexity of global construction projects they become even more critical. The degree of complexity makes it difficult to fully assess risk and maintain effective information systems. The size of financial commitment makes it impossible for one funder to take the risk on its own, so a complex finance mechanism tends to emerge involving many different parties. The reliance on large numbers of organisations makes it important to have a procurement approach that ensures an appropriate sharing of risk and rewards combined with risk offsetting mechanism such as insurance, bonds and guarantees, which have to be determined early on in the project. The key actions are:

- implement robust capital and operating-cost reduction strategies;
- during the feasibility stage to fully evaluate technical, economic, social, environmental and political viability along with early exit strategies;
- cost model and simulate future scenarios and develop adaptability strategies;
- put in place key financial sanction gates to ensure the financial viability of the project is monitored and controlled as the project progresses;

- develop a whole-life model of the project to determine the financial viability of the investment and perform a sensitivity analysis to determine the risks that need to be managed;
- adopt the most efficient and effective combination of corporate and/or project finance;
- put in place effective systems to provide accurate and up-to-date financial information along with project finance mechanisms to monitor and manage risk, thus helping to reduce the cost of borrowing;
- adopt appropriate relationship contracting and effective governance approaches that ensure a sharing of the risks and rewards and the commitment of an expert project management team;
- ensure there are effective business strategy and accounting procedures in place;
- use bonds, guarantees and insurance to offset the risk where appropriate.

9.13 Discussion questions

1. Compare and contrast bilateral and syndicated loans.
2. Discuss the main types of project finance in relation to global construction projects.
3. Compare and contrast corporate and project finance.
4. Explain the main steps within risk assessment.
5. Discuss risk offsetting mechanisms that are suitable for global construction projects.
6. How can project costs be modelled and monitored?
7. What are the main sources of project finance?
8. What is an SPV?
9. What are the main steps in completing a syndicated loan?
10. Discuss how the various sources of finance could be best combined.

9.14 References

1. Christy, G. A. and Roden, P. F. (1973). *Finance Environment Decisions*. New Delhi: Harper and Row Publishers.
2. Altunbas Y., Gambacorta L. and Marques-Ibanez, D. (2009a). Securitisation and the bank lending channel, *European Economic Review*, **53** (8), pp. 996–1009.
3. Chui, M., Domanski, D., Kugler, P. and Shek, J. (2010). The collapse of international bank finance during the crisis: Evidence from syndicated loan markets, *BIS Quarterly Review*, pp. 39–48.
4. Office of Management and Budget (1985). Impact of Offsets in Defence-related Exports.

5. Baum, W. C. and Tolbert, S. M. (1985). *Investing in Development-Lessons of World Bank Experience*. Washington, DC: Oxford University Press.
6. Schleifer, T. C. (1989). Why some contractors succeed and some don't, concrete construction. http://www.concreteconstruction.net/concrete-articles/why-some-contractors-succeed-and-some-dont.aspx [Accessed January 2012].
7. Capacity Development for CDM (2007). *Guidebook to Financing CDM Projects*. Roskilde: Risø National Laboratory.

10

Factors Affecting Perspectives on Uncertainty and Risk in Global Projects

10.1 Introduction

Global projects are essentially *dynamic* items: they require, have, or produce energy; they are functional (in having planned objectives), and they are in motion (over both time and geography). Indeed, it is arguable that projects can be regarded as living entities (as in being a thing that exists as a particular and discrete unit). In legal terms, most countries regard persons and corporations as being equivalent entities; if a corporation can be classed as equivalent to a living entity, then why not a project also? The argument for recognition of projects as entities does not only relate to the legal perspective and the dynamic nature of projects. It also relates to the recognition that projects are defined as having a life cycle (a key difference from production management). While the simplest life-cycle model consists of only three stages – beginning, middle and end – there are more complex models that can be applied if project planning requires this.

Part of the *dynamic* character of projects comes from the fact that they rely, at least in part, on humans. People of differing levels of experience and expertise are encountered within a project environment and these differing levels result in a degree of *uncertainty*. In such a situation or set of circumstances there is insufficient knowledge or information to accurately describe the current or future state of the project, or even the number of possible outcomes (more a matter of uncertainty than risk) from the project. Because of this requirement for people to be involved in projects, it can be argued that projects exhibit behaviours that reflect the behaviours of those self-same people. Even in the simplest of projects this perspective raises some difficult questions with regard to management generally, but as the complexity of the project increases there tends to be an increasing focus on risk management, and identifying and responding to areas of uncertainty within the project.

When a project grows to become a global construction project, the level of complexity (in terms of interconnections and uncertainty of outcomes) increases as factors such as differing cultures, legal requirements, performance standards and quality of materials add to the demand to manage both risk and uncertainty. In addition, the supply chain demands of a global project will need to be addressed, and factors such as the quality of a country's infrastructure starts to become a consideration. Aspects of infrastructure such as the quality of road and rail links, along with the quality of the electricity supply, can vary considerably between countries, even when they may be in the same region of the world, and can impact positively or negatively on the level of risk and uncertainty within a project. The local population will be well aware of the problems involved in moving resources around, but an incoming project manager may prove guilty of adding to the risk of project failure through simply assuming that the transport infrastructure is of a particular quality. Accurate information is, as always, the key to reducing uncertainty and determining the actual level of risk.

10.2 Learning outcomes

The specific learning outcomes for this chapter are to enable the reader to gain an understanding of:

>> risk and reward;
>> differentiating between uncertainty, risk and hazards in projects;
>> perception of risk and uncertainty;
>> responses to risk – response decisions;
>> uncertainty and success/failure ambiguities;
>> global perspectives on uncertainty;
>> tools and techniques for managing risk/uncertainty.

10.3 Human behaviours

If a piece of equipment either breaks or is not working properly, it can generally be replaced or fixed. People are a little more complex in that they can make choices; they can choose to present themselves as feeling 'satisfaction' when they are actually feeling 'dissatisfaction', and thereby create a situation where even identifying that there is an uncertainty about an individual's actual performance (a 'break') can be difficult. In such situations the 'fix' is usually not a particularly quick one and it may not even be possible, within the constraints of resources, and so on, to attempt a fix at all. Given the social duplicity that people are capable of, it is useful for a project manager to at least have an appreciation of the various behaviours that persons may exhibit.[1] In dealing with behaviours there are two aspects where caution should be exercised:

1. Individual behaviour can be modified by membership of a group (where groups are regarded as defective in terms of decision making) or a team (comprising more effective relationships between members). Group behaviour can in turn be modified by membership of an organisation.
2. Accurately predicting behaviour becomes more problematic as 'unit' size increases – prediction at the level of the individual is less problematic than prediction at the level of the group (and global construction projects comprise some very large groups).

In the context of the above 'cautions', it is understandable that the project manager feels the need to make use of 'certainties' wherever they can be found. Given that global project will typically involve large number of people from many different backgrounds and cultures, the 'behavioural' aspect of the project may be a particular focus for a project manager. It can therefore be tempting for a project manager to apply 'labels' (a form of stereotype) that appear to help predict the behaviours of individuals and groups. A general example of this type of label (very few labels of this type are construction-specific and so more general labels tend to be made use of) can be found in a brief overview of the behaviours exhibited by two classes (as identified by Reiss) of project player. This illustrates how contrasting behaviours can be present within a project environment.[2] It also has the benefit of testing how strongly behaviours can become linked (not always accurately) with particular functional specialisms (e.g. architects, quantity surveyors) in the form of stereotypes.

- *Explorers/promoters* are quite good at thinking up new ideas (but not quite as good as creator/innovators!) but are more focused on developing ideas (possibly initiated by others). However, their initial enthusiasm wears off quite quickly (resulting in a possible negative impact on energy release) with the implication that they may not be best suited to long-duration projects. They like to work informally (not good at keeping records!) and are good team motivators.
- *Controllers/inspectors* are not good team motivators as they rarely speak, preferring to focus their efforts on listening and checking detailed figures. They do not like the idea of change being imposed on the way that they work and can be almost completely inflexible. They are very good at keeping records, particularly where related to cost control. There is the implication that their level of enthusiasm does not appear to be high (due to their quiet and conservative nature) but is consistent until such time as they are faced with the threat of change.

The above examples in total represent observations of specific behaviours that could be presented by *any* individual but a single specific behaviour is of little value to a project manager when seeking to predict the overall behaviour (and therefore performance) of an individual or group. However, by combining specific behaviours in a manner that is most strongly associated with a specific 'class' of person, it then becomes temptingly easy to assume that by identifying one of the behaviours (such as explorer/promoters) within a specific combination, the total behaviours of that individual can be predicted. The label effectively becomes a stereotype of behaviours.

10.3.1 Stereotypes

Each of the preceding descriptions may have resulted in a triggering of a particular stereotype, for example, estimators (in the UK context) are stereotyped as being unimaginative, very precise, concerned with detail and lacking a sense of humour (or at least one that is recognised as such by other functional specialists). These national or even regional stereotypes are far from uncommon in that it seems every society has its own versions, although the enthusiasm with which they are applied may vary considerably. However, the ongoing process of globalisation is bringing more cultures into closer proximity with each other, even outside of the environments of global projects, and it would therefore seem reasonable to expect that stereotypes would diminish as we become increasingly aware of other lifestyles. Given that a stereotype is essentially a generalisation based on established opinion, surely being more aware of other groups would reduce the use of stereotypes, or at least change them so as to be more truly reflective of the group concerned? In fact, it seems as if the reverse is the case – the more we interact with one or more individuals who we perceive to represent a particular group, the greater our expectation that all members of that group will behave in the same manner as the individuals we have been interacting with. However, even in the most conformist of cultures and groups there will be some who differ from the 'norm'. In the global context it is important that the project manager remains aware that culture, particularly with regard to acceptable behaviours, flows to a large extent from the local environment and is thus inevitably specific to a region, resulting in a potentially wide range of cultures within a single country. One example of how culturally diverse a region can be is to consider London in 2012: commentators on the 2012 Olympics by and large agreed that one of the key factors in the success of the Olympics as an event was that the host city (London) is truly a global village. Over 300 languages are claimed to be spoken on a daily basis in London. Such is the extent of the cultural diversity that large construction projects will typically have their signage produced using the native languages of the main groups comprising their workforce.

Taking the workforce consideration one stage further, assume for one moment that you are a project manager in which the workforce comprises Mexican, Irish, Swedish and Spanish workers. You will almost immediately have conjured up your own particular stereotypes of each of those nationalities (even if you are Mexican, Irish, Swedish or Spanish yourself). In order to gauge the impact of those stereotypes on your management of the workforce, which of these four nationalities are:

- The most punctual?
- The best workers?
- The most aggressive?
- The most helpful?

Research carried out by a team from the University of Boras, Sweden found that the answers given by respondents (ten students from each of the named countries) were:[3]

- Most punctual = Swedish (all the Swedish students voted for their country, while not even the Mexican students voted for their country, resulting in '0' votes).
- Best workers = Swedish (Mexicans came bottom of the list).
- Most aggressive = Irish (the Swedes got '0' for aggression).
- Most helpful = Swedish (only the Mexican students did not identify the Swedes as being the most helpful).

It should be noted that the research was based on a small sample and perhaps should not be taken too seriously, but it does have the value of indicating how positive and negative behaviours can be expected as a result of national stereotypes. However, along with national stereotypes the project manager of a global construction project will also have to deal with what is termed 'functional' stereotypes.

Functional stereotypes can be an important part of an individual's identity and are arguably a 'normal' part of interaction between groups, particularly within a competitive environment such as a male-dominated construction project (or is this yet another stereotype of construction projects?). An Australian survey found that the industry with the highest percentage of female employees was health and community services (in line with the 'caring' stereotype for females) at 71.1 per cent, while the lowest figure was construction at 9.9 per cent (even mining achieved a higher figure: 13.3 per cent[3]). India, however, has a roughly 50:50 male–female workforce in its construction industry, but social constraints typically result in women being allocated the unskilled tasks. An Australian project manager on an Indian project could well be surprised by the labour resource available!

Various countries have made attempts to create gender-balanced construction industries, and a range of reasons are typically put forward for seeking to achieve a 'better' male–female balance in the industry. One such reason is that the industry would become more collaborative in nature, but is this based on another female stereotype? Perhaps not, as a 2011 survey found that women outscored men in 15 of 16 leadership competencies, one of which was collaboration and teamwork (men scored 49 per cent and women scored 53 per cent).[4] On that basis, the argument that women are more collaborative than men is more fact than stereotype. While stereotypes can be regarded as having usefulness in terms of a 'shorthand' for describing groups in terms of key behaviours, they can also present risks through the varying levels of certainty that they comprise. As with heuristics (rules of thumb) generally, stereotypes can have an air of certainty about them and therefore be perceived as a valid base for managing groups in complex environments, such as are found in global construction projects. However, when collaboration is preferable to competition (as is the case for global construction projects), stereotypes can be problematic in that, along with possibly not being accurate, the application of stereotypes reinforces differentiation, which in turn reinforces boundaries between groups; strong boundaries are not conducive to collaboration.

For the global project manager there is a further problem with regard to stereotypes – the stereotypes applied by groups to each other may be invisible to the said manager. Groups tend to develop their own particular culture – a combination of values and beliefs that 'guide' the actions of the group in a consistent manner; all members of group 'A' behave towards all members of group 'B' based on 'A's' stereotype of 'B' – an image of the group that may not be made more widely available. The interaction is between 'A' and 'B', therefore anyone not in either group may be completely unaware of the stereotypes being applied by each group. In such situations, there is a probability that the 'unaware' individual (the project manager) will not be able to predict accurately the behaviour of either group whenever they are required to interact – this is just a small example of the manner in which *uncertainty* can impact on a project. When this problem is multiplied by the number of groups possibly existing within a global construction project, a level of uncertainty (with regard to behaviour and thus performance) will result that is capable of crippling the project.

A final point to consider is that stereotypes are not the same thing as prejudices, in that prejudice, while generally seen as a negative (as in being prejudiced *against* a particular individual or group) can also be acted on in favour of a particular individual or group (as in prejudiced *towards*). In project management terms, both forms of prejudice are detrimental to a project, in that a project manager must remain neutral; he or she is responsible for the project and the safety and wellbeing of *all* working for the project. Good leaders are, in many cultures, those who evidence consistency in their actions. However, the expected consistency may, in some cultures, include actions that do not seem, from the project manager's perspective, to be truly even-handed to all project players: consistently reinforcing cultural hierarchies and boundaries between groups in a manner that could seem prejudicial to groups at the bottom of the hierarchy may be expected by some cultures.

The behaviour of projects does not solely result from the people involved. Projects require energy in other forms to support their dynamic nature, and there is therefore a need to also consider the non-human factors and behaviours involved in projects.

10.4 Non-human behaviours

Projects can face any number of changes, problems and areas of uncertainty. Problems tend to be bundled up together and labelled as 'risk', with the gradual result of risk consistently being regarded as a negative (in that it may be seen as contributing to project failure). However, problems may also present projects with opportunities, and it is on such occasions when it becomes apparent as to whether the project manager is focused on avoiding failure (in which case he or she will be reluctant to embrace the opportunity if it is perceived to be 'risky') or on achieving success. In other words, will the project manager focus on the 'positive' opportunity for the project that has arisen, or on any 'negative risk' that can be identified within the opportunity? In the latter case, the project manager will reject the opportunity due to the emphasis placed on the associated risk and its possible contribution to project failure. This will particularly be the case when he or she has confused 'risk' with 'uncertainty'. When the uncertainty arises due to a lack of understanding that an aspect of the project's 'behaviour' is not actually linked to a particular human behaviour (such as falling behind schedule being linked with the project workforce being lazy) the situation can be particularly difficult to deal with. The project manager has, in effect, wrongly 'created' a direct link (linear relationship) between project performance and worker performance, thereby creating a false sense of certainty with regard to the cause (and therefore the 'solution' to be applied) of the problem. Such false certainty arises when project managers apply another non-human (and inaccurate) stereotype; all projects are based on relationships of a linear nature.

Linearity is a simple but powerful tool for analysis (particularly with regard to a systems approach) and there is therefore a considerable temptation to apply it in a 'blanket' manner. In doing so, a project manager will in one respect make their life a little easier (by reducing complexity within the relationships that the project relies on) but may also introduce a significant amount of uncertainty into the planning and control of the project. The essence of a linear relationship can be represented simply as $X + Y = Z$ in all cases. A basic chemical reaction (relationship) such as can be found between water and cement powder evidences $X + Y = Z$ linearity – adding water (X) to cement powder (Y) results in a reliable reaction (Z): subject to the quality of the powder, an exothermic chemical reaction will always occur. However, even in this example, at least one level of uncertainty can be identified: is the cement powder of the quality required for the reaction to take place? Thus, even in apparently 100 per cent linear relationships it can be possible to identify some quantity of uncertainty and this results in a supposedly linear relationship actually being (perhaps only to a very small extent) non-linear. A construction example would, in order to be meaningful, actually have to be somewhat more involved than the basic $X + Y = Z$ representation; an enclosed space (building) would be floor + walls + roof, or $W + X + Y = Z$. However, even this is not going to apply in all cases, in that not all enclosed spaces (in a global context) have one or more walls. Thus it is important to be aware of the level of detail at which the linear representation is being applied. Fortunately, linearity is an essentially self-similar concept in that it can be applied from the strategic level all the way down to the task level, particularly when aided by a degree of pretence that the relationships being represented are truly and fully linear.

While the traditional approach to project planning and control has made highly effective use of 'pretending' that the project environment is based entirely on linear relationships (I supply the bricks, you build the wall), it is less effective when dealing with complex projects such as global construction projects. In such projects the uncertainty issue has to be addressed explicitly, in that seeking to continue with the traditional 'pretence' of true linearity simply adds a degree of uncertainty. Modelling a project environment composed of non-linear relationships requires the project team to add considerably to the data and information-gathering and processing activities, but brings the valuable benefit of more explicitly identifying where the areas of uncertainty are within those relationships.

In order to better appreciate the argument that will be presented for the existence of non-human behaviours in projects, a quick overview of systems thinking is required. One approach to systems thinking can be presented as ICE (import, conversion, export), which can be applied to everything involved in a project. Certainly, all production processes require imports (e.g. materials), which are then 'converted' (some manufacturing process is applied to the materials) before being 'exported' (either as a finished product in its own right, or as an import for a subsequent conversion process). This model implies two environments. First, the production or project environment: this is where the conversion process is carried out and can be referred to as the internal environment. The second environment is one from which 'raw' resources are supplied, and to which 'finished' products are exported. This can be referred to as the external environment, and it is from here that the majority of non-human behaviours originate. The potential for such behaviours can be assessed through analysis tools that break down the external environment into a number of different factors, each of which could have an impact on how the project behaves. One of the more complex but frequently cited examples is PESTLE (political, economic, social, technical, legal, environmental). However, other analysis tools that use a greater level of detail exist and it is possible to attempt a highly detailed analysis of a project's external environment. Prior to attempting a highly detailed analysis it must be noted that this will take time, and time equals money. There is therefore a balance to be achieved between identifying all possible problems (or opportunities!) within a complex external environment and expending energy on dealing with those problems.

The impact of any disorder in the external environment is evidenced in terms of non-human behaviours: a particular change comes along that causes the project to behave in a new, possibly unplanned manner. This new behaviour may be a positive or a negative with regard to the project's ability to achieve a successful outcome. In either case, it will still need to be managed and it is at this point that the human and non-human behaviours within a project will *interact*. The interaction of human and non-human behaviours in a project is therefore an important factor in achieving a successful project outcome and is worthy of further consideration by any 'good' project manager (see Box 10.1).

Box 10.1 Vignette – non-human project behaviours and unplanned change

In Chapter 4, dealing with measuring and improving project performance, two case projects were discussed: the Lake Turkana fish-processing project and the Gassinski CMFN project. Both can be used to provide evidence of non-human project behaviours and the response of the project to the concept of unplanned change. The Lake Turkana project can, in retrospect, be regarded as almost having a built-in probability of failure in that it is an example of the traditional approach to the planning and controlling of projects in action. A considerable number of 'assumed' linear relationships can be found within the project, with one example being that bringing together an impoverished social group and a relatively valuable resource was believed to result in alleviation of the group's perceived poverty (and the attendant low social status applied by other social groups in response). While there can be no doubt that the Norwegian and Kenyan aid teams involved in the project carried out their research, the methods applied to that research do suggest an overall assumption of linear relationships within the 'environment' of the project. There also seems to have been a problem with regard to identifying factors within the project environment as being 'discrete', when in fact they were not, thereby placing a boundary around the project in a manner that created a number of problems. However, both of these problem areas can be attributed to the application of a project planning and control method (based on assumptions of linear relationships between various processes) that was regarded as being 'good' in the 1960s and 70s (the project could trace its roots back to 1961). Non-linearity was not considered as being relevant to project planning or control at that time and was still a relatively new concept. The evidence for the so-called 'butterfly effect' had been identified only in 1963 and was not referred to as such until 1979. Arguably, the impact of non-linearity (particularly regarding complexity theory's connection with unpredictability) did not even become linked to project management in any meaningful manner before 1993 at the earliest.[4] The Gassinski Model Forest (MF) project commenced its scoping activity in 1994 and was thus presented with an opportunity to go about planning and controlling in a different manner to that available to the Lake Turkana project team. Whilst it must be accepted that the Gassinski MF project was not 100 per cent successful in achieving its stated objectives (the project was classed as 'closed' in 2010[5]) it is generally regarded as being an example of a successful project of its type. Much of this success can be argued as resulting from the efforts of the project team to map out the areas of uncertainty in a more non-linear manner than had happened in the Lake Turkana project. For example, while the population level within the forest area is relatively low (approximately twenty villages and two towns) the project team spent three years discussing with the population (and other players) what the objectives should be for a project situated in a landscape that is:

> ...made up mostly of plain wetlands (70 per cent) and mountains (30 per cent). Of this, the national park 'Aniuiski' covers about 49 per cent of the Gassinski MF site and includes protected territories, such as that of endangered Siberian tigers (11 per cent) and forest lands previously announced as those traditionally used by indigenous peoples (24 per cent of the park area).[6]

A traditional approach to planning and control would quite quickly have assumed linear relationships between, for example, the presence of endangered Siberian tigers and a variety of other factors such as ecotourism, conservation guidelines, illegal hunting (possibly by indigenous groups regarding the Gassinski MF area as their traditional 'home') and climate change. The more non-linear approach of the Gassinski project team resulted in strategic objectives that were actually strategic (as opposed to operational) and therefore less pre-determining of the 'solution(s)' to be applied:

- To promote the economic and social development of indigenous communities.
- To promote the conservation of biodiversity, and the protection of rare and endangered species.
- To achieve and support sustainable forest management (SFM) through decision-making processes that take into account the interests of the local people living in and around the Gassinski Model Forest, and which are based on monitoring of the condition of forest and water ecosystems.
- To promote environmental knowledge among forest specialists, students and academia.
- To conduct and support forest research for SFM.
- To monitor local level indicators (LLI) and provide the information to the public.

The Gassinski MF project not only evidences a more 'open' approach to project planning and controlling, through its greater acceptance of non-linearity in the relationships between various factors that could contribute to success or failure, it can also be argued to make more explicit the link between risk and reward than was the case with the Lake Turkana project.

10.5 Risk and reward

There is a management dictum that any given level of responsibility should come with an equal level of authority, and risk management follows this in that it seeks to achieve an appropriate reward for 'accepting' a risk. In this context it is useful to adopt a definition of risk as being the probability of an unacceptable loss arising.[7] The nature of the 'loss' arising may vary considerably between individuals and groups, and the project manager should be aware of the potential impact on a project of assuming that loss has a homogenous nature: one person's loss may well be another person's reward. A project manager should particularly avoid the assumption of it being entirely reasonable that what they personally determine as being a 'loss' will also be determined as such by all other players in the project environment. Risk can be a very personal concept but there are two 'rules' (actually heuristics – rules of thumb) that can be applied when seeking to manage it. Possibly the primary heuristic is that as the level of risk increases, so should the level of reward. This should be kept in mind when a project manager seeks to implement the second heuristic: in order to achieve a structure for managing risk, follow the risk hierarchy:

- Identify: identification allows at the least the possibility of a managed response. If a risk cannot be identified as such it remains an area of uncertainty.
- Assess: how significant is the risk? This helps to determine the appropriate response in that a 'small' risk should not generally elicit a 'major' response (essentially a matter of proportionally 'balancing' resource requirements).
- Eliminate: complete removal of the risk requires no further response and is therefore the most desirable, from a project management perspective, of the possible responses.
- Reduce: when complete removal is not possible, aim to reduce (mitigate) the risk by putting in place processes or procedures that incorporate less risk.
- Transfer: seeking to move a risk that you do not have the expertise or resources to reduce to an acceptable level is appropriate when it can be transferred to an individual, group or organisation having more expertise and therefore willing to accept the risk on your behalf. This will typically involve a risk versus reward negotiation – how much is it worth for you to transfer the risk and how much reward would a second party require for accepting the transferred risk?
- Accept: when the risk versus reward negotiation fails to achieve an agreement, then there are only two options available to the project manager – either 'accept' the risk or do not proceed with the project.

An important consideration regarding the risk hierarchy is that it is not consistently valid across all cultures, in that some cultures will, in effect, ignore all of the actions other than 'accept'. This response is covered in more detail in Section 10.11, but a brief overview is relevant at this point in that it is useful to be aware of this response when considering the material (particularly the material covering 'responses' in Section 10.8) in the intervening sections. The response of simply accepting a risk (or uncertainty) without any consideration of possible mitigation, is essentially one of either fatalism (what will be, will be) or determinism (all events are predetermined by various laws and cannot be changed). Thus if a worker believes that the result of working on a scaffold that has no hand-rail is decided before he even sees the scaffold, and he is willing to accept that outcome, then he will see no point in attempting to reduce the risk by adding a hand-rail. In essence, there is no reward that can be linked with attempting to reduce the risk, because any attempt will not change the predetermined outcome.

As far as reward is concerned, the first two stages of the hierarchy can be argued as meriting at best a low reward, in that the risk is not as yet being directly addressed. The elimination stage may merit a slight increase in reward, in that the action of eliminating may present some risk to those involved, but this is not common in that the activity involved is usually a design one (as in designing out the risk). The reduce stage may merit a slight increase in reward in that it could involve some physical interface with the risk (as opposed to purely designing the risk out). Transferring the risk is essentially the basis of all contracts, with the argument being that the risk is 'transferred' to a person or organisation with the expertise to deal with it appropriately and also the willingness to do that (for an appropriate reward). Acceptance of the risk may result when the risk is perceived by those to whom it would ideally be transferred as meriting a higher reward than the client can agree to. Alternatively, there may simply be no willingness to accept the risk and it therefore cannot be transferred.

A final aspect of risk and reward to be considered in this section relates to ethical issues, particularly regarding the definition of 'ethical' applied by any cultures active within a global construction project environment. While many countries apply the concept of the level of reward being related to the level of risk 'accepted', not all countries are consistent in this regard. Nor are all organisations and individuals consistent with the national stance on this issue. In terms of countries to avoid doing business in, a useful starting point is to consider the evidence for the most corrupt countries in the world. Various measures are applied to the assessment of corruption and, consequently, there will be variations in the order of countries listed, but the Top 10 usually commences with Venezuela in tenth place, with Haiti, Iraq, Sudan, Turkmenistan, Uzbekistan, Afghanistan, Myanmar and North Korea filling places nine down to two, and Somalia topping the list. None of these countries are generally considered safe places to do business, although the rewards flowing from activities related to resource extraction (usually oil and gas) are sufficient to tempt larger organisations into limited operations in some of the countries. This then leads on to the manner in which projects are 'awarded' in such countries, and it is not unknown for projects to be 'bought' through the giving of gifts (at the low end of unethical behaviour) through to large bribes (at the top end of the scale). While such an approach may be the norm in certain countries, there is a risk to global (transnational) construction companies related to the loss of reputation if it should emerge that projects were obtained in such a manner. There is also the risk of legal action (both by the 'host' country and the construction company's home country) and longer-term threat to the company's survival. In this context, business ethics are also a consideration for investment ethics.

Beloe (2011)[8] developed an investor tool to be used for identifying and managing business ethics risks (so-called 'bear traps') that comprised initially 50 'red flags' (later trimmed to 18 key and measurable flags) within the broad categories of:

- type of industry;
- country of operation;
- company structure and business model;
- management integrity and supervision;
- high-level financial indicators.

A key indicator with regard to the country of operation was that of having more than 30 per cent of total operations in countries having a corruption perception index lower than 3. For comparison purposes, each of the ten countries identified previously as being the most corrupt had a 2011 corruption perception index of less than 2, while the least corrupt country in 2011 was New Zealand with an index of 9.5. In terms of key emerging construction markets, the two countries typically identified as prime targets are India and China, but India has a corruption perception index of 3.1 (95th most corrupt country) and China has an index of 3.6 (75th most corrupt country). In terms of 'red flags', any construction company operating in both countries could be seen as being an investment 'bear-trap' and thus avoided by investors.

Box 10.2 Vignette – Doing business in India

An example of how business can be conducted in India was told to me by an Indian national (Mr A) who operated in several industries, of which the construction industry was one. The example involved Mr A and his attempt to secure a project carrying out some rewiring of a large government building. Mr A identified the government official (Mr B) with overall responsibility for the awarding of contracts and duly arranged to meet Mr B to discuss the project. Mr B was careful to ensure that the meeting was on site in the largely empty building (there were some shops, including a café, still operating on the ground floor) so that 'the project could be fully appreciated without being disturbed'. Mr A duly arrived at the site and the discussion went very smoothly, with Mr B indicating that Mr A was now one of the front-runners for winning the project. At this point Mr A realised that he should improve his chances by 'thanking' Mr B for his consideration, and so duly asked if he could in any way be of help to Mr B so as to show his gratitude. Mr B replied that such an offer was very generous but all he would accept would be a cup of tea. Mr A was surprised that this was all that was required for him to be awarded the project and immediately agreed to go and fetch Mr B a cup of tea. Mr B then added that the cup of tea would have to be from the café on the ground floor as it was, in his opinion, the best cup of tea to be had in the whole of the city. If Mr A was to go and see the owner of the café and ask him to bring a cup of tea with three spoons of sugar added up to Mr B, then Mr B would be most happy.

Mr A happily went in search of the café owner to place an order for a cup of tea with three sugars for Mr B. The café owner then informed Mr A that the bill for the tea was 45,000 rupees! A rather gentle and indeed almost cultured form of corruption was being engaged in. Mr B realised that the café owner was at liberty to charge whatever he wanted for a cup of tea and used this to ensure that no cash ever came directly to him, and thereby reduced the risk of being caught engaging in corruption; he simply varied the number of spoons of sugar in the cup of tea so as to indicate to the café owner the appropriate charge (bribe).

10.6 Differentiating between uncertainty, risk and hazards in projects

There is one key aspect to successful risk management. This can be summarised as knowing when a risk is not actually a risk! Hillson,[9] for example, argues that *risk* should be differentiated from *uncertainty*, particularly with regard to the result (effect) flowing from the cause. It is important to be aware of the need to place risk in a context – it does not exist in a vacuum and it should be related to particular project objectives. This is essentially the first stage of the risk hierarchy (identify) and the more specific the context that a risk can be placed in, then the more targeted the response (in terms of addressing the identified cause(s) and effect(s) as separate issues) that can be designed. These project objectives are either under threat in some way or there is an opportunity cost attributable to them. General uncertainty, as stated previously, is not the same thing as risk. There are many possible areas of uncertainty, but unless these are directly linked to specific project objectives (as either threats or opportunities), and a probability can be attached to them, they remain simply uncertainties, and it is therefore more appropriate that uncertainty management rather than risk management is implemented. As an example of this in the context of global construction projects, it was previously mentioned that such projects are demanding in terms of the infrastructure available for the movement of resources: for such complex projects the quality of various aspects of the infrastructure around the project site is an important factor and should therefore be consistently addressed, preferably in terms of a 'known' benchmark. Whilst it would seem reasonable to assume that the quality of roads, as one aspect of infrastructure, would be higher in developed than developing countries, if this remains an assumption then the project supply chain may suffer. Table 10.1 illustrates how widely aspects of infrastructure can vary and also evidences that developed countries do not always have better quality infrastructure than do developing countries.

Table 10.1 Infrastructure quality rankings and public debt levels (data, other than Corruption perception index, sourced from Global Construction 2020)[10]

Country	Infrastructure quality world ranking				corruption perception index	Ranking (in this sample)	Public debt 2009 (% of GDP)
Developed countries	Overall	Roads	Rail	Electricity supply			
US	23	19	18	23	7.1	4	53.5
Canada	13	17	16	14	8.7	3	82.5
Greece	58	57	64	65	3.4	6	113.4
Germany	9	5	5	6	8	2	73.2
Developing countries							
China	72	53	27	52	3.6	7	16.9
Singapore	3	1	6	9	9.2	1	110.0
India	91	90	23	110	3.1	8	57.3
Thailand	46	36	57	42	3.6	5	44.9

The table evidences that assumptions regarding the relative situations in developed and developing countries are not always accurate. Investment in a country's infrastructure is a key aspect of development and typically results in large projects such as road, rail and airport construction, all of which represent business opportunities for transnational construction companies. Such projects are usually funded by government, and therefore it could be tempting to regard them as being 'safe' in terms of being paid for work completed. However, hitting the top levels of the world rankings for infrastructure quality comes at a price; the level of public debt rises to the point where it becomes difficult for government to continue to pay for any large projects, infrastructure or otherwise. As a country approaches the 100 per cent mark for public debt in relation to GDP, it may be regarded as less of a viable market in the medium term as it will encounter difficulties in borrowing the money to fund large projects. At this point, any transnational construction company that has developed an operating structure in that country essentially has two choices: move on and find another developing market elsewhere, or stay put and find ways of helping the government fund projects. Chinese construction companies are, for example, able to access loans at preferential rates for projects in Africa and it is not unknown for these companies to take payment in kind (typically resources that are in short supply elsewhere) rather than cash. In effect, the companies are accepting the risk of resource values varying over time in order to secure a certainty of supply for those same resources over the medium to long term. The nature of China's system of government allows companies operating in a commercial capacity on behalf of the government to make use of a competitive benefit (being able to 'trade' probable upside opportunities and uncertain downside losses) not available to the majority of independent transnational construction organisations.

10.6.1 Impact of uncertainty

The impact of uncertainty on assessment of a situation (in terms of there being one or more unacceptable outcomes) should not be underestimated by a project manager. Where an individual or group becomes sensitised to risk by the additional presence of uncertainty (it is certain that there will be an unfavourable outcome but it is uncertain precisely what that unfavourable outcome will actually be), they become more anxious about a situation.[11] This then makes it more difficult for an individual or a group (particularly given a group's tendency towards defective decision making) to respond appropriately to the actual risk (as opposed to the perceived risk when uncertainty is added). Part of the problem in such circumstances is the tendency to be distracted by the effect rather than focusing on the cause. This becomes particularly problematic when the effect 'focus' emphasises one or more *possible* effects (usually those perceived as having the most unacceptable outcomes) and pays less attention to the most *probable* effects. If the individual can firmly label the item as 'impossible' then they cease being anxious about it, but if they continue to regard it as being 'possible' then it is a potential cause of anxiety. In circumstances where the 'possible' can be, as it were, transferred into an environment labelled as being 'probable' (in other words a specific probability of occurrence can be determined), then anxiety starts to reduce.

The need to link risks to project objectives is made difficult in situations where there are multiple possible causes and/or effects for a particular risk. In a situation where one cause leads to one risk resulting in one effect, the risk management approach can be a relatively simple one. However, when there are multiple possible causes in a project environment it can be difficult to determine which of those causes could lead to uncertainty, and which ones bring about genuine risks. Following on from a risk or an uncertainty is an effect. Effects are simply something that could happen, in the future, and are dependent upon a particular event occurring. It is therefore not certain that they will occur, and consequently we should not be seeking to deal with effects through risk management either. The argument is that risk can only be managed effectively if it is separated from both its cause(s) and its effect(s). An added confusion in this regard is the existence of one or more *hazard*s in the project environment.

When considering the requirement to provide a safe working environment within projects, confusion can sometimes arise with regard to the difference between a risk and a hazard. The HSE (UK's Health and Safety Executive) provide quite a range of information relevant to this question but, at its simplest, the following definitions can be applied when seeking to differentiate between the two:[12]

- Risk – a rating that illustrates the probability of 'harm' or an unacceptable loss. The rating may be a precise quantification based on numerical data (0.235) or simply a hierarchy of a qualitative/subjective nature (high, medium, low).
- Hazard – anything that can possibly cause harm, irrespective of the probability that it will cause.

In the construction context, there is a regular misuse of the two terms in that reference to a 'risk' should actually be reference to a 'hazard'. This generally arises due to the problem of a culture of safety and risk avoidance that results in a perception that all hazards, along with all risks, should preferably be eliminated. Certainly in the UK a number of hazards have been targeted in terms of specific requirements being mandated regarding the manner in which they are addressed (as with the Control of Substances Hazardous to Health regulations relating to certain commonly used but hazardous materials such as cement) or have been eliminated from all new construction work (such as certain forms of asbestos). For anything that can possibly cause harm, then the UK perspective is to eliminate or manage it and this is frequently done without any explicit consideration of the probability that it will actually cause harm, even when that harm is not regarded as life-threatening such as with noise-induced hearing loss (NIHL). UK statistics are available for the number of cases of NIHL identified each year since 1995/6 and so it is feasible to calculate the probability of a construction industry worker developing NIHL. As the number of cases is quite low (in 2009/10 there were around 200 cases identified[13]) in relation to the size of the industry workforce, the calculated probability will also be quite low. Nonetheless, the impact of hearing loss on an individual's life is such that even a low-probability, non-life-threatening hazard merits a response aimed at eliminating the probability of its occurrence in the workplace. For global projects, however, the majority of cultures within the project environment, along with the national culture of the country in which the project takes place, may not be particularly concerned about such issues. Such a situation can return the project manager to consider the ethics of knowingly exposing workers to specific risks and/or hazards that the culture of the workers involved does not regard as being of concern. In addition, there may be a business ethic to be considered in terms of adding to the cost of the project through being proactive regarding health and safety beyond the level legally required.

10.7 Perception of risk and uncertainty

There are a number of reasons as to why individuals may have different subjective perceptions of the extent of uncertainty or level of risk inherent in any given situation/scenario and, as with any area of human behaviour or activity, there are various theories that can be put forward as to why an individual or group behaves in a particular way. Over time, these theories are either discredited, or evolve in the light of findings by new research. With regard to *risk* perception, the currently favoured theory is that of psychological perspectives underpinning *perceptions* of risk. A wide range of personal biases that diminish or increase the perception of risk have been identified, and these are modified by various social processes attuned to the perception of risk to a community, in particular the community that an individual finds themselves a part of. This area is developing in some communities to the point where there is evidence that it is modifying the communities' perception of risk (at least insofar as risk equates to uncertainty) to the point that it is increasingly being accepted that even the rational calculation of risk becomes uncertain to a degree, related to the component of risk that is *socially* (as opposed to individually) constructed.[14]

The socially constructed component of risk perception is also referred to as *attenuation* (increased sensitivity to a particular risk as a result of processing information about the causes and effects of that risk through social structures and process). The project manager therefore needs to ensure effective communication, not only so as to learn about a particular community, but also to ensure that, for example, the appropriate tone, content, symbols, quantity of information are used when seeking to present a specific situation to that community. This is, in a sense, concerned with the reporting of risk by the project manager to the community in a manner that will not inflame that community's response to a risk through modifying their perception of it adversely (attenuation). An important factor in this regard is to be aware of possible personal biases.

Differences in risk perception amongst project stakeholders are argued to be a major cause of conflict encountered in the execution of projects. Keil *et al.* found that the differences in perception of risks between project managers and users in IT projects was sufficient to be a cause of conflict, and that this conflict could be reconciled through a process of identifying zones of *accordance* and *discordance*.[15] The previously discussed study by the University of Boras team (Section 10.3.1) provides an example of a multicultural group (Swedish, Irish, Mexican, Spanish) having at least one zone of accordance as represented by the responses to one of the questions (that the most punctual nationality was Swedish) in that not even the Mexican students voted for Mexico. However, on the other three questions listed, the responses indicate varying levels of discordance, largely (it can be assumed) because of the existence of bias (possibly on the basis of national or regional cultural values) within the decision making of each individual. This is typical of the early stages of developing a multicultural environment; this is not an easy task but cultural diversity can be managed so as to gain performance improvements and risk reduction through actions such as:

- Communicating effectively (to create awareness amongst all employees regarding the diverse values of their workmates)
- Cultivating support (all employees to acknowledge, support and encourage any employee)
- Capitalising processes and strategies (linking of cultural diversity to every business process and strategy).[16]

Examples of bias that are cultural in origin include several heuristics or *rules of thumb*. These are a widely acknowledged part of the decision-making process. An example of a heuristic common within developed countries is the 80:20 rule:

80:20 – One commonly occurring heuristic is the 80:20 'rule', which comes in a variety of forms. The construction industry in the UK, as one example, uses a heuristic which claims that 80 per cent of the cost in a project can be attributed to 20 per cent of the design. Applying this heuristic in an unquestioning manner should (!) obviously be a risky approach to project management.

An example of a heuristic that occurs in both developed and developing countries is the availability heuristic:

Availability – Relates to how easily an individual can recollect a similar incident to the one under consideration. In other words, how 'available' information on risk is within that individual's personal experience. An example of this is that, in the UK context at least, more people will have encountered someone suffering from some form of cancer (estimated at a 1 in 3 frequency in the UK population) than someone suffering from Alzheimer's (around 1 in 70 of the UK population). Individuals will therefore tend to place the risk of death from cancer higher than the risk of death from Alzheimer's, simply because the experience of cancer is more readily available to them.[17]

Understanding the factors that alter an individual or group perception of risk is useful in that it helps the project manager to both understand why a particular (inappropriate) response may have been selected by that individual or group, and also to help guide the individual or group towards a more appropriate response.

10.8 Responses to risk – response decisions

The project manager has essentially only two different responses available: to do something or to do nothing. As far as doing something is concerned, the trick is to select the 'something' that can realistically (probability again) be expected to bring about a successful result. However, many of the 'somethings' that are available are actually inappropriate. Two example responses not to implement are:

- **Avoidance syndrome:** a common response. Fear of conflict causes individuals to avoid situations and/or people. The more tense the situation, the stronger the avoidance behaviour. This results in increasing numbers of unsolved problems, favouring of extremists, and a lack of understanding of the events faced by others.
- **Frenetic syndrome:** a manager makes numerous decisions hastily, many of which later become obvious as being contradictory. The frenetic player usually believes that time is too short to carry out a full analysis, resulting in wrong decisions, or correct decisions that cannot be implemented in the time deemed to be available, thereby destabilising management systems.

As to the option of doing nothing, the perception of risk in this option would generally be seen as high (largely due to personal biases). However, within the management of risk the *actual cost* of responding should always be a mitigating factor, rather than focusing only on the *perceived cost* of not responding (effect). If the cost of responding is say £1000 but the cost of not responding is £100, then there needs to be a valid reason for incurring the cost of responding. Such decisions are normally made in the context of the requirement to reduce risk to a level that is as low as is *reasonably* practicable. This, of course, then places emphasis on defining what is deemed to be 'reasonable' within any given set of circumstances. A number of situations can be identified when it would be regarded as reasonable to do nothing:

- low risk that is within acceptable limits;
- existing controls will minimise threat or maximise opportunity;
- high probability of a risk 'disappearing';

- high probability of an opportunity increasing in value;
- high probability that taking an opportunity will prevent other, potentially greater, opportunities from arising;
- high probability that dealing with a problem will cause other problems to occur.[18]

In each of the above situations, it is reasonable to do nothing. However, it is important that the identification of a particular set of circumstances is based on realism rather than denial of the true nature of the risk or opportunity. It is also important that the nature of the situation is monitored once the decision is made, and that the project manager is willing to revise their decision in the light of changing circumstances. With regard to doing something, the response should relate to the desired outcome and, in order to guide this decision, a hierarchy of responses (in terms of outcomes) can be applied: eliminate, substitute, reduce, adapt, transfer, accept.

10.9 The 'optimal response' concept

In the context of managing uncertainty within project environments, there is a need to appreciate that the traditional approach to dealing with risk (primarily) and uncertainty (secondarily) is very much focused on engendering an overall feeling of certainty that the relevant activities are 'under control'. Achieving this very much depends upon the manager identifying and implementing what may be referred to as the 'optimal response' – the response that is going to give the optimal balance between consumption of resources and the 'creation' of certainty of outcome. However, increasingly complex projects have brought about a gradual realisation that the search for an optimal solution based on traditional project management values may actually be adding to uncertainty in a subtle and 'hidden' manner. Placing risk and uncertainty in the environment of society (as this seems to be the driver, in terms of a society's accepted perspective on both factors), suggests that possibly the most well-known approach in contemporary research on the sociology of risk is Beck's 'risk society' (Beck 1992).[19]

The risk society perspective brought about a significant and relatively rapid change in the study of risk, but as with all significant developments, has more recently had to deal with conceptual and empirical critiques. A particular focus has been a criticism that Beck's perspective has the effect of narrowing the 'boundaries' of the risk concept to being essentially a series of responses of a technical and environmental nature to unforeseen consequences of industrialisation. Given the relatively short human history of 'heavy' industrialisation it is perhaps understandable that this emphasis may emerge in a more contemporary perspective on risk and uncertainty – the populations of developed and developing societies may well have become 'attenuated' to risks and uncertainties that flow directly from the technological developments not available to their grandparents. One example of such a 'modern' take on uncertainty could be the general feeling of unease that so-called Frankenstein science (anything that the general public – wider society – does not particularly understand, such as genetically modified crops) seems to cause.

Given the technical nature of much of recent human development, an emphasis on technical (including mathematical – much of what a modern society does is dependent upon computerised systems) responses can be seen as appropriate. However, such a 'narrowing' of the range of responses available to society in general (and project managers in particular) has been criticised as resulting in an impoverished repertoire of risk and uncertainty responses, particularly when dealing with complex (highly interconnected) problems. The concern is that the focus has become more on risk and less on uncertainty, as a result of technical responses being seen as more appropriate to providing 'certainty' when dealing with areas of risk than when dealing with areas of uncertainty. The critiques of Beck's risk-society are essentially constructing an argument for a broader perspective on society's relationship with risk. Such a broader perspective can only be achieved by considering uncertainty in addition to risk.

In a perspective that includes an appreciation of uncertainty, it becomes apparent that the concept of risk as a probabilistic construct should be recognised as being in some manner 'special' with regard to its use in engendering a feeling of security through certainty of outcome. In some respects this could be argued to be turning the clock back – prior to the development of Pascal and Fermat's Laws of Probability it seems all societies had been aware of risk as a concept involving 'loss' but not in terms of being able to calculate the probability of that loss occurring. While the earliest evidence for 'gambling' (a type of die made from Deer ankle-bones) dates from around 4500 years ago, there is no evidence for there being any belief that this risk could be 'controlled' until the development of the Laws of Probability in the mid seventeenth century. The belief was in the individual gambler's 'skill' (any game of pure chance is unaffected by the skill of the player) rather than any calculation of probability.

By regarding the 'calculability' of risk as a cultural construct embedded in society it becomes valid to consider probability calculations as grounded in key aspects of a society's culture. Typically, this will place the interpretation of a probability calculation in a more subjective environment (rather than the traditionally required objective environment) that is aware of concepts of, say, science and acceptable behaviours, which are regarded as being of value to a particular society. To some extent this broader perspective requires a degree of acceptance of uncertainty due to its emphasis on a situation or context-based interpretation of a probability calculation; interpretation within a systemic environment appears precise whereas interpretation in a situational environment appears to be 'messy' – it depends! It may seem somewhat ironic that a modern society with all its technology and advances in materials, health care, computing and so on has, in its search for 'control' and 'certainty' of outcome, actually created an environment in which uncertainty is fundamental. Uncertainty has become a consistent component of modern life – we simply need to accept it as such and then change the manner in which we respond to it.

10.9.1 Secondary uncertainties

The argument for a culture-based interpretation of probability calculations is that in striving to implement certainty and order there is a risk of becoming overly focused on a single 'solution' and not appreciating the secondary uncertainties or dangers that may accompany it. It is an often-quoted truism that no one can know the unknowable. As project managers we have to accept that there are limits to our knowledge and understanding of any given environment, and that the interpretation of a probability calculation should incorporate this awareness, rather than regard a particular 'solution' as being infallible simply because it was calculated in great detail. Any action taken will possibly bring about the intended outcome but it may well also bring about unintended outcomes. What Zinn (2005) refers to as the class of 'unpreventable uncertainties' is one outcome of the shift in focus from probabilistic risk on to a wider consideration of uncertainty.[20]

The posited class of unpreventable uncertainties has similarities with the work of Perrow (1984) on the concept of complex interactions; unfamiliar sequences, or unplanned and unexpected sequences, are either not visible or not immediately comprehensible.[21] When considered in the context of contemporary technological systems with high levels of complexity, the potential for an interaction of multiple, independent failures can be identified. The degree of potential interaction between apparently independent failures (even when they can be identified in advance) is such that it is arguably unrealistic for designers or operators to fully comprehend what outcomes are uncertain but preventable, or uncertain and not preventable. Given the nature of modern global construction projects, with their emphasis on tight budgets and timescales, they can be regarded as examples of environments characterised by high levels of tightly coupled systems; one event rapidly and invariably follows another in such a manner that the traditional responses of elimination or mitigation are not available to the project management team.

The consequence of an awareness of issues such as unpredictable uncertainties and complex interactions is that the management of risk and uncertainty has increasingly recognised a need to identify uncertainties within a project environment. This is then followed by consideration of what can be most effectively managed in terms of focusing effort on responding to those uncertainties that can be classed as unpreventable. This approach differs significantly from the traditional approach of seeking to transform uncertainty into certainty that can then be controlled. An example of such an approach can be found in Loosemore *et al.* (2006) and their suggestion that risk and uncertainty should not be regarded as distinct as doing so only serves academic, rather than practical, purposes.[18] In this context, the term 'risk' also encapsulates uncertainty in that it is defined as: 'A potential future event which is uncertain in likelihood and consequence and if it occurs could affect your company's ability to achieve its project objectives'.[18]

From this point forward, unless otherwise specifically stated the term 'risk' will be used in the manner proposed by Loosemore *et al.* in that it also comprises 'uncertainty'.

10.10 Uncertainty and success/failure ambiguities

Developing the suggestion that contemporary management of risk and uncertainty should not focus on calculated probabilities interpreted in a systemised manner that isolates them from the culture in which they are to be carried out (and which will also have a say in forming decisions on success and failure), leads to an approach that Chapman and Ward (2008) describe as 'common sense captured in a manner that provides structure to guide decision-making'.[22] Possibly the key drive in such an approach is to 'minimise the time spent on relatively minor sources with simple response options, so as to spend more time on major issues involving complex response options'. As part of 'streamlining' the risk management process, a first pass approach may make use of more assessments based on qualitative criteria than those based on quantitative criteria. Such an approach is fully reasonable in that neither base has any inbuilt superiority over the other (even though it may be argued, particularly from the traditional perspective, that objective, quantitative measures are 'better' in terms of accuracy than are qualitative, subjective measures). Nonetheless, it is important for the project manager to be aware, especially when dealing with multicultural environments typical of a global construction project, that subjective measures such as the frequently used assessment of risk as being 'high', 'medium' or 'low' are open to variations of interpretation due to differing cultural perspectives. This does not present an argument for the complete removal of subjective assessments – many experts can make rapid subjective assessments with a high degree of accuracy – but it does suggest that, as with any measure, there are circumstances where their use is more appropriate than others. The observation in a previous section (10.9) that contemporary risk management increasingly recognises that probability calculations should be interpreted in the context of the society or culture that the project will be carried out in, is also a factor in the use of subjectivity within project decision-making processes.

A further consideration is that however objective an individual attempts to be in their decision making, we are all human and therefore will inevitably include some component of subjectivity within any interpretation applied to a probability calculation. It is also worth noting that a probability calculation itself may well have some values within it that are wholly or partly subjective. It is therefore argued by Chapman and Ward (2008) that it is a sensible approach to explicitly articulate the perceptions of uncertainty that have affected an assessment by individuals.[22] In doing so, a basis for dealing with uncertainty as effectively as possible can be provided, through seeking to precisely define specific issues identified by individuals, motivating individuals to communicate about uncertainty in a clearer manner and providing clarity on what is of importance (or not) to the project.

Ultimately the most important item for a project is to deliver the stated/expected objectives successfully. Previous chapters have highlighted the need for clearly stated and valid success criteria and also the need for a project manager to be aware of the constraints on creativity that can be introduced by the development of a risk management approach more focused on avoiding failure than on achieving success. Perceived high levels of uncertainty in a project environment tend to focus individuals more towards avoiding failure than achieving success (due largely to it proving difficult to plan in detail with confidence, thereby achieving the traditional objective of risk management; turning uncertainty into certainty). A further issue to consider with regard to uncertainty is ambiguity: an expression capable of more than one meaning. This is an area of concern in this chapter, in that the impact of ambiguity can be similar to that of uncertainty in either of the forms that it can be manifest:

- when an individual can interpret a statement as having more than one meaning, he or she can become uncertain as to which meaning to respond to;
- when a manager becomes aware that a statement by a more senior manager can be interpreted in a number of different ways, he or she can become uncertain as to which meaning each individual within the project environment is responding to.

One technique that can be used to address this form of ambiguity is to develop and apply SMART goals and objectives (this will be addressed in more detail in a later section). When ambiguity is present in stated success criteria for a project, an area of uncertainty is introduced by the project manager, albeit unintentionally and frequently as a result of the problem of knowledge ownership not being consistent across all affected project players. A simple example of this can be found by considering a subject that you are particularly knowledgeable about – BMW motorcycles produced from 1920 to 1940, for example! Whatever the subject may be, you are then asked to explain its key items to someone with little or no knowledge of that same subject. How do you deal with this? Do you simply download every fact that you can remember? That would not be within the set criterion of explaining the key items. But how do you know what the key items are? Perhaps a little uncertainty begins to creep in – what if the person who tasked you with this job (who you know to be an expert on the subject also) is actually testing you? Perhaps they have a 'list' of key items that you need to match; if not, then you will 'fail'. From this point on, the level of uncertainty will gradually increase as you try to second-guess what it is that you have actually been tasked to do (even though the initial statement seemed perfectly clear). The real problem in this scenario is that the person tasking you has assumed that, because they know what they want you to do, you will also know. As a result, they have been somewhat sloppy in forming their requirements and ambiguity has been built into the description of the task, thereby creating ambiguous success criteria.

When presenting objectives in a global project environment the considerable level of potential for ambiguity to occur has to be a factor that the project manager and other members of the project team are constantly aware of. The sheer diversity of interpretations that is possible within the environment of a multicultural, multinational global project can seem overwhelming. However, a good place to start dealing with this is to appreciate some of the different perspectives on uncertainty that could be present within the project environment.

10.11 Global perspectives on uncertainty

In the interests of avoiding ambiguity, the intention in this section is not to judge any particular perspective on uncertainty as being 'better' or 'worse' than any other. The starting point is simply to recognise that all individuals will bring into the project environment their own particular perspective on what uncertainty is and how it should be responded to. Given the size and diversity of the human population, seeking to present even a summary of global perspectives on uncertainty may seem somewhat daunting. However, the task can be simplified somewhat by starting with what is possibly the simplest model of the range of perspectives possible: each perspective, no matter how rare or exotic, can be placed in one of five classes, as discussed below:

1. **Fatalism**: essentially summed up as the belief that 'what will be, will be'. However, it is more accurate to regard this perspective as believing that whatever happens, happens because there was no other choice; it had to happen. On this perspective, any action that an individual decides to take has already been predetermined, usually by a particular deity. An individual cannot therefore bring about any change in his or her future; they simply do what they have to do.
2. **Free will:** there is always freedom of choice. An individual can choose between alternatives and that will affect the future. Also referred to as 'self-determinism'.
3. **Determinism:** all events are the result of a cause (one or more other prior events) and all events within a universe are governed by causal laws and are therefore predetermined by those laws. A deterministic perspective does not allow for free will; all future events are pre-existent in that the chain of events that will ultimately cause them has already started.
4. **Indeterminism:** the possibility of events that do not have any cause, either physical or psychological. The future can be changed but the problem then becomes one of selecting the appropriate action to bring about a desired future.
5. **Chance:** events simply come about as a result of luck. Flipping a coin, for example, can be seen as a chance event in terms of the outcome (heads or tails). However, the flipping itself is causal and therefore 'determined'. The outcome nevertheless remains uncertain (in this perspective) and is therefore subject to chance. This is usually accompanied by the belief that luck can be 'created' or 'lost' by the actions of the individual (as in good-luck superstitions).

The project environment may well be populated by representatives of each of these perspectives on uncertainty, and this would produce a challenging mix of perspectives for the project manager to deal with. If, for example, a construction project was commenced in Southern Tagalog, Philippines, it would be considered unlucky to begin raising any roof supports in the form of posts in any other order than commencing with the most easterly one first, followed by each post in a clockwise direction. This suggests that the people of Southern Tagalog believe that they can, to some degree, affect their future. However, a project manager of a deterministic mindset would simply want to build in the most efficient (rather than most lucky) manner.

The perception of risk may be substantially different between individuals and groups (cultures) for a variety of 'everyday' reasons that can be argued to vary in response to the specifics of a given culture or location:

- Memorable events alter perception. Media articles, lots of deaths occurring at one time, sudden deaths occurring (e.g. airline crash, student shooting spree).

- Population risk versus individual risk alters perception – uncertain as to who will be the one in a million. People still play the lottery hoping to be that one in a million or even one in ten million.
- The way information is presented affects public perception (risk of death is 32 per cent versus risk of survival is 68 per cent say the same thing but can lead to different perceptions).
- Voluntary versus involuntary risks cause different reactions from the public; it is okay for me to put myself at risk but it is not okay for government or industry to put me at risk.
- Natural versus artificial risk are often viewed differently, even if the risk is the same (a reaction to spices versus pesticides, for example).[23]

10.12 Tools and techniques for managing risk/uncertainty

A range of tools and techniques for the management of risk and uncertainty are available to those involved in managing projects and it is not essential that all of these are covered here. Rather the approach will be to take a more strategic perspective and focus on four examples of differing tools and techniques that are particularly relevant to the complexities of global projects.

10.12.1 Restructure the risk

This technique is a relatively simple one in that it is based on a series of questions concerning the risk identified within a proposed project. The steps are usually identified as select, assess, size, restructure. This is in accordance with the risk hierarchy in commencing with selection of an issue (identify the risk) and followed by assessment of whether the issue merits further investigation (evaluate the risk). The second stage acts as a filter in order to focus resources (in terms of analysis, for example.) on the 'important' issues with regard to risk. The third stage is to provide some measure of the size of the uncertainty within the issue. Typically this 'sizing' will not be especially sophisticated, but as long as all involved are clear about and agree with this then it does not represent a problem. The main concern is that the same degree of 'unsophistication' is consistently applied to all issues so that results can be compared with an acceptable level of confidence. Once the size of the issue has been determined, the restructuring question can be considered. The guidance here is that if the issue is sized as being 'small' it does not merit the effort of restructuring and it can be flagged appropriately (not requiring special attention). When an issue is sized as 'large' then it becomes eligible to be considered for restructuring. The restructuring activity is focused on decomposing the issue so as to better understand the nature of the uncertainty that is being dealt with, so that an appropriate response can be put in place.

10.12.2 Risk transfer

Transferring risk to another party has a direct and an indirect benefit. The direct benefit is that the risk should be with a specialist who can handle the 'threat' more effectively. The indirect benefit is that there is no longer an uncertainty as to whether the risk would in some way increase. The two benefits combine to remove the immediate form of the risk and also any future changes in the risk. However, there are also two disadvantages to the transference of risk: the cost of transferring may be high, and you are missing any opportunities that may result from the risk changing in a positive manner (reduction of the impact), possibly as a result of changes elsewhere. In many circumstances it is arguable that sharing of risk is more appropriate than transferring it.

10.12.3 Risk sharing

The wholesale transferring of an entire risk to another party, while having the benefits previously outlined also has one possible problem – the tendency of the individual or group from which the risk is transferred then proceeds to forget all about it. Whilst it may seem the purpose of risk transfer is that it is no longer an individual or group's responsibility, it must be kept in mind that, even within the drafting of a contract, it is rarely feasible to completely isolate an individual or organisation from a risk through transferring it in its entirety. There are also valid reasons (a collective responsibility for risk throughout the project environment can be regarded as evidence of a risk-mature organisation) for the adoption of risk sharing. Risk sharing can be accomplished through a number of different mechanisms: gain sharing (not commonly used in construction projects but widely used in other industries), insurance and retention being three of the more commonly used mechanisms.

Gain sharing is based on the sharing of opportunities as well as the traditional sharing of threats. It can only be used effectively when partners have information that could create possible opportunities, and managers trust employees to cooperate. In some countries the social and legal systems may make it difficult to achieve one or both of these requirements. Insurance risks are typically those that are of low probability in terms of occurrence but of high impact if they were to occur. Natural disasters are typical of the kind of risks that fall into this category but man-made risks can also be dealt with through insurance. The insurance company will typically balance occurrence against impact, add an overhead for operational costs and then add a profit margin. In those societies where the culture is of a more fatalistic perspective, obtaining insurance cover for the sharing of risk may prove difficult. Retention is a very traditional means of sharing risk in the construction industry and operates on the simple basis of the employer (client) in a project withholding a percentage (usually in the order of 3–5 per cent) of each payment due (typically on a monthly basis) to cover any defects or the contractor ceasing to trade.

10.12.4 SMART objectives

Objectives are the detail of any project plan in that they relate to actions that can be defined and communicated effectively to others. SMART is a structure that has been in use for some time and is, in many projects, standard. While the structure is well known it is not always well understood, in that some managers believing they are being SMART are actually far from it. SMART can be broken into:

1. **Specific**: an objective must be well defined, detailed and focused.
2. **Measurable**: uses a measurement base that allows monitoring of results in a consistent manner.
3. **Achievable**: should set the right balance between being sufficiently short-to-medium-term to retain motivation and also challenging enough to retain interest.
4. **Realistic**: achievable objectives may not be realistic in that you may not have the resources (but may have the knowledge) required or may require a fundamental change of priorities within the project.
5. **Time-bound:** deadlines need to be set for objectives to be achieved within and allow for the required focus, incentivisation and setting of priorities.

By using the SMART structure to ask pertinent questions, areas of uncertainty can be identified and assessed, in a subjective manner initially, before decisions are made regarding appropriate priorities within the planning and execution of the project.

10.13 Chapter summary

Uncertainty and risk within project environments are in a complex relationship with all aspects of a project's planning and execution. A fuller understanding of the nature of risk (particularly when it is viewed as also comprising uncertainty) is relevant to the project manager of a global project being effective in responding in an appropriate manner to the actual (as opposed to supposed) risk within the constraints of the available resources, and while acknowledging the societal and cultural context of risk perception. In project management terms, a range of responses are available to the project manager, of which four (restructuring the risk, transferring the risk, sharing the risk, and setting SMART objectives) are relevant as an indication of what the global project manager may choose to implement.

10.14 Discussion questions

1. Outline the nature of the problem faced by project managers with regard to intergroup stereotypes.
2. Discuss the nature of the suggested need to link risk to project objectives and how this differs from the concept of uncertainty.
3. What is the impact of uncertainty with regard to 'sensitising' an individual or group to a particular risk?
4. Outline the manner in which 'availability' modifies an individual's perspective on a specific risk.
5. Discuss the difference between fatalism and determinism with regard to perspectives on risk.

10.15 References

1. Hadley, B. J. (2009). *Duplicity/complicity: Misperforming the social drama of disability*. In Proceedings of the PSI 15: Misperformance: Misfitting, Misfiring, Misreading, Performance Studies International Conference No.15, 24–28 June, University of Zagreb, Zagreb, Croatia.
2. Reiss, G. (1992). *Project Management Demystified*. London: E and F. N. Spon.
3. Milan, A. G, Russell. D., Ronan, P., Spring D. and Martinovski B. Globalization of stereotypes. Sweden: School of Business and Informatics, University of Borås. http://nic.hb.se/assets/media/globalization_of_stereotypes.pdf [Accessed June 2012].
4. Zenger, J. and Folkman, J. (2012). Are women better leaders than men? http://blogs.hbr.org/cs/2012/03/a_study_in_leadership_women_do.html [Accessed August 2012].
5. Cooke-Davis *et al.* (2007). Red Orbit. Mapping the strange landscape of complexity theory and its relationship to project management. http://www.redorbit.com/news/science/1015233/mapping_the_strange_landscape_of_complexity_theory_and_its_relationship/ [Accessed June 2011].
6. Gassinski Model Forest. http://www.mtnforum.org/sites/default/files/pub/1387.pdf [Accessed August 2011].
7. Project Stakeholders. http://www.projectstakeholder.com/ [Accessed July 2011].
8. Beloe, S. (2011). Avoiding bear trapsa: An investor tool for identifying and managing business ethics. In *Risk and Reward: Shared Perspective*, ACCA.
9. Hillson, D. (2005). When is a risk not a risk? http://www.risk-doctor.com/ [Accessed January 2013].
10. Global Construction Perspectives (2013) *Global Construction 2020*. Oxford Economics, http://www.globalconstruction2020.com/
11. 2009 *Future angst? Brain scans show uncertainty fuels anxiety*. http://www.physorg.com/news169755202.html [Accessed June 2011].
12. http://www.hse.gov.uk/ [Accessed August 2011].
13. Noise-Induced Hearing Loss (NIHL) in Great Britain. http://www.hse.gov.uk/statistics/causdis/deafness/index.htm [Accessed July 2011].
14. Zinn, O. J. (2006). Recent developments in sociology of risk and uncertainty. http://www.qualitative-research.net/fqs-texte/1-06/06-1-30-e.htm#g3 [Accessed January 2013].
15. Keil, M., Tiwara, A. and Bush, A. (2002). Reconciling user and project manager perceptions of IT project risk: A Delphi study. http://www.blackwell-synergy.com/doi/pdf/10.1046/j.1365-2575.2002.00121.x?cookieSet=1 [Accessed June 2012].
16. Cascio, W. F. (1995). *Managing Human Resources*. New York McGraw-Hill.
17. Stephens, M. (2007). *Early Detection of Alzheimer's Disease: The Influence of the Availability Heuristic and Person Perception*. University of Colorado.
18. Loosemore, M., Rafferty, J., Reilly, C. and Higgon, D. (2006). *Risk Management in Projects*. 2nd edn. London: Taylor and Francis.
19. Beck, U. (1992). *Risk Society: Towards a New Modernity*. London: Sage.
20. Zinn, O. J. (2006). Recent developments in sociology of risk and uncertainty. http://www.qualitative-research.net/fqs-texte/1-06/06-1-30-e.htm#g3 [Accessed July 2012].
21. Perrow, C. (1984). *Normal Accidents: Living with High-Risk Technologies*. New York: Basic Books.
22. Chapman, C. and Ward. S. (2008). *Project Risk Management*. 2nd edn. Chichester: John Wiley and Sons.
23. *Risk Perception*. Carnegie Mellon. http://telstar.ote.cmu.edu/environ/m3/s6/08perception.shtml [Accessed June 2011].

11

Global Sustainable Development and Construction

11.1 Introduction

The output of the global construction industry has a major impact on our environment. If we cannot meet our agreed targets for the environmental impact of infrastructure and buildings, then we have to change the way we design and build. The main aim of this chapter is to explore how to implement and manage sustainability in global construction projects. In addition, we will set out specific actions by industry that will contribute to the achievement of overarching targets in the next 20 years.

11.2 Learning outcomes

> The specific learning outcomes are to enable the reader to gain an understanding of:
>
> >> sustainable development;
> >> sustainable construction;
> >> corporate social responsibility;
> >> best sustainable practice.

11.3 Background to sustainable development

11.3.1 Origins and definitions of sustainable development

As the pace of economic development and industrialisation has accelerated there has been growing concern about the impact this has on the long-term viability of the planet's natural resources, not only through the use of the resources such as raw material but also due to increased global warming mainly associated with CO_2 emissions. The increased pressure on the Earth's resources is of major concern, particularly in urban settings that host many social, economic and environmental problems. However, the concept of sustainable development is not a new phenomenon and there are many early examples of the concept being applied, for example within forestry where there is evidence of the balancing of consumption and reproduction from as early as the twelfth century. The concept of sustainability has even earlier roots when in 400 BC, Aristotle described the emerging concept that the household had to be self-sustaining. Although the environment is the foundation upon which sustainable developments are built, the social and economic challenges that arise are equally important. In the 1970s, the term sustainability started to be more frequently used to describe the need to have equilibrium between the economy, environment and social wellbeing.

Short-term economic pressures, mainly driven by the demands of shareholders for early returns on their investments, have led many organisations to focus on making a profit with little regard for the negative environmental and social impacts that often have to be borne by others. Deforestation of the rain forest, the use of child labour to provide cheap goods, and industrial disasters such as Bhopal, Chernobyl and British Petroleum Deepwater Horizon drilling rig explosion are only a few of the many examples. Sustainability principles can be applied to a wide range of activities such as urban planning, construction, manufacturing, agriculture and banking. Governments worldwide have thus recognised the importance of sustainability to national security and prosperity. Although more firms are recognising the economic savings and improved reputation that can result from more sustainable approaches to the social and environmental impacts of their projects, it has taken extensive government legislation and initiatives to drive the required change in many organisations, some of which are discussed below:

- At the United Nations Conference (1972),[1] the principles behind sustainable development resulted in the following definition: 'the interdependence of human beings and the natural environment; the links between economic development, social development, and environmental protection; and the need for a global vision and common principles'.
  The United Nations published the Brundtland Report in 1987.[2] This was based on the work of the World Commission on Environment and Development, which highlighted the underlying principles behind sustainable development (i.e. equity, viability and liveability) and defined sustainable development as:

  > Development that meets the needs of the present without compromising the ability of future generations to meet their own needs ... sustainable development is a process of change in which exploitation of resources, the direction of investments, the orientation of technological development, and institutional change are all in harmony and enhance both current and future potential to meet human needs and aspirations.

- Sustainable development has received much attention globally. Various definitions have been developed out of which two have been found to be more relevant to this subject. The first is adopted from Caring for the Earth, IUCN/UNEP 1991;[3] which defines sustainable construction as 'improving the quality of human life while living within the carrying capacity of supporting ecosystems'. Sustainable development will have little meaning especially in developing nations if it is not seen to improve the quality of human life.
- The International Council for Local Environmental Initiatives[4] (ICLEI 1996) defined sustainable development as being 'development that delivers basic

environmental, social and economic services to all residences of a community without threatening the viability of natural, built and social systems upon which the delivery of those systems depends'.

- The Kyoto protocol was first established in 1997 and involved 55 industrial countries responsible for 55 per cent of the world's greenhouse gas emissions.[5] The protocol's aim was to establish ways of dealing with global warming. Today, 156 countries have signed up to the protocol. However, this does not include the US and Australia.

- The Millennium Summit (2000) was attended by the largest gathering of world leaders who ratified the United Nations Millennium Declaration,[6] including the following agreed millennium development goals aimed at encouraging development by improving the social and economic situation in the world's poorest countries: Goal 1: eradicate extreme poverty and hunger; Goal 2: achieve universal primary education; Goal 3: promote gender equality and empower women; Goal 4: reduce child mortality rates; Goal 5: improve maternal health; Goal 6: combat HIV/AIDS, malaria and other diseases; Goal 7: ensure environmental sustainability; Goal 8: develop a global partnership for development.

- The United Nations 2005 World Summit,[7] a follow-up to the Millennium Summit, highlighted the 'interdependent and mutually reinforcing pillars' of sustainable development as economic development, social development and environmental protection.

11.3.2 Sustainable construction

Sustainable construction is the way of providing the necessary built environment and associated infrastructure whilst improving the inhabitants' quality of living or working. The close relationship between the terms 'sustainable development' and 'sustainable construction' has resulted in the terms being highly interchangeably in practice. Kibert defined sustainable construction as 'the creation and responsible management of a healthy built environment based on resource efficient and ecological principles'.[7] Since the Brundtland Report, there has been growing recognition of construction's important role within sustainable development. The design, construction and use of the built environment has direct economic benefits but also impacts on the rate at which we consume our natural resources. Whatever methods, material or technology, are adopted during the construction phase to mitigate their environmental impact, it is unlikely that there will be no irreversible impact on the natural environment. According to the Global Construction Report (2010),[8] 'buildings are responsible for almost half of the country's carbon emissions, half of our water consumption, about one-third of landfill waste and one-quarter of all raw materials used in the economy'.

Sustainable construction should not only take account of environmental impacts on the natural and built environments, but due consideration also needs to be given to the impact on social factors such as health, education and crime. Many of the problems associated with sustainable development are often attributed to the construction sector. However, construction projects are generally in response to a pressing social need. Despite construction's potential impact on the natural environment, it has a vital role to play in sustainable growth and development. There are many societal issues, for example population growth, that need serious attention if the goal of sustainable development is to be achieved. The construction of new road infrastructure will invariably cause some damage to the natural environment, for example, through the extraction of raw materials; resulting changes in water run-offs; changes in human and wildlife habitats. However, better transportation provides many lasting benefits, including access to vital services such as health and education. Achieving, measuring and demonstrating sustainable development and sustainable construction is complex and involves the consideration of many highly interlinked issues connected through social, environment and economic systems, which in their own right have a high degree of complexity. Adetunji *et al.*[9] considered the issues that contribute to sustainable construction and concluded that: from an environmental and social standpoint, there needs to more emphasis on whole-life costs and the consumption of energy, carbon, waste and materials – and although the construction sector has sufficient expertise to deal with these issues, it has yet to fully understand many of the associated social needs and impacts such as health, wellbeing and equity.[10]

The joint UK government/industry *Strategy for Sustainable Construction* was published in 2008 and presents a shared vision of 'construction as a competitive sector which plays a central role in delivering sustainability and prosperity across the economy'. The strategy aims to 'promote leadership and behavioural change, as well as delivering substantial benefits to both the construction industry and the wider economy'. Although it complements the 2007 Action Plan for Civil Engineering it does not include some of the broader issues such as planning, the management of existing assets and transport policy.[11]

11.3.3 Sustainable development and the EU

The European Union (EU) is committed to sustainable development, whether it be maintaining and increasing long-term prosperity, addressing climate change or working towards a safe, healthy and socially inclusive society.[12] This is reflected through the EU's green papers, white papers, directives, decision papers and communication instruments, of which there are many. It is neither practical nor necessary to discuss each of these instruments in detail. However, of particular interest to the construction sector are the EU's ambitious and comprehensive Sustainable Development Strategy and Construction Products Directive. The first Sustainable Development Strategy (SDS) was produced in 2001 and reviewed in 2004 and 2006. As part of this strategy, all EU institutions were encouraged to ensure that major policy decisions had been supported by a high-quality Impact Assessment (IA) on the social, environmental and economic issues compared to the costs associated with inaction.[13] However, in some areas, environmental assessment is not simply encouraged but mandatory for plans and programmes likely to have significant environmental effects or those for agriculture, forestry, fisheries, energy, industry, transport, waste management, water management, telecommunications, tourism, town and country planning, or land use.[14] The Construction Products Directive (89/106/EEC) is also an important EU policy instrument, which requires that construction products bear the Conformité Européene (CE) marking,[15] thus demonstrating that the construction product meets the required technical specifications (standard or approval) and has properly gone through a product assessment.[16] Many of these directives and policies have been translated into corresponding sustainable development strategies by individual EU member states.

11.3.4 The triple bottom line

Sustainable development requires the management of economic, social and environmental capital, frequently referred to as the triple bottom line, as presented in Figure 11.1 and defined below:

- **Economic capital:** financial capital and goods that have been manufactured (or constructed) and tend to be of a physical nature such as buildings, vehicles, plant and machinery.
- **Social capital:** the institutions, relationships and norms that shape the quality and quantity of connections among individuals, and society's social interactions, such as social cohesion, mutual benefits, reciprocity and fellowship.
- **Environmental (natural) capital:** the natural environment in the form of natural assets, such as water, land, air, minerals, natural forests that provide a flow of useful goods or services.

The consumption of these three types of capital are usually non-substitutable and irreversible. For example, it may not be appropriate to substitute natural capital with economic capital (although it may be possible to manufacture material to replace some types of natural resource, it is unlikely that they will ever be able to replace the ozone layer or the rainforest). However, the triple bottom line can be considered from a different perspective, as suggested in the Brundtland Commission Report, in which the interlinkages between economic development, environmental degradation and population pressure were emphasised. Rather than having three types of capital, the triple bottom line can be expressed in terms of four capitals with the addition of:

- **Human capital:** properties of individuals usually expressed in terms of health, education, skills and knowledge.

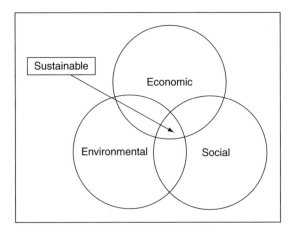

Figure 11.1 Triple bottom line

Sustainable development principles have been well accepted within the construction industry, with many companies integrating sustainable strategies and policies. The emphasis has tended to focus on environmental and economic issues, however, to fulfil the requirements of the social dimension of sustainability there needs to be a better understanding of associated issues such as social capital and corporate social responsibility (CSR).

11.3.5 Resilience of the global construction projects

The frequency and severity of recent disasters has resulted in greater recognition of the need to improve the resilience of the built environment and its critical infrastructures such as utilities, transport and communication networks and hospitals. United Nations Development Programme in India defined a disaster as 'a serious disruption of the functioning of society, causing widespread human, material or environmental losses which exceed the ability of affected society to cope using only its own resources'.[17] Although the characteristics of disasters vary considerably depending on their origins, they can usually be classified as either natural or technological.[18] However, most disasters 'are not always singular or isolated events',[19] and many potential hazards have to be considered when assessing resilience and developing appropriate response strategies. For example, the 2011 Japanese disaster comprised an earthquake, tsunami and nuclear leaks. The 2010 Port au Prince earthquake (Haiti) and the 2004 Indian Ocean tsunami have illustrated the different types of risks and their impact on health, with many health-care services not being available when they were most needed.

> There are countless examples of health infrastructure – from sophisticated hospitals to small but vital health centres – that have suffered this fate. One such case occurred in the Hospital Juarez in Mexico. In 1985, almost 600 patients and staff lost their lives when this modern (for its time) and well-equipped hospital collapsed in the wake of an earthquake.[20]

Natural disasters affect most parts of the world, with a recent dramatic increase in frequency: threefold between 1945 and 1975; sixfold between 1975 and 2010. According to United Nations International Strategy for Disaster Reduction (2009) there were 3501 disasters between 2000 and 2008 (excluding epidemics, insect infestations and technological disasters) affecting over two billion people, and resulting in over 770,000 deaths and damage of £570 billion.[21]

Some of these events, such as flooding, can directly be linked to climate change, others such as earthquakes can have a greater impact because of the effects that climate change has already had on the resilience of the natural environment. Increased population growth and urban development, urban drift and the quality of the built environment all magnify the impact of natural disasters. Recent floods in the UK highlighted the fragility of the built environment and supporting infrastructure. The 2007 floods proved to be a major turning point for UK strategy, resulting in first UK National Risk Registry and National Security Strategy aimed at spreading knowledge and improving resilience.[22] Resilience is thus being recognised as an important sustainable development consideration and, although the number of people affected by disasters has increased since the start of the twentieth century, the number of fatalities has, encouragingly, decreased.

11.4 Social capital

11.4.1 Background to social capital

The concept of social capital has become a dominant paradigm in the quest for sustainability and is increasingly being seen as a powerful instrument towards: the improvement of human capital such as health and happiness; increased economic development; better functioning schools; safer neighbourhoods; more responsive governments.[23] The World Bank's view of the role of social capital as part of sustainable economic development features high in contemporary literature on development economics. Regions or countries with relatively higher stocks of social capital (measured in terms of generalised trust and widespread civic engagement) tend to have higher levels of growth compared to those with lower levels.[24,25,26] The design and form of cities, neighbourhoods and individual buildings will significantly impact upon the way in which people interact and bond with each. Some urban designs encourage social ties and informal contact among residents, while others do not. It is therefore important to understand the key determinants of social capital within an urban context.

11.4.2 Definition(s) of social capital

The concept of social capital draws on a wide range of disciplines and there are many definitions, which can be categorised under the four broad subject areas of anthropology, sociology, economics and political science, as illustrated in Figure 11.2 and summarised below:

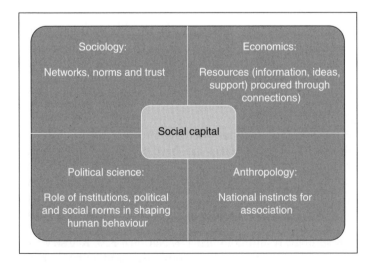

Figure 11.2 Definitions of social capital

- **Anthropology:** social capital is embedded in the belief that humans have a natural instinct for association and social order.[27]
- **Sociology:** emphasises the features of social organisation such as trust, honesty, reciprocity and civic engagement.[28]
- **Political science:** emphasises the role of institutions, political and social norms, trust and networks in shaping human behaviour.[27]
- **Economics:** emphasises the time-investment strategies of individuals assuming that they endeavour to maximise their personal utility and decide to interact and draw on social capital resources to perform group activities.[29]

11.4.3 Types of social capital

There are three main types of social capital: social bonds, bridging and linkages, as illustrated in Figure 11.3 and expanded upon below:[30]

- **Bonding social capital:** the relations among members of families and ethnic groups that are effective in sustaining group solidarity and support for individual group members.[31]
- **Bridging social capital:** tends to be more outward looking and is usually associated with relationships with distant friends, associates and colleagues, thus bringing together individuals from different social groups and encouraging cross-cultural understanding and tolerance.[32]
- **Linking social capital:** the relationships between different social hierarchy levels, leading to open and accountable relationships between citizens and their representatives.[33,34]

Given the wide range in perceptions relating to social capital, definitions vary depending upon the viewpoint of the analyst. However, sociological definitions appear to be more relevant to the concerns of urban sustainability. Social capital

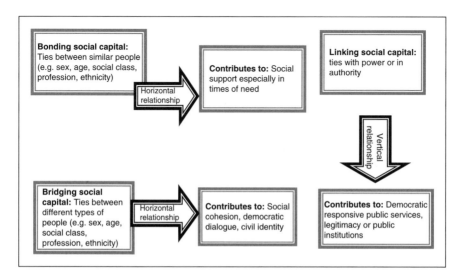

Figure 11.3 Types of social capital

is thus best considered as 'the collective value of all social networks and the inclinations that arise from these networks to do things for each other, i.e. social networks and the norms of reciprocity and trustworthiness that arise from them'.[35]

11.4.4 Why social capital in urban sustainability development?

Economic studies suggest that social capital can help to make workers more productive, firms more competitive and nations more prosperous.[36] It also has positive external value in so much that it is not limited to those within the networks, but extends to those outside the system, thus having a ripple effect on a wider cross-section of a community, including those not practically participating in the network. There is clear evidence that demonstrates how social capital can impact on the wellbeing of individuals, organisations and nations.[37] This impact can be best explored by drawing upon the four schools of thought (i.e. anthropology, sociology, economics and political science) as highlighted above.

Social capital has many practical benefits to individuals and communities, which need to be given appropriate attention as contributors to achieving sustainable urban development. It can have a significant impact on health and wellbeing, as demonstrated by Durkheim[38] and Wasserman,[39] who identified a relationship between the level of social integration and suicide rates. It can also have many other impacts, as demonstrated by sociology research, which indicates that improved social capital can enhance relationships and problem solving among members of groups.[40,41] There is a growing body of evidence that suggests that the size and density of social networks, and the nature of interpersonal interactions, are significant determinants of the sustainability of development projects.[42]

The role of social capital within urban sustainability can be examined by referring to the UK government sustainable development policy, particularly the sustainable communities' agenda, which defines sustainable communities as:

> Places where people want to live and work, now and in the future. They meet the diverse needs of existing and future residents, are sensitive to their environment, and contribute to a high quality of life. They are safe and inclusive, well planned, built and run, and offer equality of opportunity and good services for all.[43]

Although it is relatively easy to explain the contribution of social capital to sustainable urban development, it is more challenging to identify how, or even if, social capital is being created, the determinants of social capital in an urban development context, and how much social capital is needed to ensure sustainable urban development.

11.4.5 Social capital and the physical urban environment

Physical and social urban environments are intertwined, and few would argue that society should be treated in isolation from its physical environment. The urban environment, as a physical setting where people live, is both a determinant for and a consequence of social relationships.[44] Despite this seemingly obvious symbiotic correspondence, there is relatively little research evidence that defines the relationship between the physical urban environment and social capital. The most significant work in this area is that of Putnam *et al.* who helped to improve understanding of what constitutes social capital and devised strategies aimed at increasing civic engagement.[36] Putnam's work identified '150 things you can do to build social capital' and made recommendations on methods to replenish it in five categories: the workplace, the arts, politics and government, religion, and youth and education. However, further work is needed in order to better understand what individuals and groups need to do to generate social capital.

11.4.6 Physical determinants of social capital

If the debate is to move on, there needs to be a focus not on what the individuals and groups can do, but what the physical urban environment needs to be in order to help in the creation of social capital. Table 11.1 thus summarises some of the key physical determinants of social capital within an urban setting.

Although the above list provides an excellent starting point for an analytical framework, it is by no means a comprehensive list of (physical) determinants of social capital within an urban setting. Many of the identified factors relate more to residential areas than other forms of urban environments as most social capital manifests itself in places where people live, although workplaces are also responsible for a considerable amount of social capital generation. Given the complex multidimensional characteristics of social capital, a clear under-standing of its role in urban sustainability assessment is best achieved based on a selection of the key determinants. Although a holistic approach has its advantages, Glasson *et al.* highlighted the impracticalities of comprehensiveness

Table 11.1   Physical determinants of social capital

Determinants	Explanations
Pedestrian-oriented designs	Decline of daily walking and cycling associated with lower social capital
Mixed-use and clustered developments	Limited household variety and mix discouraging social capital. Clustered developments maximise number of people within walking distance. Social polarisation is identified with large estates in outer suburbs, and a particular social class
Proximity to public transport	Increases physical interaction
Effective lighting	Safety and security issues
Public spaces	Increase in social interaction
Houses with front porches	Increases in social interaction
Sidewalks	Increases permeability and therefore interaction
Open space designs	As opposed to gatedness
Proximity to local amenities and infrastructure	Local tavern, local coffee shop, post office, schools, police station, resource centres, etc. within walking distance
Mixed-use recreational facilities	Recreational facilities meeting the requirements of all social classes have the potential of enhancing interaction
Children's play areas	Both the children and their parents/guardians will have a chance for physical interaction

when dealing with sustainable urban development.[45] By focusing on a carefully thought-out shortlist of the social capital determinants it is possible, however, to develop a predictive model. Although there are very few effective predictive models of social capital, as stated, the physical factors summarised above provide an ideal starting point for the construction of such a model that enables the physical urban environment's ability to generate social capital. Complexity theory considers most social and physical systems as complex and adaptive, and comprising large numbers of interacting agents. According to Waldrop, any system coherence (order) starts with the interaction between the individual agents, which is critical to self-organisation of the system as a whole.[46]

11.5 Corporate social responsibility

11.5.1 Globalisation and corporate growth

Increased globalisation and the drive for improved performance through increased size has resulted in many firms becoming so large and powerful that it has been necessary to reflect on and reassess the role of the firm in society. Margolis and Walsh emphasised that the role of the firm needs to change in response to increasing environmental, economic and social concerns, but also new roles emerge from the reality of having to take on new responsibilities imposed by clients or government legislation.[47] Increased globalisation has also resulted in many firms operating in countries with fewer laws, regulations and welfare rights than their own countries. As a result, many global players, at least in the recent past, became used to applying different standards depending upon in which country they were operating. The Bhopal disaster highlighted this issue. However, it is becoming increasingly difficult to apply such inequalities and double standards, especially when developing an organisation's strategy, ethical standards and international reputation. Many organisations, including those operating in the construction sector, need to assess and report on the performance of their supply chain for compliance with the standards they aspire to. The recently published *ISO 26000* provides voluntary guidance on social responsibility worldwide (more on this in Section 11.5.8). It informs public and private organisations on how to operate in a socially responsible way, based on a distillation of international experience. The standard includes: terms and definitions, social responsibility principles, core guidance on social responsibility, and guidance on implementation.

11.5.2 Corporate social responsibility and corporate responsibility

Corporate social responsibility (CSR) was defined by Hopkins as being: 'concerned with treating the stakeholders of the firm ethically or in a responsible manner'.[48] The World Business Council for Sustainable Development provided an alternative definition and defined CSR as: 'the continuing commitment by business to behave ethically and contribute to economic development while improving the quality of life of the workforce and their families as well as of the local community and society at large'.[48] The UK government stated that by 'adopting socially and environmentally responsible behaviour businesses can make a significant contribution to boosting wealth creation and employment, fostering social justice and protecting the environment.'[49] It has been more widely accepted that, as with many other sustainability practices, CSR should not be treated as an optional 'add-on' but needs to be well integrated into normal business practice if it is to be meaningful and accepted by a firm's shareholders, directors, employees and the wider community.

The Department for Business Innovation and Skills (BIS) leads the UK government's interest in corporate responsibility (CR), defined as: 'how companies address the social, environmental and economic impacts of their operations and so help to meet our sustainable development goals'. This means firms taking into account the economic, social and environmental impacts that arise from their operations. To ensure a holistic sustainability approach and avoid overemphasis on the social aspects, the UK government have adopted the term corporate responsibility, rather than corporate social responsibility, when referring to the voluntary actions a firm undertakes towards sustainability rather than an approach that is purely aimed at compliance with current legislation. As part of the UK government's strategy for achieving corporate responsibility, they have developed a policy framework that: provides minimum performance levels relating to health and safety, environmental impact and employment practices; encourages voluntary performance above the minimum standards; sets out to encourage and support innovation and the application of best practice. The government has also recognised the role it has to play in creating the right climate and strengthening the business case for corporate responsibility. The BIS Corporate Responsibility Report builds upon the 2004 publication of *Corporate Social Responsibility – A Government Update.*[49]

11.5.3 Corporate responsibility and profitability

As with many social issues, it is difficult to place a monetary value on, and business case for, corporate responsibility (CR), since many of the benefits are subjective and perceived differently, depending on individual and group perceptions of value.[50] Porter and Kramer (2003) in the 1970s',[51] along with many others, argued that there was a case for organisations pursuing CR with a focus on profits rather than fully engaging in social responsibility; however, this view is not widely held today. This may in part be to avoid expensive litigation but also to protect brand image. However, there has been increased understanding of how to value a wide range of environmental and social impacts and how these can be embedded with business case development. Also, many social factors associated with CR can easily be quantified, such as high staff turnover measured in terms of loss of human capital with the subsequent loss of tacit knowledge, lost productivity, recruitment costs and additional training costs. Many construction firms have started to put considerable effort into changing the image of construction, in part to attract and retain the best staff and customers. CR has been an excellent starting point in this culture-change process. An organisation's approach to CR will impact on how it conducts itself within the communities it serves, whether they be local, national or global. Good CR can help to improve: staff recruitment, retention and productivity; stakeholder and shareholder loyalty; working practices. It also can ensure that the organisation stays ahead of future legislation and competition. However, these will only be achieved if CR is embodied wholly within the business and not merely as a risk-minimisation technique.

11.5.4 Developing a social conscience: business ethics

The terms business ethics, corporate responsibility, social responsibility and stakeholder engagement have become commonplace throughout construction. However, in today's society, organisations are expected not only to behave according to certain moral standards, but they also need to demonstrate a level of social responsibility and accountability to a range of stakeholders that includes employees, customers, suppliers and the wider community.[50]

11.5.5 Construction ethics

When dealing with complex ethical issues in what can be stressful work environments, codes of practice, vision statements, policy documents and value statements are good starting points. However, just having a code of ethics does not guarantee good ethical practice within an organisation or company. According to McNamara many ethicists believe that developing and continuing dialogue around an organisation's values is critical in this regard, with the code being a live document frequently updated based on reflection and dialogue, thereby creating a consensus.[53] The initial stages in the development of a 'code of ethics' are not easy, especially for construction companies that operate in a sector that is mainly client-led, with procurement trends often dictating the pace of change within the industry. Many construction organisations have in the past waited for their client to demand change before taking the appropriate action. An example of this is the introduction of quality assurance and partnering, which has tended to be client led. Such a reactive approach is not ideal and can create problems with materiality or contextualisation, as clients tend to have different stakeholders and priorities. Construction has traditionally based its responsibilities and reporting procedures on operational and financial parameters; however, over the past two decades there has been a shift towards a broader range of these responsibilities and investment in the broader development agenda.[51] The transient nature of the workforce and the way in which major projects are undertaken has been a barrier to CR adoption and subsequent integration. The fact that most construction projects differ due to the construction technique, design or simply the location in which it is to be built, creates additional challenges to the introduction of CR. However, this can to some extent be mitigated by establishing long-term relationships throughout the supply chain.

11.5.6 Shareholders and stakeholders

There are often conflicting requirements of shareholder and stakeholder groups. However, CR is being increasingly used as a measure of company performance. Many financial indices have been developed to enable shareholders and stakeholders to make informed decisions regarding where to invest. Many organisations report performance in this regard on the Internet, thus reaching a wide audience, although this is not as easy as it might initially appear as such indicators have to fit in with existing project management structures, whilst at the same time being meaningful to a diverse range of stakeholders.

11.5.7 CR steps

CR has been achieved to varying degrees in different organisations and sectors. Adopting the following steps should help firms to start the process of embedding CR concepts within their mission statements, culture and business practices:

1. Determine where the organisation currently is with regard to its approach to internal and external CR.
2. Appoint, as a champion, a senior director who is passionate about how CR can improve the business.
3. Perform a stakeholder analysis determining the key stakeholders, what information they require, how the business affects them.
4. Align the business objectives with the results from the stakeholder analysis.
5. Develop the business case for adopting CR based on the need for 'sustained competitiveness'.
6. Address internal CR issues first in order to gain credibility and acceptance of the CR principles being embedded within the organisations.
7. Address external CR issues throughout the organisation's supply chain.
8. Provide the CR information in a transparent manner that is accessible to all stakeholders and shareholders.

11.5.8 Reporting corporate and social responsibility

There is increased recognition of the need to promote CR internationally and the UK government has been working closely with business to develop frameworks and guidance. The UN's Special Representative on Business and Human Rights (SRSG) proposed a framework based on three core principles:[55]

- the state duty to protect against human rights abuses by business;
- the corporate responsibility to respect human rights;
- the need for more effective access to remedies.

Over 100 developing and developed countries were involved in drafting the Global Social Responsibility Guidance Standard, ISO 26000, which was published in November 2010. The standard covers voluntary guidance on social responsibility for public and private organisations. It is intended to inform and advise organisations on how to operate in a socially responsible way and provides guidance on implementation. According to ISO 26000 Clause 7.5 Box 15 – Reporting on Social Responsibility: 'an organisation should at appropriate intervals report about its performance on social responsibility to the stakeholders affected'. Social responsibility was defined as:[56]

- [the] responsibility of an organisation for the impacts of its decisions and activities on society and the environment, through transparent and ethical behaviour; that
- contributes to sustainable development, including health and the welfare of society;
- takes into account the expectations of stakeholders; is in compliance with applicable law and consistent with international norms of behaviour; and
- is integrated throughout the organisation and practiced in its relationships.

The Global Reporting Initiative (GRI) is a network-based non-governmental organisation that aims to drive sustainability and environmental, social and governance (ESG) reporting. GRI produced a sustainability reporting framework that provides principles and indicators that organisations can use to report their economic, environmental and social performance. This GRI and ISO 26000 (2011) publication is intended to relate the social responsibility (SR) guidance in ISO 26000 to reporting guidance provided by GRI (GRI and ISO 26000: How to use the GRI Guidelines in conjunction with ISO 26000).

11.6 Sustainability assessment

11.6.1 Need for incentives in sustainability assessment

Sustainable urban development needs effective tools and metrics for measuring, assessing and monitoring systems to support key decision-making processes and make sure they are consistent with sustainability principles. Incentives to undertake sustainability assessment and the adoption of tools are essentially embedded within a cocktail of legislative instruments at various levels of governance and are channelled through the agency of different users. Policymakers, local authorities, planners, developers, consultants and quasi-non-governmental organisations (quangos) all make key decisions affecting the urban environment, which require the use of effective sustainable assessment tools. However, their needs will change depending on the type of sector or profession, the stage in the project life cycle at which the decision is being made, and the spatial scale of urban development that is being assessed. The project life cycle includes: planning, design, manufacture, construction, operation and decommissioning. Consequently, there are many different sectors or professions involved in the delivery and use of the built assets, for example construction, manufacturing, transport and utilities. Although there are different approaches, the spatial scale can be categorised as an urban development in terms of: building element (materials), project, neighbourhood, and city, regional, national or global levels.

The increase in the prominence of sustainable development has driven the need for holistic assessment mechanisms that take into account economic, environmental and social impacts of urban development, and it has resulted in a burgeoning of tools and metrics for assessing sustainable development. The complex nature of sustainable development and the prevalence of a multifarious array of assessment tools created a need for better integrated tool kits. However, there is a lack of tools to holistically predict, monitor or evaluate social sustainability impacts on urban development, which may in part be attributed to how social sustainability is perceived and translated into development criteria.

11.6.2 Barriers to sustainability assessment

Sustainability assessment is an intellectually challenging task due to: the need for transparent assessment of complex issues (i.e. environmental, economic and social); the multiplicity of stakeholders with diverse values and differing perceptions about the concept of sustainable development; the need to develop solutions that integrate different systems and stakeholders; the dynamic nature of the urban environment. According to Meppem and Gill, the complexity of the environmental, sociocultural and economic systems can hinder conventional processes of scientific verification.[54] Many sustainability assessment issues are deeply rooted in societal structures and institutions. They have multiple causal mechanisms, cover multiple fields and are intertwined to such an extent that they cannot be solved in isolation.[55] It is only after a considerable period of time that the performance of an urban development in terms of sustainability can be assessed, as many issues may take several decades to manifest, for example: most major construction projects such as bridges and dams are designed to be used for many decades, but the demand and use of these projects is very difficult to accurately predict as there are many competing external factors that can have a major impact. This lack of evidence can be a major barrier as users may not be able to define the real benefits of using a specific form of assessment. The broad sets of barriers are explored below: perceptual/behavioural; institutional; economic; those that are specific to the software and hardware associated with the tools. The main enablers are associated with the various policies and legislative instruments that have emerged.

11.6.3 Perceptual and behavioural barriers

There are many perceptual and behavioural barriers that mainly relate to individuals and groups, for example, lack of knowledge and understanding, apprehension and uncertainty, and resistance to change. This human dimension further complicates sustainability assessment. People have different value judgements and are interconnected through a complex web of family, professional organisations, workplaces, community, belief systems and political groupings.[57,58] Although individual and societal perceptions and priorities change with time, there is also an inherent resistance to change and a preference for the status quo to take precedence over rational judgement. Many social sustainability tools and approaches attempt to impose rational quantification by adopting approaches imported from the natural and/or physical sciences; however, Therivel suggested that this treatment of social issues could lead to inappropriate sustainability assessment.[59] Lack of knowledge and understanding of sustainability and the available assessment tools is a frequent barrier and can manifest itself in the development and implementation of best practice, metrics and tools. Until recently, the lack of formal guidance on sustainability assessment mechanisms has further compounded the problem, but this has to some extent been addressed through the emergence of new publications and tools. The inherent complexity of sustainability assessment due to many cause-and-effect relationships compounds the problem of lack of understanding and the process becomes very difficult to understand and difficult to address in terms of decision making.[59,60]

11.6.4 Institutional barriers

The main institutional barriers relating to the adoption and use of sustainability assessment tools relate to the lack of cooperation among tool users, resulting in some tools being incompatible and thereby reducing their effectiveness. Many organisations operate in competitive environments rather than working cooperatively towards a common goal. This constraint, combined with the weak links between policymakers and practitioners, can result in a lack of information sharing. Many policies are so broad and vague that translation into specific sustainability assessment mechanisms is open to interpretation by individual tool developers and users. Most public and private sector institutions traditionally undertake a discrete function rather than supporting integration between functions.[61,62] However, industry is working towards relationship contracting and many governments are recognising that an integrated approach to service provision is the way forward. These trends have led to increased awareness and a more open and cooperative approach to many issues, especially those such as sustainability assessment and improvement. Political and institutional approval can be a barrier as many decision-support tools may need to be approved by elected members or senior management before being adopted. Also, with the increased focus on equity in many policy and legislative instruments, some assessment approaches may not be politically acceptable.

11.6.5 Economic barriers

Economic barriers are the most tangible impediments and revolve around monetary or resource constraints. To some extent, sustainability assessment tools are unlikely to be used if they adversely affect the economic wellbeing of the organisation.[63] Lack of sufficient funds to support the adoption and use of appropriate tools can also be a major barrier to such an extent that more conventional but no longer appropriate cheaper tools would be adopted. Moreover, it may not be easy to justify the cost of shifting to a new approach, even when the new system is more appropriate. These short-term economic pressures often detract from long-term sustainability interests. The lack of compatibility between different software tools and hardware systems is in part inevitable given the competition between tool developers. Also, tools developed to suit certain contexts with specific sustainability problems may not be easily integrated with the emerging, more generic tools developed to assess triple bottom line.[64] Some of these technical barriers have reduced with the increased use of building information modelling (BIM).

11.6.6 Social return on investment (SROI)

Social return on investment (SROI) analysis is one of several emerging methods for quantifying, measuring and accounting for the environmental and social value created in addition to any financial value generated. It aims not only to maximise financial return but to quantify a broader set of 'blended returns' relating to economic, social and environmental impacts[65] thus providing a broader concept of value whilst reducing inequalities and environmental degradation, and improving wellbeing.[66] As with traditional return on investment (ROI) analysis, SROI also uses a cost-benefit approach. SROI is used to determine the monetised net present value of the social (and environmental) costs and benefits.[67] According to SROI Network,[68] SROI can be performed retrospectively based on actual outcomes, or to predict the social value that will be created if the planned outcomes materialise. Benefits include economic, social and socio-economic parameters. Some tradable benefits are relatively easy to quantify in economic terms, however, non-traded impacts such as population health and wellbeing can be difficult to monetise. SROI encompasses a range of techniques that can be used to overcome this difficulty, such as:

- The Contingent Valuation Method (CVM) based on a willingness to pay (WTP) established by asking consumers to identify the maximum that they are willing to pay (WTP) for specified improvements; or
- Revealed Preference or Stated Preference Techniques, which infer valuations from the prices of related market-traded goods and actual values for complementary effects.[68]

11.7 Promotion of best sustainable practice

11.7.1 Managing sustainable projects

There are five critical actions for achieving sustainable global construction projects:

- Reliable evidence-based approaches that relate the economic, social and environmental impact of government policy and industry practice.
- Science-based approaches, performance indicators and accounting systems for measuring and monitoring environmental, economic and social impacts.
- Whole-life analytical techniques and management practices that effectively integrate environmental, social and economic objectives, such as cost-benefit analysis, multi-criteria, foresight analysis, change management, simulation and modelling.
- Appropriate community consultation and stakeholder engagement especially during the planning stages.
- Proactive measures that support and complement government and policy guidelines.

11.7.2 Business case for sustainable development

Early concepts around sustainable development were very often considered in terms of trade-off, mainly driven by short-termism. Environmental and social sustainability were usually seen to come at a financial cost. However, the business case for sustainable development has strengthened, in part due to increased sustainability-related legislation but mainly due to the fact that the performance of urban developments and construction projects is more frequently being measured in terms of their economic, environmental and social impacts. Increased community engagement and consultation has also strengthened the business case. The UK government/industry Strategy for Sustainable Construction supports the business case for sustainable construction achieved by:

- increasing profitability by using resources more efficiently;
- firms securing opportunities offered by sustainable products or ways of working;
- enhancing company image and profile in the marketplace by addressing issues relating to corporate and social responsibility.

Dyllick and Hockerts emphasised that the traditional economic aspects of a business case rarely represent sustainable development, however, other criteria such as eco-effectiveness, socio-effectiveness, sufficiency and eco-equity need to be satisfied.[69]

11.7.3 CR is good business

CR policies need to be embedded within the organisation's business strategy and this needs to be done in a way that makes good business sense and can result in sustained competitive advantage, improved working conditions for employees, and improved project output for other stakeholders. For example, CR policies and most construction organisations have health and safety as an important strategic goal, with the aim of reducing accidents at work and providing healthier working environments. The health and safety performance of a construction organisation can be easily measured and used to develop competitive advantage within the marketplace. Involving the local community within the planning, design and construction process can help to improve the chances of project success, thus demonstrating how good CR makes good business sense.

11.7.4 Sustainable supply chain management (SSCM)

As discussed earlier, the perceived recurrent problems of poor quality of work, underperformance, low productivity and low profit margins fuelled adversarial relationships and resulted in various government reports and initiatives suggesting that construction would benefit from adopting better SSCM. This has since emerged as an important strategy in the improvement of construction performance. According to Adetunji *et al.* this resulted in:

- an increase in the different types of construction supply chain management (SCM) and procurement;
- the iron triangle performance measures (i.e. time, quality and cost) being expanded to consider cost and value from a whole-life perspective, and quality measures to take account of environmental and social impacts;
- suppliers' previous innovations and sustainability track records becoming important selection criteria;
- a burgeoning of the policies, strategies and tools aimed at improving, measuring and demonstrating performance.[70]

Organisations with genuine sustainability ambition should not only focus on their internal operations but address the organisation's whole supply chain, working towards SSCM and forming partnerships with key members of the supply chain to collectively develop products and services based on environmental integrity, social equity and commercial viability.[70] The implementation of supply chain management (SCM) and sustainability are complex undertakings in their own right and there are conditions and strategies for achieving successful SSCM on global construction projects. To achieve an integrated approach to sustainable construction and SSCM, public bodies can lead the change process through their spending power and the introduction of new legislation, standards and guidance. There is still much debate about how best to integrate SCM and sustainability. Adetunji defined SSCM as:

> Identification of problematic economic, social and environmental issues throughout the supply chain; assessment of their potential impact and risks; and development of measures to enhance impact and mitigate risk.[9]

Until recently, sustainability and SCM implementations were confined to a few proactive large construction companies (notably those providing services to the public sector or organisations with a sector-specific focus, such as road maintenance, with clearly identifiable supply chains). For some time there has been a lack of integration of the spectrum of sustainability issues within SCM with many organisations, with much more emphasis on environmental impact rather than, for example, the social impacts of projects. This has in part been due to the concerns regarding increased CO_2 emissions but also due to the fact that environmental impacts can be more easily legislated for, visualised, measured and quantified than the social impacts. Also, measures introduced to reduce environmental impact when done well can also result in financial savings, not least in that major negative environmental impacts can severely damage a company's reputation and result in hefty fines and large compensation claims. Substantial cost savings can result from well-crafted and successful integration of environmental/sustainability best practice throughout the supply chain, for example: better waste management to reduce landfill; recycling initiatives; material innovation and product stewardship; pollution avoidance. However, although not immediately obvious, the same could be said for some social issues such as health and safety. Based on well-articulated business cases, more construction firms are proactively implementing sustainability objectives. However, there needs to be improved awareness of the need to involve the whole supply chain in a company's sustainability agenda.

11.7.5 Successful SCM: two schools of thought

There is much debate as to how SCM can be best achieved in the construction sector, but most literature relating to the conditions for its success can be divided into two main schools of thought:

1. School A: operational efficiency and effectiveness via collaboration based on equitable relationships;
2. School B: strategic efficiency and effectiveness via collaboration based on power relations.

Both schools of thought agree on some of the basic requirements that make collaborative arrangements successful, such as trust openness, adaptability, coordinated teams, cross-functional integration and commitment to common goals. However, SCM relationships within the construction sector tend to be based more on dominance and power regimes. Where clients have a large and regular workload, and extensive knowledge of the process; the main contractors have a high revenue dependence on that client. There is a tendency to adopt an extended structural dominance approach to controlling its supply chain through long-term strategic partnering arrangements with its contractors. The main contractor selection criteria, apart from cost, tend to require policy and previous performance evidence relating to: environmental policy, innovation, supply chain management and health and safety. With a highly proactive approach, clients are able to directly manage the upstream and indirectly the downstream supply chains thus keeping its goals at the top of the agenda. Increased requests from clients for better environmental, and health and safety performance have been influential in obtaining board support and has been the main catalyst for achieving environmental management accreditation to ISO 14001.

11.7.6 Barriers to integrating sustainability issues in SCM

The major barriers are briefly described below:

- **Low risk culture:** many clients are resistant to change, are still unwilling to take risks or reluctant to share rewards and tend to select tried and tested solutions, which often lack innovation and include poor sustainability practices.
- **Restrictive procurement legislation:** for example, the 'open tendering' and non-discrimination procurement practices mandated within the World Trade Organization's (WTO) Government Procurement Agreement; and the European Union (EU) Treaty.
- **Inappropriate public sector procurement policy:** although mainly based on the concept of 'value for money', an increasingly devolved approach to procurement results in the effectiveness of its practical application, depending on local approaches.
- **Minimum standards:** the reliance of standards and guidance encourages pre-selection systems, which tend to be based on minimum standards and accept 'just enough' as the entry point prior to moving towards a tender process mainly based on a lowest cost.
- **Cost of innovative solutions:** many clients are unwilling to share the cost or pay a premium for innovate sustainable construction solutions, even with government incentives. Those that have commercialised innovate solutions tend to need a payback period and are naturally unwilling to freely share their ideas with others without long-term relationships in place.

11.7.7 Corruption

Successful trading nations cannot afford the entry barriers created by international corruption, and many are committed to dealing with the global 'demand' and 'supply' sides of corruption. Recent government anti-corruption actions have tended to focus on: the investigation and prosecution of bribery overseas; money laundering; the recovery of stolen assets; international anti-corruption initiatives; the promotion of responsible business conduct overseas. Organisations must not underestimate the risks associated with international corruption, and governments need to encourage good business practice and effective risk management strategies. The recent UK Bribery Act (2010) has reformed criminal law, helping courts and prosecutors to deal more effectively with bribery at home or abroad. The Act replaces common law and the Prevention of Corruption Acts 1889–1916, and mainly creates two new offences: the promising or giving of an advantage, and the requesting, agreeing to receive or accepting of an advantage. Anti-corruption practices form an important part of CR and sustainable development, especially in developing countries.

11.8 Infrastructure projects in developing countries

11.8.1 Background

Until recent years there was little or no effort made towards achieving sustainable development within many developing countries. Many projects with funds provided by external agencies for new capital investment in public infrastructure and buildings were provided without sufficient consideration of the associated recurrent costs, or how to develop the local capacity and skills required to sustainably manage and maintain the new infrastructure. In many cases there was a lack of local community engagement, thus depriving that community of the opportunity to clearly identify their priorities. Consequently, many centrally controlled maintenance programmes have completely failed.

The all too often lack of policy direction for maintenance of infrastructure and many global construction projects has compounded the problem. There are many examples where major development projects have not only failed to deliver the anticipated financial return, but there has also been considerable decay of key social infrastructures mainly as a result of poor design, poor construction and poor materials, and from many years of indifference in attitudes towards maintenance. Sustainability in construction has also been compounded by the planning process and decisions made outside beneficiary communities. A high degree of central control has meant that planners have often had little or nothing to do with the intended beneficiaries' most urgent problems affecting their quality of life. Infrastructure sustainability for developing countries is a way of improving quality of life long after the project has been completed; consequently, the long-term continuation of a project following withdrawal of external support is an important sustainability concept for global construction projects in most developing countries.

11.8.2 The rationale

The UN's action plan for sustainable development, Agenda 21, provides an important foundation upon which action plans can be developed, based upon goals and actions to be taken by the stakeholders within a strategic programme. Even more relevant to the construction industry is the Habitat Agenda, which focuses on human settlement and shelter (United Nations Commission on Human Settlements).[71] Paragraph 25 of Habitat Agenda II states that governments should encourage the construction sector to promote:

> locally available, appropriate, safe, efficient and environmentally sound construction methods and technologies in all countries particularly in developing countries, at local, national, regional, and sub-regional levels to emphasise optimal use of human resources and to encourage energy saving methods that are protective of human health.

Two decades ago, the main emphasis was technical issues relating to construction methods such as materials, building components, construction technologies and on energy-related design concepts. In recent years, the shift has been towards the non-technical 'soft' issues that are so crucial to sustainable development. Sustainable construction has different priorities and approaches throughout the world. Although the environmental, economic and social issues are critical to sustainable construction, there has been a tendency by developing nations to place greater emphasis on the social aspects of sustainable development, and addressing problems associated with poverty and underdevelopment are sometimes included as objectives within the definitions of sustainable development. Social equity is now much higher on the agenda than environmental concerns, an issue that has not been fully grasped by those providing financial assistance to developing countries. Tools for social analysis are not fully understood and are infrequently used.

Poor communities can be characterised by scarce, inaccessible and unsafe water; isolation, with poor roads and inadequate transport infrastructure; precarious shelter; lack of energy for cooking and heating; poor sanitation. New appropriate infrastructure can stimulate investment in other sectors and has a direct impact on poverty alleviation: water supply and sanitation, and rural transportation infrastructure projects make major contributions in the process of pro-poor growth (i.e. growth that reduces poverty). Sustainable transport infrastructure is crucial to urban and rural development. It can help to provide access to jobs, health, education and other amenities, without which quality of life suffers. Without the appropriate access to resources and markets, growth stagnates and poverty reduction is not addressed.

11.8.3 External funding issues

Most developing countries lack the financial resources required for the development of vital infrastructure. They are often left with no option but to seek financial assistance from external funding bodies, who often have predetermined procedures and policies with regard to how projects are funded. Although sustainable development/construction can be high on the agenda of all parties, its perception in developing countries has to be balanced with that of developed countries. This often results in the process of approving funds for development/infrastructure projects being long and laborious, frequently leading to communities being deprived of much-needed service. Developed and developing countries often have very different environmental, economic and social priorities, due to the difference in the levels of development. Economic prerequisites of the project are drawn from the prevailing situations in the developed countries, thus emphasising impacts to the environment rather than the more pressing social issues. The relationships of funders are mainly limited to a local government agency or non-government organisation (NGO); consequently, the cultural values and community priorities unique to developing nations are given insufficient consideration, leading to the problems of poverty, underdevelopment or social equity. Also, the local community's long-term benefits and community expectations are often not understood, with the result that the potential long-term benefits of the project are not fully realised.

11.8.4 Stakeholder and community participation

The challenges that arise from sustainable infrastructure development are too large for a single sector to deal with. New forms of partnership are emerging with private sector participation, and the roles of public sector bodies, local communities and international donors are being redefined. However, there is concern as to whether or not current donor-imposed conditions help or hinder developing countries to develop sustainably. There is also concern at the failure of some guidelines and manuals to help create synergy between social, economic and environmental sustainability rather than the all too frequent trade-off in favour of economic sustainability. Global construction projects that are centrally planned and executed without input from end users and local communities have a higher probability of failure and are usually poorly maintained. A proactive approach to local participation increases efficiency and strengthens a sense of community ownership, provides transparency, and enables the projects to meet the needs of the end users, which encourages them to use the facilities provided and contribute to project operations and maintenance costs. As broad a range of stakeholders as possible should be involved in the consultation process. World Bank guidelines for community-based investments state that local community-based investment can be a good substitute for external control. Governments in many developing countries are encouraging local communities to contribute towards construction and maintenance of infrastructure. This contribution may be in kind, as many rural communities are unable to contribute in monetary terms. If the quality of life of poor communities is to be significantly improved, the priorities of these communities need to be clearly identified through a process of proactive engagement.

11.8.5 Economic sustainability

Investment approaches such as net present value and economic rate of return are widely used for cost-benefit analysis and during project design and selection. Project benefits tend to be assessed in term of willingness-to-pay as demonstrated in the market, however, this often conflicts with social equity and poverty reduction objectives. To obtain a more balanced approach, economic and social issues need to be considered together. Social issues, such as poverty, can still be considered from an economic viewpoint by determining which cost options can best achieve the desired policy objective. The economic dimension will only have any real value if the development generates local employment for the rural communities, both in the long term but also during the construction phase by the use of local labour-intensive approaches, which can also provide cost savings. However, it is important not to generate too much schedule growth as this will increase borrowing requirements and reduce project-generated income. Training and capacity building are also key components of social and economic impact.

11.8.6 Environmental sustainability

There is generally a low level of environmental and hygiene awareness in many developing communities, especially regarding the problems associated with environmental conservation/protection, water use and sanitation. These communities do not possess the technical information or knowledge required to address the problems and there needs to be greater effort directed towards educating local people on how to better maintain their environment. Many government bodies, such as the Zambia Social Investment Fund (ZAMSIF), have provided guidelines for communities on ways of enhancing the environmental sustainability of projects. The Fund recommends environmental assessments (EA) to be used as part of a systematic review conducted (using a variety of tools) in order to integrate environmental factors into the sustainability of construction projects. The EA should not be seen as an activity performed at the end of a project, but as an ongoing process integrated throughout the project's cycle.

11.8.7 Social sustainability – quality of life aspects

Poor transport and communication systems have a very high negative impact on the health and wellbeing of women, who are traditionally used as transporters of heavy loads such as firewood, water and other products. There needs to be a deliberate policy to factor and mainstream gender issues into development activities. Health and quality of life benefits result from the sustainable development in the construction of road and water infrastructure. Better infrastructure for transporting agricultural product will encourage individuals to produce more than that which just feeds the family, with the surplus produce being sold for much needed cash. The increase in agricultural production will subsequently help to alleviate poverty and improve quality of life. Improved transport to rural areas also improves the sustainability of local communities and enhances the socio-economic status of the community. Water supply infrastructure development is perceived by rural communities as directly increasing their financial security, as the poorest spend up to one-third of their income on inadequate water sources. It also improves health, reduces medical expenses and increases time available for productive work.

11.9 Governments taking actions

According to the *Global Construction Report*,[8] more and more governments are making plans for new policies to promote or force clients to build new energy-efficient buildings. The European Union's revised Energy Building Performance Directive, the Obama administration's recently introduced Better Building Initiative and China's Building Energy Efficiency legislation are a small number of examples of this trend. It is worth noting that the need for sustainable buildings goes further than climate change and energy efficiency. The construction industry is accountable for more than one-third of global resource consumption, including 12 per cent of all fresh water use, and contributes to the production of solid waste (estimated at 40 per cent).[8]

Evidence shows that while the rationale for sustainable buildings is shared by Mexico, India, South Africa, South East Asia and other nations, the culture to affect change and to achieve benefits from sustainable buildings is deeply rooted at the regional level.[8] The role of government policymakers, construction clients, architects and planners, will eventually decide the shape of upcoming construction markets. What is clear in this chapter, is that clients and contractors cannot afford to pay insufficient attention to environmental prerequisites increasingly placed on the performance of construction projects. The message is clear: there is a need to integrate sustainability policies and procedures at all levels. UNEP's Sustainable Buildings and Climate Initiative (SBCI) is collaborating with both private and public sectors globally to find ways of moving the building and construction industry towards sustainable buildings.[8] Financial institutions and green building rating schemes such as LEED, Building Research Establishment Environmental Assessment Method (BREAM) and Green Star are supporting this move.

11.10   Case study: London Olympics 2012

The London 2012 Olympic programme had sustainability as a major priority, guided by the World Wide Fund for Nature/BioRegional concept of 'One Planet Living' with a commitment to minimising the use of the planet's resources through the Sustainability Plan, which covers three key phases:

- **Preparation:** the design and construction of venues and infrastructure.
- **Event staging:** the Games.
- **Legacy:** the post-Games economic, social, health and environmental benefits.

The second edition of the London 2012 (2007) Sustainability Plan details the approach to and vision for sustainability. The plan is broken down into considerable detail, with very specific performance targets where performance can be measured. Its main themes are:

- **Climate change:** vision – 'To deliver low-carbon Games and showcase how we are adapting to a world increasingly affected by climate change.'
- **Waste:** vision – 'To deliver a zero waste Games, demonstrate exemplary resource management practices and promote long-term behavioural change.'
- **Biodiversity:** vision – 'To conserve biodiversity, create new urban green spaces and bring people closer to nature through sport and culture.'
- **Inclusion:** vision – 'To host the most inclusive Games to date by promoting access, celebrating diversity and facilitating the physical, economic and social regeneration of the Lower Lea Valley and surrounding communities.'
- **Healthy living:** vision – 'To inspire people across the UK to take up sport and develop more active, healthy and sustainable lifestyles.'

11.11   Chapter summary

Sustainability is a major consideration on global construction projects, which tend to be large, multicultural and have the potential to make significant economic, social and environmental impact. They will tend to include the investment of large sums of money not only in terms of initial capital expenditure but also in terms of operating costs. The economic sustainability of such projects can have considerable long-term impact on the economy and the wellbeing of the workforce, especially in developing countries where investment in global construction projects represents significant national investment, and which has to be undertaken in a sustainable way. The size and resource-intensive nature of such projects can result in considerable environmental impact and will need to be well managed to ensure a positive outcome through practices that include responsible sourcing and SSCM. The long-term social impact of global construction projects is often underestimated, in part due to their complexity and the wide-ranging types of impact that are difficult to measure and quantify.

11.12 Discussion questions

1. Discuss the emergence of sustainable development within the context of global construction projects.
2. Discuss the importance of the triple bottom line to the sustainability of global construction projects.
3. Discuss the four main types of capital and how these relate to project performance.
4. Discuss the physical determinants of social capital that could be used on global construction projects.
5. Explain the role of corporate governance.
6. Discuss the role of construction ethics within sustainable development.
7. How can the barriers to sustainable assessment be overcome?
8. Discuss how social return on investment can be used to assess the feasibility of a global construction project.
9. Discuss the business case for sustainable development.

11.13 References

1. United Nations (1972). Declaration of the United Nations conference on the human environment. http://www.unep.org/Documents.Multilingual/Default. asp?documentid=97&articleid=1503 [Accessed February 2012].
2. United Nations (1987). Brundtland Report: Report of the World Commission on Environment and Development, General Assembly Resolution 42/187 General Assembly Resolution 42/187. http://www.un-documents.net/a42r187. htm [Accessed November 2011].
3. IUCN/UNEP/WWF (1991). *Caring for the earth: A strategy for sustainable living.* Gland, Switzerland: IUCN/UNEP/WWF.
4. International Council foe Local Environment Initiatives (1996). *The Local Agenda 21 Planning Guide.* Toronto: ICLEI.
5. United Nations Framework Convention on Climate Change (1997). Kyoto Protocol to the United Nations Framework Convention on Climate Change. http:// unfccc.int/resource/docs/convkp/kpeng.html [Accessed January 2012].
6. UN (2005). World Summit Kyoto protocol: Resolution adopted by the General Assembly. http://unpan1.un.org/intradoc/groups/public/documents/un/ unpan021752.pdf [Acceesed January 2012].

7. Kibert, C. (1994). *Establishing principles and a model for sustainable construction.* In Proceedings of the First International Conference on Sustainable Construction, 6–9 November, Tampa, University of Florida.

8. Betts, M., Robinson, G., Blake, N., Burton, C. and Godden, D (2011). Global Construction Report 2020: A global forecast for the construction industry over the next decade to 2020. Oxford Economics, 3 March 2011. London.

9. Adetunji, I., Price, A. D. F., Fleming, P. and Kemp, P. (2003). *Sustainability and the UK construction industry: A review.* In Proceedings of ICE: Engineering Sustainability, 156, pp. 185–199, Paper 13472.

10. Davis Langdon Consulting (2003). Investing in Sustainable Developments Key Players Workshop, 11 November, London.

11. Pepper, C. (2007). Sustainable development strategy and action plan for civil engineering. London: Institution of Civil Engineers. http://www.ice.org.uk/get attachment/276bd7f7-e28c-4473-8aec-1d804902e022/Sustainable-development-strategy-and-action-plan-f.aspx [Accessed November 2011].

12. EU (2006). Renewed EU Sustainable Development Strategy, EU, Brussels, 9 June.

13. Smith, S. P. and Sheate, W. R. (2001). Sustainability appraisal of English regional plans: Incorporating the requirements of the EU Strategic Environmental Assessment Directive, Impact Assessment and Project Appraisal, Vol. 19 (4), pp. 263–276.

14. Barker, A. and Wood, C. (2001). Environmental assessment in the European Union: Perspectives, past, present and strategies, *European Planning Studies,* 9, pp. 243–254.

15. DCLG (2006). CE making-the gateway into the European single market, Department for Communities and Local Government, London. http://www.communities. gov.uk/ [Accessed June 2012].

16. Sjöström, C. (2001). Approaches to sustainability in building construction, *Thomas Telford Journals,* 2, pp. 111–119.

17. UNDP India (2008). Guidelines for Hospital Emergency Preparedness Planning: GOI-UNDP DRM Programme (2002–2008), Government of India (GOI) and United Nation Development Programme (UNDP), New Delhi.

18. Centre for Research on the Epidemiology of Disasters (2008). Disaster Data: A Balanced Perspective, Issue No. 13, Brussels.

19. EEA (European Environment Agency) (2003). Mapping the impacts of recent natural disasters and technological accidents in Europe, EEA, Copenhagen. http:// mmediu.ro/RO-EEA-EIONET/Publications/Mapping%20the%20impacts%20of %20the%20natural%20hazards%20low%20res.pdf [Accessed September 2011].

20. Achour, N., Miyajima, M., Kitaura, M. and Price, A. D. F. (2011). Earthquake induced structural and non-structural damage in hospitals, *Earthquake Spectra,* 27 (3), pp. 617–634.

21. United Nations International Strategy for Disaster Reduction (2009). CRED disaster figures: Deaths and economic losses jump in 2008, Report No. UNISDR 2009/1, Geneva.

22. Cabinet Office (2008). National risk registry of the United Kingdom, *Security in an Interdependent World,* Cabinet Office, TSO London.

23. Sander, T. H. and Lowney, K. (2003). Social capital building toolkit, version 1.0, John F. Kennedy School of Government, The Saguaro Seminar, Harvard University.

24. Brown, D. L. and Ashman, D. (1996). Participation, social capital, and inter-sectoral problem solving: African and Asian cases, *World Development,* 24, pp. 1467–1479.

25. Knack, S. and Keefer, P. (1997). Does social capital have an economic payoff? A cross-country investigation, *Quarterly Journal of Economics,* 122, pp. 1251–1288.

26. Krishna, A. and Uphoff, N. (1999). *Mapping and measuring social capital: A conceptual and empirical study of collective action for conserving and developing watersheds*

in Rajasthan, India, Social Capital Initiative Working Paper No. 13. Washington, DC: World Bank Publications.

27. OECD (2001). *The well-being of nations: The role of human and social capital.* Centre for Educational Research and Innovation, Paris: OECD.

28. Putnam, R. D. (2000). *Bowling alone: The collapse and revival of American community.* London: Simon and Schuster.

29. Glaeser, E. L., Laibson, D. and Sacerdote, B. (2002). An economic approach to social capital, *Economic Journal,* **483**, pp. 437–458.

30. Woolcock, M. (1998). Social capital and economic development: Towards a theoretical synthesis and policy framework, *Theory and Society,* **27**, pp. 151–208.

31. Jochun, V., Pattern, B. and Wilding, K. (2005). Civil renewal and active citizenship: A guide to the debate, National Council for Voluntary Organisations, London.

32. Moobela, C., Price, A. D. F, Taylor, P. J. and Mathur, V. N. (2007). International Conference on Whole Life Urban Sustainability and its Assessment.

33. Roberts, J. and Chada, R. (2005). What's the big deal about social capital? In Khan, H. and Muir, R. (eds), *Social Capital and Local Government: The Results and Implications of the Camden Social Capital Surveys 2002 and 2005.* London: Borough of Camden.

34. Halpern, D. (2005). *Social Capital.* Cambridge: Polity Press.

35. Putnam, R. D. (2000). *Bowling Alone: The Collapse and Revival of American Community.* London: Simon and Schuster.

36. Putnam, R., Feldstein, L. and Cohen, D. J. (2004). *Better Together: Restoring the American Community.* London: Simon and Schuster.

37. Portes, A. (1998). Social capital: Its origins and applications in modern sociology, *Annual Review of Sociology,* **24**, pp. 1–24.

38. Durkheim, E. (1997 [1893]). *The Division of Labor in Society.* London: Free Press.

39. Wasserman, I. M. (1984). Political crisis, social integration and suicide: A reply to Boor and Fleming, *American Sociological Review,* **49**, pp. 708–709.

40. Putnam, R. (2002). *Democracies in Flux.* Oxford: Oxford University Press.

41. Savage, M. (2001). *Class Analysis and Social Transformation.* London: Open University Press.

42. Simpson, L. (2005). Community informatics and sustainability: Why social capital matters, *The Journal of Community Informatics,* **1** (2), pp. 102–119.

43. ODPM (2003). Sustainable communities: Building for the future, Office of the Deputy Prime Minister, London.

44. Hillier, B. and Hanson, J. (1984). *The Social Logic of Space.* Cambridge, MA: Cambridge University Press.

45. Glasson, J., Therivel, R. and Chadwick, A. (2005). *Introduction to Environmental Impact Assessment.* London: Taylor and Francis.

46. Waldrop, M. M. (1992). *Complexity: The Emerging Science at the Edge of Order and Chaos.* London: Penguin.

47. Margolis, J. and Walsh, J. (2001). People and Profits? *The Search for a Link between a Company's Social and Financial Performance.* Mahwah, NJ: Lawrence Erlbaum Associates, Inc.

48. Hopkins, M. (2003). *The Planetary Bargain. Corporate Social Responsibility Matters.* London: Earthscan.

49. BIS (2009). Corporate Sustainability Report, Department for Business, Enterprise and Regulatory Reform, London. http://www.bis.gov.uk/files/file50312.pdf [Accessed August 2012].

50. BIS (2012). Corporate Sustainability Report, Department for Business, Enterprise and Regulatory Reform, London.

51. Porter, M. and Kramer, M. (2002). The competitive advantage of corporate philanthropy. *Harvard Business Review on Corporate Responsibility.* HBS **80** (12), pp. 56–68.

52. Moon, C. and Hagan, J. (2001). *Business Ethics.* London: The Economist Books.
53. McNamara (1999). Complete guide to ethics management: An ethics toolkit for managers. http://www.infra.kth.se/courses/1H1146/Files/ethicsmanagement. pdf [Accessed July 2011].
54. Meppem, T. and Gill, R. (1998). Planning for sustainability as a learning concept. *Journal of Ecological Economics*, 26 (2), pp. 121–137.
55. Shaughnessy, H. (2004). *Implementing CSR Communications for Business Results.* Conference Report. 24–25 Feb 2004. London: CSR Datanetworks.
56. UN Special Representative on Business and Human Rights. http://www.business-humanrights.org/SpecialRepPortal/Home [Accessed June 2011].
57. ISO 26000: GRI and 26000: How to use the GRI Guidelines in Conjunction with ISO 26000. https://www.globalreporting.org/resourcelibrary/How-To-Use-the-GRI-Guidelines-In-Conjunction-With-ISO26000.pdf [Accessed June 2012].
58. Moobela, C., Price, A. D. F. and Taylor, P. J. (2006). The case for social capital in the quest for holistic urban sustainability assessment, Paper presented at the 1st International Conference on Sustainability Measurement and Modelling, 16–17 November, Terrassa, Spain.
59. Therivel, R. (2004). Analysis of sustainability/social tools: Results of a research conducted on behalf of the Metrics, Models and Toolkits for Whole life Urban Sustainability Assessment consortium of the UK.
60. Van Rees, W. (1991). Neighbourhoods, the state and collective action, *Community Development Journal*, **26**, pp. 96–102.
61. Sayer, J. and Campbell, B. (2003). *The Science of Sustainable Development: Local Livelihoods and the Global Environment.* Cambridge: Cambridge University Press.
62. Moore, J. L. (1994). What's stopping sustainability? Examining the barriers to implementing clouds of change. MA Thesis, University of British Columbia.
63. Mittler, D. (1999). Environmental space and barriers to local sustainability: Evidence from Edinburgh, Scotland, *Local Environment*, **4** (3), pp. 353–365.
64. Bebbington, J. and Gray, R. (1996). Incentives and disincentives for the adoption of sustainable development by transnational corporation. Geneva: UNCTAD Report TD/B/ITNC/AC1/3.
65. Pope, J., Annandale, D. and Saunders, A. (2003). Conceptualising sustainability assessment, *Environmental Impact Assessment Review*, **24**, pp. 595–616.
66. Carleton Centre for Community Innovation. (2008). Social return on investment.
67. Nicholls, J. (2004). *Social Return on Investment: Valuing What Matters.* London: New Economics Foundation.
68. SROI Network. (2009). A guide to social return on investment: Cabinet Office.
69. Dyllick, T. and Hockerts, K. (2002). Beyond the business case for corporate sustainability, *Business Strategy and the Environment*, **11** (2), pp. 130–141.
70. Adetunji, I., Price, A. D. F., Fleming, P. and Kemp, P. (2003a). *Trends in the conceptualisation of corporate sustainability.* In Proceedings of the Joint International Symposium of CIB Working Commissions W55, W65 and W107, 23–24 October, Singapore.
71. UNCHS (1996). *The Habitat Agenda: Goals, Principles, Commitments and Global Plan of Action*, 3 –14 June, Istanbul, Turkey.

12

Conclusions: Global Construction Project Management

12.1 Introduction

Globalisation of the construction industry has brought unique problems and challenges, such as integration of project teams from different countries. The problems have become greater in extent and severity in recent years. In this conclusion, we focus on lessons drawn from a global perspective to highlight factors that influence project delivery. We consider some of the more important differences in project delivery and the implications for practical global construction organisations, as set out in the following sections:

- implementation of global construction projects;
- tackling bribery and corruption during implementation;
- comparison of identified project practices;
- information and communication technology (ICT) in global construction projects;
- barriers, benefits and attributes of inter-organisational ICT;
- implementation of ICT on construction projects: the way forward;
- the value of project management learning and higher education.

12.2 Learning outcomes

> The chapter's specific learning outcomes are to enable the reader to gain an understanding of:
>
> >> how to implement global construction projects;
> >> how to address bribery and corruption during implementation of global construction projects;
> >> factors to consider when implementing information communication technology on global construction projects;
> >> different project management standards

12.3 Implementation of global construction projects

As has been discussed, each country in which construction organisations operate has a unique economic, political, legal, cultural and competitive context that contractors, suppliers, subcontractors and designers have to deal with.[1] Every country has its own distinct values that are not quite the same as others. As revealed in the previous chapters, changes in the global economy are presenting construction organisations with both new opportunities and challenges. Rapid progress in innovation, increasing international trade and investment, and growing wealth and prosperity across the global economy are all compelling clients to invest in new emerging economies. Essentially, the new global environment is forcing organisations to acclimatise to a new global order.[2] Clearly, the global trends are restructuring the competitive dimensions in economies and driving the need for increased application of global project management tools. To implement global construction projects successfully, governments, clients and contractors must be prepared to overcome distinctive project challenges. The challenges will arise from differences in cultures, tradition, values, legal frameworks, languages of the contractors, time differences, uncertainty in the market, currency fluctuations, unexpected foreign taxes, bribery, ethical conflicts, bureaucracy and the use of organisational processes and project frameworks.

In implementing and managing global construction projects, clients and contractors will have to methodically address the above global challenges as well as learn how to address global risks. The main challenge faced by a global construction project manager is to sustain, implement and control project tasks and cultures while simultaneously addressing harsher global economic conditions and pursuing productivity, efficiency and sustainability targets set by the client and governments. To implement and deliver projects successfully, contractors and subcontractors will have to be able to affect change and transform their organisational processes in order to capitalise on both local and international opportunities. Uncertainty in the global market has increasingly made things more difficult. For instance, contractors and subcontractors are under increasing pressure from clients to deliver projects that are of a high quality over shorter timescales. The current trends towards fast-tracking projects are increasing the importance of effective implementation strategies. Two of the core problems facing a global project manager lie in the coordination of dispersed multiple tasks and integration of ethical standards. Most global construction projects involve complex arrangements across numerous contractors and subcontractors. In order to address these two issues, organisations need to introduce a holistic planning cycle that creates a clear process of priority setting and resource allocation, whilst balancing the interests of stakeholders.[3]

When different sets of cultures are brought together in global construction projects, organisations have to incorporate a specific project code of conduct that reflects ethical standards in a diverse team. This will minimise ethical conflicts in a multicultural global environment.[4] The choice of implementation strategy in a global environment is as important a decision as the choice of management strategy, which identifies the soft and hard issues of the project. In order for the implementation strategy to be effective, project managers need to ensure that individual values and goals are fully aligned at the onset of the project. Failure to do this at the inception of a global construction project will result in project teams having different priorities and ways of working, thus leading to confrontational situations.[4] Given the generally dynamic nature of team affiliations in multicultural project teams, the client and project manager must determine a common team culture from the outset. What needs to be well understood is that the effective structure of multicultural team working depends on a well-balanced communication system between the client, project manager and the project team. Contractors and subcontractors need to re-examine the services they provide in order to correspond to the changing global economic and social demands of clients. During implementation of global construction projects care must be exercised with regard to contract agreements. Clients, designers, subcontractors and contractors have to ensure fairness and balance. In essence, from the onset of the project, stakeholders have to fully amalgamate dispute resolution processes and integrate sound business practices at strategic, operational and project level. Like most projects, global construction projects require commitment from the client, project manager, senior managers and a team of (in this case) multicultural participants committed to the same objective and meticulous planning. As noted in Chapter 7, to be successful with multicultural integration the project managers in construction organisations must consider cultural differences and variations in legal prerequisites. Respect of values and cultures is essential, and the project manager has to consider the team view when making critical decisions. The project manager has to ensure that team members are allowed to contribute in a style that they are happy with.

12.3.1 Tackling bribery and corruption during implementation

It is evident that senior management in construction organisations need to be aware of the risks to their global operations. Bribery and corruption is particularly applicable to those contractors and subcontractors dealing with global construction government projects. According to Deloitte,[5] bribery and corruption offences can be classified as transgression relating to the giving or receiving of consideration in order to gain undue influence over a person in a position of trust. In the new global economy, bribery and corruption are increasingly featuring as a major concern for a number of suppliers, contractors and subcontractors. Anyone finding themselves answering '*yes*' to the following set of questions should be aware that bribery and corruption could be an issue that needs addressing at the strategic level of an organisation:[5]

- Do you run your operations through the use of agents, joint ventures and business partners?
- Do you acquire joint venture projects in foreign environments without carrying out appropriate meticulous assessment designed to identify potential corruption risk?
- As an organisation are you delivering construction projects in countries with a high perceived risk of corruption?
- Do you carry out business with government ministries or bodies in which a foreign government has an ownership interest, for example when delivering large public construction projects?

Prosecutors and regulators in the world over are becoming ever more active in introducing anti-corruption legislation. Evidence shows that, over the last few years, the number of enforcements and jurisdictions have increased significantly.[5] As a result, Transparency International noted that amongst the Organisation for Economic Co-Operation and Development (OECD) signatory nations, there has been significant enforcement that has been introduced by governments. The US government has led the way in introducing 33 enforcement actions under the Foreign Corrupt Practices Act, which forbids payment to a foreign government official in order to obtain, retain or otherwise gain an improper advantage in the conduct of business.[5]

In 2009, the UK government introduced its Bribery Bill. Under the new Bill, it is an offence to bribe another person or request to receive a bribe. A new corporate offence of negligency has also been introduced. Organisations that fail to prevent bribery by an employee or agent will be prosecuted. In order to minimise bribery and corruption, construction organisations have to invest significant resources into introducing and reinforcing their control methods, as well as encouraging openness and transparency through discussions with stakeholders about their obligation to ethical business. When operating in a global environment, contractors, subcontractors and suppliers ought to integrate risk assessment into their operations relating to interactions and negotiations with clients and business partners. Additionally, they need to incorporate anti-bribery and corruption criteria into their performance management models of personnel, and develop guidance and training schemes to deal with the treatment of intermediaries and business partners, including joint venture associates.[5]

12.4 Comparison of identified project practices

The process of internationalisation has increased the level of cultural complexity and diversity in construction organisations. Global project managers are facing challenges very different from those of the recent past. This has led to managerial challenges at strategic, operational and project level in these organisations. In order to incorporate effective global project management techniques, organisations will have to introduce shared cultural values among suppliers, contractors and subcontractors. For instance, suppliers, contractors and subcontractors operating in the Middle East, North Africa and Sub-Saharan Africa will have to align their project management models with the cultural values utilised in these two continents. Cultural differences in global construction projects derive from practice, processes and values. Countries have unique patterns of organisation styles and micro-level behaviours. As noted in Chapter 1, global construction projects involve collaboration between suppliers, contractors and subcontractors from different countries. Differences in project management practices, construction requirements, technical requirements, environmental regulations, international finance, legal regulations and procurement methods has contributed to the increased cost of project delivery in most continents.

The impact of economic, political and social globalisation on construction projects is significant. Economically, increasing international competitiveness is restricting construction organisation budgets. Interestingly, with the adoption of open project management models, contractors and subcontractors are becoming more aware of how their counterparts in other countries are dealing with economic and social issues. In China and Japan, it was found that the close working relationship between governments and contractors has contributed to the success of construction project management practices.[6] Emerging economies can only benefit if they review their industrial policies. It is essential for countries to devise industrial policies adjacent to their monetary policies. In the UK and US, a part of the government's role in industrial policy has been to link fiscal and legal enticements to support research and development. Part of the Chinese and Japanese success is attributable to construction contracts utilised and the contractual affiliation clients have with subcontractors, contractors and suppliers. To ensure that key features of global construction project management practices are fully incorporated, modern construction organisations will have to adopt open project delivery models. These have strong client focus. Open project management models focus on the technical and non-technical elements of the project. As illustrated in Figure 12.1, construction organisations have to ensure the full integration of global project terms, global project management requirements, global project manager technical ability and global project management guidance.[7]

Despite the fact that project management being practised in Europe may not appear to be much different to that practised in the US, Asia, the Middle East and Africa, there are a number of additional attributes to be considered:

- an overall appreciation of the construction industry in the country where contractors, subcontractors and suppliers are operating;
- understanding of the values of the local society;
- proficiency in the local language, including technical terminology;
- willingness to enlighten as well as manage;
- the project model utilised provides an all-inclusive task.

When planning to implement a global construction project, it is fundamental to the success that both regional and local suppliers, contractors and

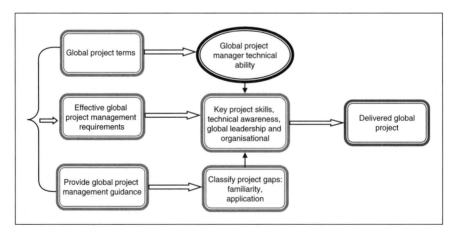

Figure 12.1 An approach to effective global project management requirements

subcontractors, and other stakeholders are aware of the issues and challenges raised at strategic, operational and project level. If not well planned-out, implementing a global construction project can be risky, expensive and complex. Success can only be achieved if a coherent strategy is developed, understood and agreed upon by stakeholders and senior managers.

12.5 Information and communication technology in global construction projects

In the new global economy, GCOs have been obligated to explore all possible options for improving project delivery. Due to the high level of uncertainty in the global market, clients are expecting contractors to deliver a better service and projects that meet their prerequisites more closely. This has challenged the sector to become more economical, integrated and more appealing, both in the eyes of society and its prospective labour force.[8] The integration of information communications technology (ICT) in the delivery of global construction projects has allowed a more standardised communication between different stakeholders in construction. The use of inter-organisational 3D CAD in construction project delivery has been a driving force for technological and organisational modernisation.[9]

However, the application of ICT in construction projects still seems quite less effective and limited than in manufacturing.[10] The main problem is the complex nature of the global environment in which global construction projects are delivered. Clients and contractors have to deal with organisational, technical, political, social and economical factors. Central steering in a global environment can be difficult because some contractors tend to act primarily according to their own business objectives. A number of large ICT modernisation projects, aiming to substitute long-established but increasingly outdated systems, are delivering all the promised business enhancements at budgeted cost and within the predicted schedule[12] but a number of the projects have resulted in total and expensive failure. Most have been moderately successful if judged against initial estimates of schedule, budget and promised benefits. As an example, the Dutch government experienced severe difficulties in delivering ICT projects. For instance, the failure of a project aimed to develop an ICT system to support a future human resources shared services centre. Because of unconstructive relationship between government and the contractor, the contractor decided to terminate the contract. The Dutch government identified three factors of complexity that affected its ICT projects, namely political, organisational and technical.[12] In 2011, the coalition government in the United Kingdom decided to put a halt on a £12 billion National Health Service (NHS) computer system project. The failure of the project was due to poor organisational alignment. Following a countrywide review, the coalition government decided to substitute a one-size-fits-all information technology (IT) project with a low-cost regional initiative, and to form a new supervisory body to ensure huge sums of capital can never be wasted on uncosted IT projects.

There has been some increase in the use of ICT in parts of Africa, but the penetration of ICT technologies is still very low compared with Western economies. In order to bridge the gap, governments need to invest in ICT infrastructures and work closely with organisations and institutions. For instance, in Kenya there is growing hope that technology, particularly ICT technologies, can help achieve vision 2030 goals and spur progress in the economy. Kenya vision 2030 is a long-term development programme covering the period 2008–30. Its aim is to help to create a globally competitive and prosperous nation to provide a high quality of life to all its citizens by the year 2030. Studies have confirmed that ICT development projects have a low success rate, in part because of poor project design and management.[13,14] Government leaders must recognise the procedures of managing projects and be aware of the infrastructures, technologies and tools available to raise project success. The success of an ICT system in global construction projects is strongly related to cultural, governance, political, technical, organisational and social factors. In this, global construction projects are complex because several contractors are involved in the delivery.

Present trends in ICT on construction projects are yielding a number of new computer-based tools to assist in the managing and integration of project phases. The application of building information modelling (BIM) tools in construction projects has increased the effectiveness and efficiency of designing and managing, but is still a problematic task in practice.[13,14] At present, there is little experience that can be used to assist practitioners with the configuration and alignment of BIM-based tools and project work processes. BIM applications are typified by the use of multitude frameworks that are supported by different sets of BIM-based tools.[15] The emergence of ICT on construction projects can be classified into three phases. The first phase focused on introducing stand-alone tools to assist designers, quantity surveyors and architects with structural analysis, estimating and architectural drawings. The second phase focused on computer-supported communications such as e-mail, the Internet and document management systems. In this phase, there are still new tools and key features emerging. The third phase has mainly focused on incorporating individual applications. This emerging theme has experienced some innovative application in the construction industry but has yet to be fully amalgamated. The third phase is characterised by technologies such as BIM, virtual design and construction.[15]

For instance, the application of ICT on global construction projects requires information schemes that are increasingly complex, central to the delivery of the project and specialised skills. In the construction sector, the opportunity to incorporate the data sets that underlie a number of computer applications has considerably increased. The ability to assimilate project data in a project environment must continue to improve to the degree that the collective project set captures much of the intrinsic interdependencies of the real world.[13] The introduction of reasonably priced mobile technologies such as handheld computers, Smartphones and Tablet PCs alongside mobile network communications infrastructure (Wi-Fi, Bluetooth, 3G, WLAN and GPRS) have provided the missing link to help address the ongoing drive of ICT in construction projects.[9] In the last two decades, construction organisations have invested in data transmission networks. New approaches to data transmission provide contractors with more affordable online data processing in real time, thus allowing better solutions to be realised. More affordable and reliable data transmission methods will enhance utilisation of client relationship management techniques.

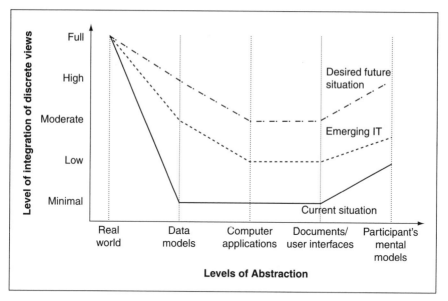

Figure 12.2   An illustration of the level of integration between views within various levels of abstraction of construction project information
Source: Adapted from Froese 2009[13]

As shown in Figure 12.2, there are several levels of abstraction of a construction project. For each level, there is an integration that exists between discrete views.

These are described as the current situation (level 1), the effects of emerging IT (level 2) and the desired situation (level 3). In all the three levels, the project constituents within the real world are mutually dependent. In the case of level 1, usually there exists a one-to-one relationship between documents. The computer application utilised to generate documents and data sets that these applications employ are capable of little or no integration.[13]

12.6   Barriers, benefits and attributes of inter-organisational ICT

Results from a recent study carried out in the US suggested that clients would benefit greatly from the use of inter-organisational ICT on construction projects. Some of the benefits were identified as:

- single data source;
- better coordination;
- higher speed of communication.[11]

Barriers to inter-organisational were identified as:

- organisations have to come up with upfront investment (costs of deployment of hardware, software, training and coordination costs);
- implementation of inter-organisational of ICT is allied with risks;
- some organisations view increased transparency as a barrier;

- alignment between ICT and working practices;
- lack of technical skills in construction organisations.

Ten primary attributes of ICT project management have been identified as:

- project definition;
- project manager;
- stakeholders participation;
- communication strategy;
- training;
- planning and managing human resources issues;
- ICT project management and risk management;
- technology;
- project control and monitoring;
- assessing project progress-independent review.[4]

The above attributes are relevant in both developed and developing countries. Global construction projects have a project scope that requires project managers to ensure that the right technology has been selected. This can be achieved by matching technological ambitions to local conditions and ensuring that there is active participation of project leaders at each level of the project. The above benefits and barriers will influence the attitudes of multinational organisations intending to adopt inter-organisational ICT on global construction projects. It is apparent that the integration of ICT on global construction projects is reliant on the structure of the organisation, size, delivery frameworks, complexity, contract types, government policies and stakeholders involved. To ensure that the implemented technology will work correctly, a complete testing strategy is required at the initial phase of the project. There are five different levels of testing that can be performed:

- Does the technology work at all in a technical sense?
- Does the technology work correctly (logic testing)?
- How will the technology perform in the real work (usability)?
- How will it cope with expected volumes and peaks (scalability)?
- Integration testing.[11]

Regardless of the size, location and nature of the project, an all-inclusive testing regime should be carried out. It is the responsibility of the project manager to set up the testing with the assistance of contractors and subcontractors. The implementation of an ICT technology requires a strategic vision in which the path towards achieving the objectives is clear.[16] To press forward this vision, the client and project manager have to formulate ways of organising the project deliverables.

12.7 Implementation of ICT on construction projects: the way forward

The global construction industry is faced with the ongoing challenge of altering and improving current project delivery and work practices in order to become more efficient. As noted in previous chapters, these challenges are due to the emergence of new global economies, greater performance expectations from clients, continued refinement of construction methods and clients demanding shorter delivery times (*fast-track projects*). Some construction organisations are now coming to grips with the transition from traditional designs to contemporary approaches using ICT on construction projects, whilst others have only started the journey. Organisations located in developed and developing countries are weighing up broadly similar technology and development techniques as they prepare for the future. In emerging economies and developing nations, they will have to establish appropriate balances between creating administrative efficiency and forming a solid model for more ICT applications.[11] For ICT to be implemented successfully on construction projects, organisations will have to use methodical approaches to project management. One important element where global construction organisations have to be mindful is the inter-organisational relations and contractual arrangements. Tendering in combination with the one-off nature of project work means that numerous joint contracts will have to be negotiated between participating organisations (clients, contractors, subcontractors, suppliers, consultants).[10] As illustrated in Figure 12.3, the distinctive nature of the global construction industry requires designers, quantity surveyors, construction managers, architects, consultants, project managers and suppliers to be fully integrated.

A significant challenge facing multinational global construction organisations intending to work in emerging economies and developing countries is how to minimise inaccurate and untimely data communications amongst primary stakeholders. In recent years, researchers have carried out several road-mapping research projects to determine how ICT could influence the future of the construction sector. The following visions were devised, taking account of client's needs and the opportunities offered by ICT:[8]

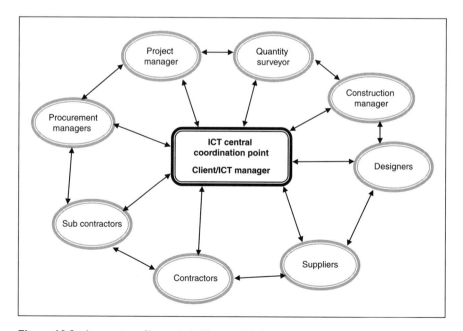

Figure 12.3 Integration of key stakeholders in global project

1. **CIB W78**

   In 1983, the International Council for Research and Innovation in Building and Construction (CIB), introduced the W78 Working Commission focusing on computer aided design, but this has now been redesigned to IT in construction.

2. **European Union's framework programme**

   The European Commission funded a number of road-mapping research projects under their fifth and sixth framework within information society technologies (ISTs). Under their initiative, some of the themes that were identified are:

   • value orientation and sustainability;
   • virtual organisations;
   • life-cycle integration;
   • reuse of information and knowledge;
   • IFC-based or model-based ICT;
   • advanced Internet technologies (based on XML, SOAP, etc.);
   • legal and contractual aspects of ICT in construction;
   • human and organisational aspects of ICT in construction.

   For example, as illustrated in Figures 12.4 and 12.5, the ROADCON (Strategic Roadmap towards Knowledge Driven Sustainable Construction) project aim was to develop a vision for agile, model-based/knowledge-driven construction.

   The vision for future ICT in construction was identified as 'the construction sector is driven by total product life performance and supported by knowledge-intensive and model based ICT enabling holistic support and decision making throughout the various business process and the whole product life by all stakeholders'.[8] As exemplified in Figures 12.4 and 12.5, the key areas of particular relevance are: ambient access and digital site.

3. **FIATECH**

   In 2004, FIATECH, a non-profit association in the US, developed the capital projects technology roadmap, which was a cooperative effort by stakeholders from government agencies and industry, working together to hasten the integration of new technologies that will enhance the capabilities of capital projects in the industry. As shown in Figure 12.6, the roadmap focused on 'Intelligent and Automated Construction Job Site' and the 'Integrated Automated Procurement and Supply Network', which amalgamates the following themes of the future:

   • location and status of resources;
   • automation of construction processes;
   • wirelessly managed construction job sites;
   • asset life-cycle information systems;
   • site monitoring systems and tracking systems that will compare daily construction progress against the baseline.

12.8 The value of project management learning and higher education

Globally, project management teaching is facing several challenges. The incorporation of real case studies through the provision of suitable learning environments, and the need for learners to reflect on their own skills and

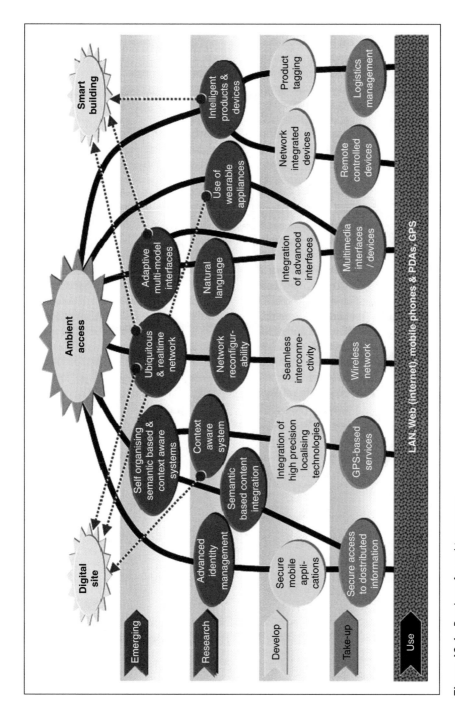

Figure 12.4   Roadmap for ambient access

Source: Adapted from Capital Projects Technology Roadmap Report[17]

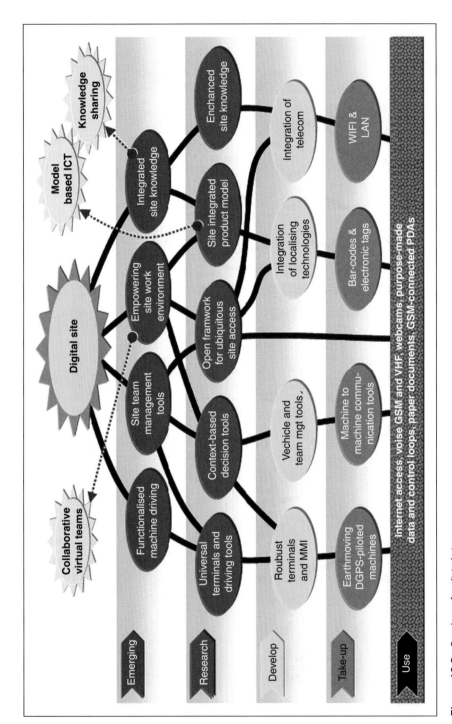

Figure 12.5 Roadmap for digital site

Source: Adapted from Capital Projects Technology Roadmap Report[17]

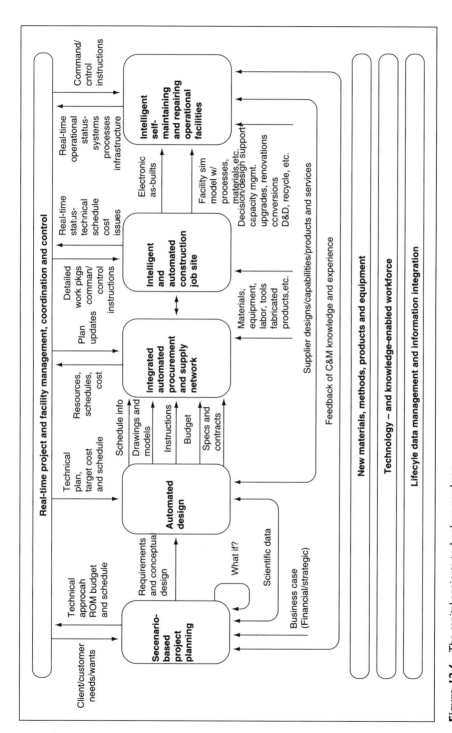

Figure 12.6 The capital projects technology roadmap

Source: Adapted from Capital Projects Technology Roadmap Report[17]

attitudes to projects, has been identified as an essential approach to promote more sensible and sufficient responses to the current complexities we face in managing projects.[18] In an attempt to learn from projects, a number of UK institutions have incorporated corporate learning in their programme portfolios. Corporate learning extends the margins of project management to include:

- consistent delivery of projects with excellence;
- alignment with client expectations; through
- understanding and definition of the project life cycle;
- service delivery to produce a quality product.[19]

These objectives are attainable through delivery that is dispersed, captured and facilitated. As noted in Chapter 2, the global economy is now complex and uncertain. The current Eurozone crisis took us all by surprise and still has colossal unexpected effects. In the Eurozone, expectations regarding economic developments are currently being reviewed and modified on a regular basis. In the UK for example, a number of clients have scaled down their operations while others have ring-fenced or prioritised their projects. In the new global economy, project managers and teams have to learn to deal with uncertainty and complexity. Cultural integration and complexity in global construction has generated an increasing interest in understanding how a global project manager can deal with pressures and demand in a global environment. The interest in understanding how global construction projects can be managed has transcended into the higher education curriculum. At present, a number of undergraduate and postgraduate courses integrate insights on project typologies, groups and management skills in order to encourage learners to think how they could better utilise their project management skills.[19] Any project management curriculum developed by higher education institutions and project management bodies needs to take into account the universal nature of the new global economy. Cordoba and Piki suggested that a possibility for improving project management can be achieved by offering generic project management in educational institutions whilst practical project management can be delivered in practice-related settings.[16] However, this should be carried out without disengaging the two.

12.8.1   Nurturing global project management in higher education

Global project management is a phenomenon; it is no more than what those involved with the discipline and profession say it is. The fact that there is no agreed single definition for global project management could be in part because the subject base needs to adapt to constant changes, a feature that could prove to be its enduring strength. Second, and arguably just as important, there is the need to provide the most appropriate learning and assessment environments to enable novices to achieve their potential. This area has to date received less attention than the drive to define 'how to?' and 'what is?' global project management. Third, as with most global project management knowledge, success in global project delivery depends on integration of the good use of tools, techniques and soft issues.[18] These complexities have emerged from the use of sophisticated technology, challenges associated in multicultural project teams, and increased coordination requirements. The global project manager has a central role in ensuring that multicultural project teams are fully integrated and that the technology used fully supports global delivery. The challenge facing professional project management bodies is to prepare learners with an in-depth extensive skill set including:

- good interpersonal skills;
- basic global business and management skills;
- knowledge and awareness of sustainability;
- excellent communication skills;
- cultural awareness;
- technical ability;
- ability to manage the project organisation to enable multicultural integration.[18,20]

In the UK, higher education institutions are required by the Quality Assurance Agency for Higher Education (QAAHE) to state the learning outcomes for modules and programmes at college, undergraduate and postgraduate level. The learning outcomes are classified into four key areas:

- transferable skills;
- subject practical skills;
- intellectual skills;
- academic knowledge.[19]

In a global environment, project management can become an enabler of competitive advantage if its function incorporates continuing learning and enhancement. Project management offers a logical model to manage organisational change or delivery of products and services.[20,21] Unlike habitual operations – processes that are repeatedly inflexible, inflated and slow to respond to changes – project management approaches are supple, opportune, adaptive and team-based.[22] In many cases, project management provides the best substitute or exceptional way to get business done. Teaching and learning project management has become an influential tool for both educators and learners to master knowledge, skills and attitudes. By placing the emphasis on both the learner and teacher, learning project management becomes a dynamic, interactive process of doing rather than only listening. A good project management programme should have a balance of three learning domains: knowledge, skill and personal development. It is usually agreed that these domains are delivered by a combination of both teaching and learning. The Higher Education Academy asserted that the last domain 'personal development' is non-existent in many higher education curricula,[22] although some courses that aim at personal awareness are becoming increasingly available in the form of executive development programmes. It is essential to note that higher education has become part of a global shift in a new way of generating and utilising knowledge. According to Ghoshal, this contemporary way is centred on solving problems, and is receptive of students needs. It strives for both quality and quantity.[23]

12.8.2 Differences between APM, PMI and PRINCE2

Over the last two decades project management has become firmly recognised as a management discipline represented by professional bodies across the globe. As tutors in well-established UK institutions, we (the book's authors) frequently get asked the question 'should I take a Project Management Institute (PMI), Projects in Controlled Environments (PRINCE2) exam or an Association of Project Management (APM) exam?' This question mirrors the state of affairs that, for many learners, there is a choice to make that is more than 'Which one should I take up first?' The question also indicates that individuals do not comprehend the absolute difference between the three. It seems to be the case that a number of people cannot differentiate PRINCE2, APM and PMI qualifications. In this section, the authors have simplified the differences so as to make it easier to prospective applicants and organisations. There is no generic conformity between bodies of knowledge and what academia, industry, associations and professional bodies believe to be the right learning outcomes, knowledge set and competencies. To fully appraise the differences we have examined different project management qualifications and certifications provided by APM, PMI and the Office of Government Commerce (OGC).

12.8.3 Association of Project Management

The association is the largest self-governing project management professional body in Europe with over 19,000 members and 500 corporate members. APM is based in the UK. Their aim is to develop and promote professionalism through five themes: breadth, depth, achievement, commitment and accountability. Examinations and training are administered through APM Accredited Training Personnel (ATP). At present the association is seeking a Royal Charter to have project management recognised as a profession.[24] As noted on their website, their mission statement is 'to develop and promote the professional disciplines of project and programme management for public benefit'. At the core of APM philosophy is the *APM Body of Knowledge (APM Bok)*, which consists of 52 knowledge areas. APM offers four qualifications that are an essential part of the APM five dimensions of professionalism. The four qualifications are:

1. **APM introductory certificate in project management:** for basic awareness of project management terminology. This course is designed to give learners some basic principles of project management. It entails a diverse range of topics, including risk management, teamwork, communications and quality management.
2. **Associate project management professional (APMP):** is for individuals with some project management background. It has been designed to give learners some foundation knowledge that will enable them to work efficiently on projects in their companies. The APMP syllabus entails a range of topics that are essential to successful project management. Some of the topics include earned value management, procurement, leadership, cost management and conflict management. The topics can be found in the *APM Body of Knowledge*, fifth edition.

3. **Practitioner qualification:** the next level available from APM, this course is for experienced practitioners who can demonstrate an ability to manage a non-complex project.
4. **APM project risk management certificate:** the next level, this course is designed to build on knowledge acquired in the APMP qualification.[24]

12.8.4 Association for Project Management Group (APMG)

APMG is based in the UK. It manages certifications, qualifications and accreditations on behalf of the Office of Government Commerce (OGC). OGC are the owners of PRINCE2, which is a structured project delivery methodology made up of four integrated components:

1. **Principles:** these are the key guiding ingredients that determine whether the project is being managed using PRINCE2 methodology. The seven principles can be summarised as: continued business justification, learn from experience, defined roles and responsibilities, manage by phases, manage by exception, focus on deliverables and tailor to suit the project.
2. **Themes:** these illustrate features of project management that must be fulfilled recurrently and in parallel throughout the project. The seven topics are: business case, organisation, quality, plans, risk, change and progress.
3. **Processes:** these exemplify the stepwise sequence through the project life cycle. Each process provides practitioners with a checklist of recommended tasks and responsibilities. The seven processes are: starting up a project, directing a project, initiating a project, controlling a stage, managing product delivery, managing a phase boundary and closing a project.
4. **Tailoring:** this theme addresses the need to modify PRINCE2 to the specific context of the project.

In the UK, OGC provides guidance on best project management and procurement policy standards for government departments. OGC also promotes collaborative procurement across the public sector in the UK, and monitors and challenges government departments on the policy standards they set out. This is to ensure that government departments integrate innovative project delivery methods and deliver better value for money to society.[25] OGC has now been incorporated into the new Efficiency and Reform Group (ERG) within the Cabinet Office.

12.8.5 Project Management Institute (PMI)

Globally, PMI is the largest project management organisation, with over 600,000 members in more than 185 countries. PMI core values are driven by a clear mission and an underlying set of standards that determine how they act and influence the expectations of stakeholders. They believe in project management impact, professionalism, volunteerism, community and engagement.[26] At the centre of the PMI ethos is *A Guide to the Project Management Body of Knowledge (PMBOK Guide)*. The PMI handbook consists of professional responsibility, project management processes and techniques. PMI offers six qualifications that acknowledge knowledge and competence. The six certifications can be classified as:

1. **Certified associate in project management (CAPM):** this qualification is designed for project team members. The course is aimed at entry-level project managers and learners.
2. **Project management professional (PMP):** this course is for people who have been managing projects for a number of years.
3. **Programme management professional (PgMPSM):** this course is aimed at people who have been managing projects and programmes for some years.
4. **PMI agile certified practitioner (PMI-ACP):** this course is designed for people working in organisations that utilise agile approaches to manage projects.
5. **PMI risk management professional (PMI-RMP):** this qualification is for individuals who are looking to fill a risk management role in their organisation.
6. **PMI scheduling professional (PMI-SP):** this qualification is aimed at people who are intending to fill a scheduling role in their organisation.[26]

PMI training is managed by registered education personnel (REP). Examinations are administered through certified PMI centres, with their head office based in the US.

The selection of project management qualifications and certifications is a complex problem for modern organisations. Distinct challenges faced by modern organisations are to identify a qualification that:

- gives details of project complexity and uncertainty;
- can be widely used by stakeholders in well-established economies and emerging economies;
- addresses global economic, social and environmental issues.

12.8.6 Choosing your project management qualification

In order to help practitioners relate the concepts to their organisations, professional bodies need to incorporate the following topics in their texts:

- strategic issues in global projects;
- managing cultural complexity in a global project;
- partnering and alliancing in global projects;
- selection of appropriate finance mechanisms in global projects;
- project sustainability management;
- management of uncertainty in global projects.

APM, PMI and PRINCE2 do not put adequate emphasis on the above topic areas. The three project methodologies have been found to be effective, but they do have limitations, especially if they are used to manage change that entails primary stakeholders changing their behaviour or working patterns. As examined in Chapter 6, project management entails initiation, planning, execution, monitoring and control. It is about achieving project objectives through a series of deliverables, arranged in a logical sequence. PRINCE2 and PMI offer little practical guidance to learners and practitioners in how to develop commitment, manage resistance to change, maximise team-working, identify stakeholders, and establish effective communication between stakeholders and the project teams.[27] The three methodologies offer little practical direction to practitioners and learners on how to:

- select appropriate finance mechanisms in global projects;
- deal with strategic issues in global projects;
- integrate sustainability into projects;
- manage cultural complexity;
- manage uncertainty in projects;
- partnering and alliancing in global projects.

It is vitally important that global construction organisations integrate project management methodologies that will incorporate the above six components. In addition, project managers need methodologies that will incorporate technological innovations. As global projects have become more complex, organisations need project management programmes that will address complexity at strategic, operational, technical and project level.

12.8.7 Amalgamating APM, PMI and PRINCE2 qualifications

By doing APMP and PMP a practitioner can cover knowledge areas that are not included in the PRINCE2 handbook. It is worth noting that there are number of key areas in the PRINCE2 handbook that are not well covered. These are:

- **Investment appraisal:** in this section practitioners are told to integrate an investment appraisal in their business case, but they have not been told how to carry out an investment appraisal. APMP and PMP explain how different types of investment appraisal can be utilised. In particular, they give a description of internal rate of return (IRR) and net present value (NPV) as investment appraisal techniques.
- **Roles and responsibilities:** PRINCE2 gives a description of roles and responsibilities in a project management team but does not illustrate how they relate to the organisation's own structure. It highlights some of the challenges faced in working with project teams. APMP and PMP compare different organisation structures and exemplify some of the merits and demerits of project teams.
- **Planning quality control:** in this section practitioners are being advised to use quality review technique, but are there any other quality techniques that practitioners could use in the new global economy? APMP and PMP provide additional quality techniques and also give a description of how to perform quality control.
- **Risk management:** on this theme, practitioners have been provided with a risk management framework, but other areas of risk such as health and safety have not been explored. In addition to the above, APMP and PMP explore tools and techniques for risk identification and the application of probability and impact grid.

There are challenges to be faced for anyone intending to apply APMP and PMP to PRINCE2 surroundings. The APM, PRINCE2 and PMI *Project Management Body of Knowledge (PMBoK)* have been written by people with different backgrounds, and as a result there is dissimilarity in some of the words used. Out of the three project management handbooks, *PMBOK* is the most used, but a number of students have misconstrued *PMBOK* as a process guide. In some sections of *PMBOK* and APM *Body of Knowledge* they lack rigour and governance models, which are accurately defined in PRINCE2. There are some organisations that offer APM, PMI and PRINCE2 training qualifications. These qualifications are not equally exclusive, and are not competing methodologies of project management. There is a misapprehension in academic circles that if someone grasps the fundamentals of *PMBOK*, PRINCE2 and APM *Body of Knowledge*, then they can manage any project and are guaranteed a senior well-paid job.

PRINCE2, APM and PMI qualifications will certainly boost skills and confidence when managing global construction projects and will greatly enhance any employment experience. The three professional qualifications will not, however, unlock potential in themselves. So what will unlock potential? It is advisable to couple professional qualification with a postgraduate qualification or an undergraduate degree in project management. Major recruiters all over the world require applicants to have a first degree or a second degree. Any meaningful global project management curriculum at any reputable higher education institution will surely help to get started in project management. Table 12.1 contains key lectures that need to be taught in a global project management curriculum.

Given that recruiters seem to trust the grounding and training provided by the standard project management programme, getting a job can become easier with a project management qualification on your curriculum vitae. New entrants to

Table 12.1   The way forward: global project management curriculum

Key lectures	Discussion areas
✓ Cultural complexity in projects	• Building multicultural and virtual construction teams
✓ Partnering and alliancing	• Drivers for and conditions of partnering and alliancing
✓ Project sustainability and performance improvement	• Environmental, social, economic, political
✓ Project delivery complexity	• Managing of project life cycles
✓ Controlling projects	• KPI for projects
✓ Global project leadership	• Managerial and technical aspects of projects
✓ Global communication strategies	• Defining a global communication matrix
✓ Stakeholders in projects	• Defining strategies for managing stakeholders and shareholders
✓ Effects of the global economy on construction projects and sources of project finance	• Conditions for joint venture funding
✓ Project success and project failure	• Defining project success
✓ Global risks and constraints	• Risk identification and mitigation strategies
✓ The reality of global project management	• Aligning global business objectives and project objectives

the profession should consider a project management degree or Masters degree in project management only if they truly think project management is a strong point and if they want to spend 20 or more years solving soft and hard issues in global construction projects. As illustrated above, APM qualifications are designed to ensure that learners have grasped generic knowledge in key areas of project management, whereas PRINCE2 is a structured methodology that is aimed at organisations delivering projects. PMI training and qualifications are appropriate for anyone involved in project management. There are two deliberations that individuals and organisations have to explore when deciding on which professional qualifications to go for:

1. Do I/we need a professional qualification that is globally recognised? If you need a global professional qualification then you can start with PMP.
2. Which professional membership do I/we need? It is important to ensure that you choose a professional membership that is aligned to your personal or organisational needs.

Table 12.2 highlights some of the factors that potential applicants have to consider when applying for a PRINCE2, PMI or an APM qualification.

Table 12.2   A summary of factors to consider

Factors to consider	APM	PRINCE2	PMI
• Syllabus based upon	• APM Body of Knowledge	• OGC PRINCE2 manual 2005 edition	• PMBOK 5th edition 2009
• Geography	• Mainly UK and other parts of Europe	• Mainly across Europe	• US-based. Most globally recognised
• Professional membership available	• Yes (but not linked to qualification)	• No	• Yes (and requires PMP qualification)
• Best-practice sharing/forums amongst members	• Yes (through specific interest groups and annual conference)	• No	• Yes (regional chapters)
• Qualifications	• APM Introductory Certificate in Project Management: • Associate Project Management Professional (APMP) • Practitioner Qualification • APM Project Risk Management Certificate	• PRINCE2	• Certified Associate in Project Management (CAPM) • Project Management Professional (PMP) • Program Management Professional (PgMPSM) • PMI Agile Certified Practitioner (PMI-ACP) • PMI Risk Management Professional (PMI-RMP) • PMI Scheduling Professional (PMI-SP)

Table 12.2 (Continued)

Factors to consider	APM	PRINCE2	PMI
• Other factors to consider	• Qualifications fully aligned with IPMA (International Project Management Association levels A to D) • APM progressing towards Chartered Status and register of Chartered Project Professionals (ChPP)	• Re-registration exam to maintain practitioner status required after three to five years	• Eligibility requirements to satisfy prior to sitting any exams based on education attainment, relevant training hours and project experience. • To retain certification status requires evidence of continuous professional development – CAPM every five years; PMP every three years

Sources: Office of Government Commerce (n.d.)[25], Project Management Institute (2012)[26], Miller (2006)[27], Provek (2012)[28].

12.9 Chapter and book summary

It has been shown that most global construction projects involve complex arrangements across numerous contractors and subcontractors. The choice of implementation strategy in global environments is as important a decision as the choice of management strategy that identifies soft and hard issues of the project. In order for the implementation strategy to be effective, project managers need to ensure that individual values and goals are fully aligned at the onset of the project. It is worth noting that in a global construction project, the integration of ICT is reliant on the structure of the organisation, size, delivery frameworks, complexity, contract types, governments and stakeholders involved. As noted in this chapter, it is relatively simple to produce a combined curriculum that meets the requirements of APM, PMI and PRINCE2. A detailed appraisal of APM and PMI showed that they have much in common. The challenge for institutions and project management associations is to develop graduates and practitioners with skill sets that are culturally competent and technically aware. Some of the benefits of achieving a project management qualification or certification include best practice knowledge, and attainment of a more holistic approach to project management.

A number of noteworthy issues have been identified that have not been discussed in the global construction management literature. The issues that have been identified relate to global construction project management. As outlined in this book, the rapid globalisation of the world's economy has had significant impact on the way construction project managers work, bringing them frequently into contact with clients, suppliers and peers who they have never worked with before. In an era of globalisation, projects in the construction industry will face unique challenges in coordinating among clients, financiers, developers, designers and contractors from different countries. In addition, clients and project managers will need to cope with the complexities of both local institutions and physical environments. Throughout this book we have discussed the challenges facing global construction organisations who are intending to work effectively across borders; we have identified the major challenges as being able to develop business models that balance global competitiveness, multinational flexibility and the building of a global learning capability. Achieving this balance will require GCOs to develop the cultural sensitivity and ability to manage and leverage learning in order to build future capabilities.

While offering opportunities, globalisation also poses significant challenges for construction project managers, especially when different cultures are involved as a team. What does this mean for clients, project leaders and international construction organisations? They must actively promote multicultural team working as the means of addressing poor performance on people management and cultural issues on construction projects. In particular, if organisational change is to be effectively introduced in developing countries, multinational construction organisations will have to ensure that their key decisions are being informed by the knowledge and experience of local or indigenous managers. This will require construction project leaders to have a better understanding of cultural change processes and procedures in developing countries.

It is worth noting that the global construction industry has been under pressure to evolve into a sector that is constantly changing to fit the needs of the broader context in which the operations are executed. Attitudes towards working have changed dramatically in recent years and there is currently much more emphasis on multicultural team working. As global construction organisations define more of their activities as projects, the demand for multicultural team working grows, and there is increasing interest in reforming the project delivery process. For GCOs there is an increasing need to refine their business models. Now that construction companies are able to move resources to almost any location worldwide and have the capacity to work on a global scale, for many organisations future opportunities will entail thinking more clearly about cross-cultural issues and more overtly and systematically an understanding of risk management, project governance and global strategic issues.

With an ongoing increase of globalisation, multinational construction organisations must be aware of social, political, environmental, economic and legal policy issues in order to function effectively in a global project environment. The proposed strategies in this book present a better way of optimising the performance of project-based operations, thus enabling GCOs to reform their poor performance on projects and empower them to better manage emerging cultural challenges in their future projects. In spite of the current difficulties the industry faces, there is an increasing need to get contractors from different nationalities to work together effectively. Many international construction organisations have found that international integration can be problematic, and at times performance is not always at the level required or expected. It is crucial for the construction research community to strengthen the debatable assumption that globalisation is an organisational variable, which is subject to conscious manipulation. A more nuanced understanding of global project complexity and structure are required if globalisation in construction is to be accurately understood and responded to.

12.9.1   Case study: Itaipu Dam (Brazil and Paraguay)

The concept behind the Itaipu Dam project emerged in the 1960s as a result of negotiations between Brazil and Paraguay. The Iguacu Act was signed on 22 July 1966 by the Paraguayan and Brazilian foreign ministers. The joint declaration of interest from the two countries was to allow exploration to be carried out in the section of the Parana River from, and including, the Salto de Sete Quedas, to the Iguazu River watershed. In 1970, a consortium formed by corporations (from the United States) and ELC Electroconsult S.p.A (from Italy) won the global bidding for the realisation of the feasibility studies and for the assessment of the construction work. After the successful completion of the feasibility study, Brazil and Paraguay signed the Itaipu Treaty in 1973, which led to a legally binding agreement being completed by the two countries. In 1974, the Itaipu Binacional entity was introduced to oversee the hydroelectric power plant construction. The construction work began in 1975. When the construction of the project commenced, approximately 10,000 families living beside the Parana River were relocated. For the first three years, contractors and subcontractors worked on moving the Parana River around the construction site. After the first three years, contractors managed to complete the diversion channel, which was 1.3 miles long, 300 feet deep and 490 feet wide. The original riverbed began to dry out, and this allowed contractors and subcontractors to start constructing the dam. In 1979, an important agreement was signed by Brazil, Paraguay and Argentina. The agreement was used to establish the allowed river levels and how much the river level could change as a result of various hydroelectrical undertakings in the watershed. Argentina felt that, in the event of a conflict, Brazil could unlock the floodgates, raising the water level in the Rio de la Plata, which would lead to flooding the capital city Buenos Aires. In 1982, the construction of the project was completed and contractors created the 125-mile-long reservoir. This led to the formation of Itaipu Lake, which within two weeks the water level rose to 100 metres to reach the spillway. During the creation of the reservoir, conservationists with the project travelled through the flooded project environment in vessels in order to save local animals. In 1984, the first generator of the project began running. The remaining 18 units were installed at a rate of two to three

Figure 12.7 Panoramic view of the Itaipu Dam, with the spillways (closed at the time of the photo) on the left

a year. The last two of the 20 units started operations in September 2006 and March 2007. With 20 generators installed, the hydroelectric plant generates up to 14 gigawatts of electricity. Due to a clause signed by the three countries, the maximum number of generating units allowed concurrently cannot exceed 18 units. The project cost US$19.6 billion. In 2009, transmission from the plant was interrupted. This led to power failure in Brazil for two hours and in Paraguay for fifteen minutes. In 1994 the American Society of Civil Engineers nominated the Itaipu Project as one of the Top 7 modern Wonders of the World Project. Figure 12.7 illustrates a panoramic view of the project.

12.9.2 Discussion questions

1. Why did this project take so long to complete? Identify some of the challenges that Brazil, Paraguay and Argentina had to overcome.
2. What were some of the strategic, operational, technical and project challenges that the two nations had to overcome?
3. What are some of the major opportunities for improvement in the management of the implementation phase?
4. Cultural dynamics causes challenges in global construction projects. What were some of the cultural issues that the two corporations had to overcome when carrying out the feasibility study in 1970?
5. Do you think the building of this project had major sociological, economical and environmental effects to society in Brazil and Paraguay?

6. Using the following grid, complete the evaluation of the following variables on a rating scale of: 1 – *very poor*, 2 – *poor*, 3 – *good*, 4 – *very good*, 5 – *excellent*.

Variables	Rating scale	Explanation
Ethical conflicts		
Construction management		
Technical management		
Environmental management		
Legal regulations		
Site management		

12.10 References

1. Mahalingam, A. and Levitt, R. E. (2007). Institutional theory as a framework for analysing conflicts on global projects, *Journal of Construction Engineering and Management*, **133** (7), pp. 517–528.
2. Young, R. B. and Javalgi, R. (2007). International marketing research: A global project management perspective, *Business Horizons*, **50**, pp. 113–122.
3. Dooley, L., Lupton, G. and O'Sullivan, D. (2005). Multiple project management: A modern competitive necessity, *Journal of Manufacturing Technology Management*, **16** (5), pp. 466–482.
4. Kang, B. G., Price, A. D. F., Thorpe, T. and Edum-Fotwe, F. (n.d.). Ethics training on multicultural construction projects. *Construction Information Quarterly*, **8** (2), pp. 85–91.
5. Thornhill, R. (2009). Real estate executive Report, Deloitte.
6. Pheng, L. S. and Leong, H. Y. (2000). Cross cultural project management for international construction in China, *International Journal of Project Management*, **18**, pp. 307–316.
7. Yasin, M. M., Martin, J. and Czuchry, A. (2000). An empirical investigation of international project management practices: The role of international experience, *Project Management Institute*, **31** (2), pp. 20–30.
8. Bowden, S., Dorr, A., Thorpe, T. and Anumba, C. (2006). Mobile ICT support for construction process improvement. *Journal of Automation in Construction*, **15** (5), pp. 664–676.
9. Adriaanse, A., Voordijk, H. and Dewulf, G. (2010). The use of interorganisational ICT in United States construction projects. *Journal of Automation in Construction*, **19** (1), pp. 73–83.
10. Nitithamyong, P. and Skibniewski, M. (2006). Success/failure factors and performance measures of web-based construction project management systems: Professionals' viewpoint, *Journal of Construction Engineering and Management*, **132** (1), pp. 80–87.
11. IBM (2004). Information and communications technology (ICT) social security project management: Ten issues on ICT management in social security organisations.
12. Leydesdorff, E. and Wijsman, T. (n.d.). Why government ICT projects run into problems (Netherlands).

13. Froese, M. T. (2010). The impact of emerging information technology on project management for construction, *Journal of Automation in Construction*, **19** (5) pp. 531–538.
14. Hartmann, T., Meerveld, H. V., Vossebeld, N. and Adriaanse, A. (2012). Aligning building information model tools and construction management methods, *Journal of Automation in Construction*, **22**, pp. 605–613.
15. Rozendal, R. (2002). The cultural and political environment of ICT projects in developing countries. International Institute for Communication and Development Research Brief-No.3
16. Cordoba, J. R. and Piki, A. (2011). Facilitating project management education through groups as systems, *International Journal of Project Management*, **30** (1), pp. 83–93.
17. Fully Integrated and Automated Technologies for Construction (2004). Capital projects technology roadmapping initiative. http://fiatech.org/images/stories/techroadmap/deliverables/oct2004/CPTR_AcknowIntro_Download_Oct2004.pdf [Accessed 15th June 2012].
18. Levene, R. J. (2003). The future directions for project management through learning. http://www.law.bournemouth.ac.uk/pm/2003/pdf/RalphLevene.pdf [Accessed July 2012].
19. Gale, A. and Brown, M. (2003). Project management professional development: An industry led programme, *Journal of Management Development*, **22** (5), pp. 410–425
20. Kezsbom, D. S. and Edward, K. A. (2001). *Organisational and Interpersonal Project Communication: Spanning across Boundaries*. 2nd edn. New Jersey: Wiley and Sons.
21. Meredith, J. R. and Mantel, S. J. (2000). *Project Management: A Managerial Approach*. New York: John Wiley and Sons.
22. The Higher Education Academy (2006). Personal development planning and employability. http://www.heacademy.ac.uk/assets/documents/tla/employability_enterprise/web0368_learning_and_employability_series2_pdp_and_employability.pdf [Accessed June 2012].
23. Ghoshal, S. (2005). Bad management theories are destroying good management practices, *Academy of Management Learning and Education*, **4**, pp. 75–91.
24. Association of Project Management (2012). APM qualification information. http://www.apm.org.uk/ [Accessed June 2012].
25. Office of Government Commerce (n.d.) PRINCE2 Maturity Model (P2MM), version 2.1.
26. Project Management Institute (2012). What are PMI certifications? *http://www.pmi.org/Certification/What-are-PMI-Certifications.aspx* [Accessed June 2012].
27. Miller, D. (2006). Putting a people focus into project management, *Project Manager Today, June*. http://www.changefirst.com/uploads/documents/Putting_a_People_Focus_into_Project_Management.pdf [Accessed July 2012].
28. Provek (2012). Project management training. www.Provek.co.uk [Accessed July 2012].

Index

Printed and bound by CPI Group UK Ltd, Croydon, CR0 4YY